Antio☠idants
FAILURES
and
DANGERS

The consequent fall of free radical theory

316 failed study reports, containing
90 reports showing harmful effects

A Selective Review for the Medical
Professional and Informed Consumer

BY

PROF. HON. RANDOLPH M. HOWES, M.D., Ph.D.

Physician, Surgeon and Scientist (Biochemist)

Free Radical Publishing Co.
27439 Hwy 441
Kentwood, Louisiana 70444

Address for communication: 27439 Highway 441, Kentwood, Louisiana
70444-8152, USA. Email: rhowesmd@hughes.net

Randolph M. Howes, M.D., Ph.D.

Antio✖️idants FAILURES and DANGERS

The consequent fall of free radical theory

BY
PROF. HON. RANDOLPH M. HOWES, M.D., Ph.D.
Physician, Surgeon and Scientist (Biochemist)

Adjunct Assistant Professor of Plastic Surgery,
The Johns Hopkins Hospital, Baltimore, MD USA

Espaldon Professor of Plastic and Reconstructive Surgery,
University of Santo Tomas, Manila, Philippines

Adjunct Professor of Biological Sciences,
Southeastern Louisiana University

Founder, Director and Chairman of the Scientific Advisory Board;
U.S. Medical Scientific Research Foundation, Inc.

ACKNOWLEDGEMENTS

Special thanks Don Neale Piatt, Sr. for proof reading.
Also, special thanks to Michael R. Root, M.S. for his unwavering
encouragement.

NOTICE TO USERS

It is understood that medicine is an ever-changing science. As new research and clinical experience broaden our knowledge, changes in treatment and drug therapy are required. The author and the publisher of this work have checked with sources believed to be reliable in their efforts to provide information that is complete and generally in accord with the standards accepted at the time of publication. However, in view of the possibility of human error or changes in the medical sciences, neither the authors nor the publisher nor any other party who has been involved in the preparation of publication of this work warrants that the information contained herein is in every respect accurate or complete, and they disclaim all responsibility to any errors or omissions or for the results obtained from use of the information contained in the work. Readers should confirm the information contained herein with other sources. For example and in particular, readers are advised to check the product information sheets (or labels) included in the package of each drug they plan to administer to be certain that the information contained in this work is accurate and that changes have not been made in the recommended dose or in the contraindications for administration. This recommendation is of particular importance in connection with new or infrequently used drugs, additives or supplements.

Disclaimers:

Please note: only your personal physician or other health professional you consult can best advise you on matters of your health based on your medical history, your family medical history, your medication history, and how information from any of these databases may apply to you. Neither Dr. Howes nor any party involved in creating, producing or delivering this web site shall be liable for any damages arising out of access to or use of this material or web site, or any errors or omissions in the content thereof.

The information given herein is not intended as medical advice. Always consult with your doctor for underlying illness. Before beginning dietary investigation, consult a dietician or a physician with an interest in nutrition. Information is drawn from the scientific literature, web research, and personal enquiry; while all care is taken, information is not warranted as accurate and the author cannot be held liable for any errors and omissions.

Financial disclosure:

Dr. Howes has no financial conflicts of interest and is not involved in the sale of dietary supplements or fitness equipment. The author holds no stocks or interests in companies in the food additive or antioxidant supplement business.

**The story of antioxidants,
as they relate to disease
prevention and cure,
is the story of ....
FAILURE!**

R. M. Howes, M.D., Ph.D.
6/2/11

ABOUT THE AUTHOR
Dr. Randolph M. Howes M.D., Ph.D.

As a champion of the people, Dr. Howes anticipates and hopes for the active involvement of all connected parties (patients, caregivers, healthcare professionals, etc.) as an integral approach to educating consumers and the public about the potential dangers of excessive antioxidant-containing supplements.

Some people are born with a silver spoon in their mouth but Dr. Howes had to earn his. Even as a child, Dr. Howes could think with adult clarity. He could envision his future but it would require "decades of dedication" to make it a reality. From childhood, Dr. Howes was motivated to become a medical doctor and scientist. Assuredly, having been born on a small strawberry farm in rural Louisiana, his journey to the top has proved to be arduous and demanding.

However, he was fortunate to acquire the confidence of Sister Elizabeth at St. Joseph's school and went on to gain the support of his high school speech teacher, Mrs. Iris Brann, who also had strong beliefs in his abilities and potential. Ultimately, with the help of his guitar and his singing ability, he defeated the star quarter back of the high school football team to become the president of the student body. With the aid of a $25 dollar legislative scholarship, he went on to Southeastern Louisiana College (SLC).

At SLC, he was selected for honors chemistry, made the Dean's list, worked at the Psychology Research Lab forty hours a week, maintained a premed study load, and was elected president of the Junior Class and the Interfraternity Council. To earn badly needed funds, he played music on weekends in a small combo, The Three Blind Mice. Next, he matriculated to Tulane University School of Medicine.

His initial dream was to try to combine both medicine and science. In that regard, he began work as a technician with Dr. Andrew Schally at the Endocrine Polypeptide Lab in the isolation of thyrotropin releasing factor. This work led to a Nobel Prize for Dr. Schally. Dr. Howes had been highly impressed with the enthusiasm of biochemist, Dr. Richard H. Steele, who accepted him as a doctoral candidate under his tutelage. Dr. Howes graduated in the top 10 of his class, won the Louisiana Pathology Association Award, was elected to the Sigma Xi honor fraternity and was the first in the history of Tulane to become a Doctor of Medicine and a Ph.D. in biochemistry concurrently. Next, he was selected to pursue a career in surgery at the prestigious Johns Hopkins Hospital.

Unbelievably, at Dr. Howes' urging, he was allowed to operate his own research lab during his surgical internship and residency training. He worked hand in hand with the greats in American medicine and surgery. Independently, he garnered grants, trained lab techs, wrote papers, slept on the cold floor, proudly served as a Captain in the U.S. Army Reserves Medical Corp and finished with board eligibility in both general and plastic surgery in an unheard of six year period. In another first, he was appointed as an Adjunct Assistant Professor of Plastic Surgery at Johns Hopkins Hospital. For decades, Dr. Howes gave unselfishly to pro bono medical missions in the Philippines and he holds the Ernesto Espaldon Chair as Professor of Plastic Surgery at the University of Santo Tomas. Upon retirement from a career in cosmetic plastic surgery, he is living his dream of trying to revolutionize the treatment of cancer, heart disease, HIV/AIDS and malaria, with his in depth knowledge of the arcane biochemistry of oxygen metabolism. He is a work in progress! Dedicated and passionate, he is on a mission for mankind.

Dr. Howes was the first in the history of Tulane School of Medicine to be awarded a Doctorate of Medicine degree and a Ph.D. in Biochemistry at the same time. He was trained as a General surgeon and a Plastic surgeon at the prestigious Johns Hopkins Hospital, in

Baltimore, Maryland. He was the first in the history of Johns Hopkins Hospital to obtain board eligibility in both general and plastic surgery in a six year period. Also, he proudly served as a Captain in the U.S. Army Reserves Medical Corp.

Dr. Howes invented the triple lumen venous catheter, which has been credited with helping save the lives of over 20 million critically ill patients worldwide. His catheter is the number one venous catheter in the world today and his name is well recognized in over 100 countries. He has been recognized as a humanitarian, visionary, entrepreneur, singer, songwriter, inventor and author.

He received the Harper Award for innovative research from the American College for Advancement in Medicine, served as their keynote speaker and his peers refer to him as "a walking encyclopedia on oxygen metabolism."

He is a Dr. Norman Vincent Peale Unsung Hero award winner, which recognized his awesome versatility. Additionally, even though he is humble and does not like talking about it, he is a self made multi-millionaire.

He is currently doing extensive research on cures for cancer and heart disease and development of revolutionary treatment modalities. He has written 16 books over the past 8 years on the subject of oxygen metabolism, as it relates to protection from cancer, heart disease, diabetes, malaria, HIV/AIDS, Alzheimer's disease, aging and arthritis. He has written many scientific and medical papers and has lectured nationally and internationally. He has written over 230 medical letters to the editor on popular topics.

His research has shown that currently common antioxidant vitamins, such as vitamins A & E, (and vitamin C to a lesser extent) can be harmful and that oxygen free radicals protect us from bacterial, fungal and viral infections and they help to control cancer growth. He has developed an effective, inexpensive singlet oxygen generating system, from orthomolecular agents, for the treatment of cancer and heart disease. He is passionate about his research and hopes to have his discoveries at the patient's bedside in his lifetime.

There are over 6,000 pages in his magnum opus and at the Howes World Selective Library on Oxygen Metabolism.

Over 3,000 pages of his opus are available online in a searchable format www.iwillfindthecure.org © 2011 by R.M. Howes

**To put one's faith in people
is to seek disappointment.
To put one's faith in ideas of discovery
is to be pestered by never-ending curiosity.**

R. M. Howes, M.D., Ph.D.
9/7/09

"It is difficult to free fools from the chains they revere." ---- VOLTAIRE

**The mind is fortified by learning
and the spirit is strengthened by doing.
Combined, they become exponential.**

R. M. Howes, M.D., Ph.D.
9/7/09

**If you believe the implausible,
you will accept the indefensible and
not recognize the inexcusable.**

R. M. Howes, M.D., Ph.D.
6/5/11

The scientific method demands that we change our beliefs or theories to fit the factual data. I believe that this applies directly to the Free Radi-Crap theory. Again, I say, "The free radical theory has fallen."

Companion Papers:

Citation: R. Howes: Mythology of Antioxidant Vitamins?. *The Journal of Evidence-Based Alternative and Complimentary Medicine*. April, 2011. 16(2): 149-189.

Citation: R. Howes: Cancer Therapy: A Review with Scientific Validation for the Role of Electronically Modified Oxygen Derivatives in Oncologic Treatment Modalities. *The Internet Journal of Alternative Medicine*. 2010 Volume 8 Number 1.

Citation: R. Howes: Hydrogen Peroxide: A review of a scientifically verifiable omnipresent ubiquitous essentiality of obligate, aerobic, carbon-based life forms. *The Internet Journal of Plastic Surgery*. 2010 Volume 7 Number 1.

Howes M.D., PhD., R. (2009). Dangers of Antioxidants in Cancer Patients: A Review. *PHILICA.COM Article number 153*. Published 7th February, 2009. (20 pages)

Howes M.D., PhD., R. (2008). Aging and anti-aging claims: a review on antioxidant vitamins A, C & E. *PHILICA.COM Article number 116*. Published on 12th January, 2008. (16 pages)

Howes M.D., PhD., R. (2007). Sleep: An original "radical" proposal. *PHILICA.COM Observation number 42.* Published on 5th October, 2007. (1 page)

Howes M.D., PhD., R. (2007). Antioxidant Vitamins A, C & E; Death in Small Doses and Legal Liability? *PHILICA.COM Article number 89.* Published on 5th April, 2007. (23 pages)

Howes M.D., PhD., R. (2007). Cancer, Apoptosis and Reactive Oxygen Species: A New Paradigm. *PHILICA.COM Article number 86.* Published on 26th February, 2007. (11 pages)

Howes M.D., PhD., R. (2007). Antioxidant Vitamins A, C and E: Assessing Potential for Harm. *PHILICA.COM Article number 83.* Published on 15th February, 2007. (14 pages)

Howes M.D., PhD., R. (2007). The Consequent Downfall of the Free Radical Theory. *PHILICA.COM Article number 75.* Published on 22nd January, 2007. (9 pages)

Howes, R.M.: "The Free Radical Fantasy," The Annals of New York Academy of Sciences, 2006, Vol. 1067, pp. 22-26.

Available at:
www.philica.com
www.medi.philica.com
www.iwillfindthecure.org

DEDICATION

To the "impossible dreamer"….ME!

"Futures's shape is sculpted by the persistent kneading hands of the impossible dreamer."

R. M. Howes, M.D., Ph.D.
5/2/04

TABLE OF CONTENTS

DISCLAIMERS 2
FINANCIAL DISCLOSURE 3
ABOUT THE AUTHOR 5
COMPANION PAPERS 9
DEDICATION 11

CHAPTER ONE: 17

- Here is the "pitch line of the vitamaniacs"
- No deaths from vitamins - none at all in 27 years

CHAPTER TWO: 21

- Fundamental value of creative ideas
- Here is "the scientific evidence-based truth" exposed by Howes
- Bias, chance and cause
- EMOD exposure
- EMOD control
- EMOD sources
- A great biochemical wonder, EMODs
- Deficiency and overexpression studies
- Quantifying antioxidant activity
- Oxidative stress measurement (biomarkers)

CHAPTER THREE: 31

HARMFUL EFFECTS OF ANTIOXIDANTS by category
- MORTALITY (14 study reports)
- CANCER (34 study reports)

- HEART DISEASE (19 study reports)
- STROKE (5 study reports)
- LUNG FUNCTION (3 study reports)
- DIABETES RELATED (6 study reports)
- PRETERM BIRTH & Pregnancy (7 study reports)
- INFECTIONS (2 study reports)
- EXERCISE (3 study reports)
- EYE DISEASES (3 study reports)
- SKELETAL (5 study reports)
- MISCELLANEOUS (9 study reports)
- Summary

CHAPTER FOUR: 53

- Organizations which do not recommend the use of antioxidant vitamins

CHAPTER FIVE: 57

- Prologue
- Cancer survivors
- Cancer cells types protected by antioxidants
- "Antioxidant" is the buzz word

CHAPTER SIX: 63

- "Disappointing" Results
- Consensus statements on the fall of the free radical theory

CHAPTER SEVEN: 75

- The epiphany of 4-4-11
- Human cancer cell types (27 human & 9 murine) shielded by antioxidants
- Five common antioxidant killers
- TEN common harmful antioxidants

- 13 ineffective antioxidants
- 8 potentially harmful endogenous antioxidants

CHAPTER EIGHT: 79

- Sale of children's vitamins on the rise
- Use of dietary supplements keeps climbing: CDC
- Variables and covariants
- Thirty variables in antioxidant and EMOD (redox) studies
- Safety and cautionary statements

CHAPTER NINE: 91

- Cocoa, chocolate and flavonoids
- Dr. Frei discounts the antioxidant properties of flavonoids
- Omega-3 fish oil
- Fish Oil Does Not Prevent Alzheimer's Disease
- Fish Oils: Is There A Downside?
- The Fish Oil Gold Rush
- Fish oil does not accelerate weight loss
- Canola oil and omega-3s (fish oil)
- Omega-3 fatty acids fail glowing predictions
- Omega 3 (fish oil): Summary
- REFERENCES: Cocoa, chocolate and flavonoids and Omega-3 and fish oil section

CHAPTER TEN: 111

- 316 Antioxidant Intervention Trial & Analysis Study Reports
- Antioxidant vitamin failures and dangers
- References

CHAPTER ELEVEN: 185

- Expanded 316 study reports

CHAPTER TWELVE: 387

- 90 Study reports showing harmful effects of antioxidants (1-92)
- Harmful effects of antioxidants - Summary
- Adverse consequences of antioxidant vitamins and increased risk of death

**We are all children of the radical,
the descendants of the di-radical,
the di-radical, OXYGEN.
Three and one half billion years ago,
we were the asexual spawn of
an illicit evolutionary paramour,
who is still the love of our lives, O_2.**

R. M. Howes, M.D., Ph.D.
6/5/11

CHAPTER ONE

Here is the "pitch line of the vitamaniacs"

FOR IMMEDIATE RELEASE - (reproduced in its entirety).

Orthomolecular Medicine News Service, June 14, 2011

No Deaths from Vitamins - None at All in 27 Years

Commentary by Andrew W. Saul and Jagan N. Vaman, M.D.

(OMNS, June 14, 2011) Over a twenty-seven year period, vitamin supplements have been alleged to have caused the deaths of a total of eleven people in the United States. A new analysis of US poison control center annual report data indicates that there have, in fact, been no deaths whatsoever from vitamins . . . none at all, in the 27 years that such reports have been available.

The American Association of Poison Control Centers (AAPCC) attributes annual deaths to vitamins as:

2009: zero	2000: zero	1991: two
2008: zero	1999: zero	1990: one
2007: zero	1998: zero	1989: zero
2006: one	1997: zero	1988: zero
2005: zero	1996: zero	1987: one
2004: two	1995: zero	1986: zero
2003: two	1994: zero	1985: zero
2002: one	1993: one	1984: zero
2001: zero	1992: zero	1983: zero

Even if these figures are taken as correct, and even if they include intentional and accidental misuse, the number of alleged vitamin fatalities is strikingly low, averaging less than one death per year for over two and a half decades. In 19 of those 27 years, AAPCC reports that there was not one single death due to vitamins. [1]

Still, the Orthomolecular Medicine News Service Editorial Board was curious: Did eleven people really die from vitamins? And if so, how?

Vitamins Not *THE* Cause of Death

In determining cause of death, AAPCC uses a four-point scale called Relative Contribution to Fatality (RCF). A rating of 1 means "Undoubtedly Responsible"; 2 means "Probably Responsible"; 3 means "Contributory"; and 4 means "Probably Not Responsible." In examining poison control data for the year 2006, listing one vitamin death, it was seen that the vitamin's Relative Contribution to Fatality (RCF) was a 4. Since a score of "4" means "Probably Not Responsible," it quite negates the claim that a person died from a vitamin in 2006.

Vitamins Not *A* Cause of Death

In the other seven years reporting one or more of the remaining ten alleged vitamin fatalities, studying the AAPCC reports reveals an absence of any RCF rating for vitamins in any of those years. If there is no Relative Contribution to Fatality at all, then the substance did not contribute to death at all.

Furthermore, in each of those remaining seven years, there is no substantiation provided to demonstrate that any vitamin was a cause of death.

If there is insufficient information about the cause of death to make a clear-cut declaration of cause, then subsequent assertions that vitamins cause deaths are not evidence-based. Although vitamin supplements have often been blamed for causing fatalities, there is no evidence to back up this allegation.

References:

1. Download any Annual Report of the American Association of Poison Control Centers from 1983-2009 free of charge at http://www.aapcc.org/dnn/NPDSPoisonData/NPDSAnnualReports.aspx The "Vitamin" category is usually near the very end of the report.

Most recent year: Bronstein AC, Spyker DA, Cantilena LR Jr, Green JL, Rumack BH, Giffin SL. 2009 Annual Report of the American Association of Poison Control Centers' National Poison Data System (NPDS): 27th Annual Report. Clinical Toxicology (2010). 48, 979-1178. The full text article is available for free download at http://www.aapcc.org/dnn/Portals/0/2009%20AR.pdf

The peer-reviewed Orthomolecular Medicine News Service is a non-profit and non-commercial informational resource.

Editorial Review Board:

Jorge R. Miranda-Massari, Pharm.D. (Puerto Rico)
Erik Paterson, M.D. (Canada)
W. Todd Penberthy, Ph.D. (USA)
Gert E. Shuitemaker, Ph.D. (Netherlands)
Jagan Nathan Vamanan, M.D. (India)

**Andrew W. Saul, Ph.D. (USA), Editor and contact person.
Email: omns@orthomolecular.org**

CHAPTER TWO

FUNDAMENTAL VALUE OF CREATIVE IDEAS

In evaluating the fundamental value of an individual's creative ideas, one must not rely solely on the preconceived acceptance of the so called good-old-boy network of experts, how many meetings the author has attended, how many organizations he belongs to, how many degrees he has accumulated, how many articles he has published or his demonstrated skills as a medico/scientific politician and con artist. Instead, I suggest that before accepting or rejecting another's new and creative ideas, there is an absolute requirement for the unbiased evaluation of the intellectual integrity of the ideas themselves as regards their inherent honesty, ascertaining their persuasive strength based on the magnitude of prior established data and the courage to judge for yourself. Anything less is cowardly pandering and appeasement of peers, which is likely to stall scientific progress, just as it has, so many times in the past. Do not stick with a popular but erroneous notion just because it is the convenient and comfortable thing to do. Do not be afraid to open your mind and to step into the future of scientific discovery.

(R.M. Howes MD, PhD, 2008)

BIAS, CHANCE & CAUSE

Epidemiology teaches that every statistical association has only 3 possible explanations: bias, chance, and cause.

However, I impugn this statement and would add to this 1) partial fabrication of data, 2) unscrupulous, self-serving manipulation of data and 3) completely made up data, i.e., bald-faced liars.

EMOD EXPOSURE

EMOD (electronically modified oxygen derivative)

Every day, our bodies take 630 quadrillion damaging oxygen hits per day

Each cell is exposed to 4 lbs superoxide/year; and up to 10-15 fold that amount (40-60 lbs/year) with exercise

According to Balz Frei, normally each cell is exposed to 10^{10} molecules of superoxide each day

(figures can vary)

$O_2^{.-}$ generated by a mitochondrion has been estimated to be as high as 10^7 $O_2^{.-}$ radicals per day. This calculates to 2.3 X 10^{16} superoxide anions produced by the body every second of every day. This huge amount of superoxide and H_2O_2 begs the question, "Just how toxic are these EMODs?" Obviously, they are not highly toxic or we would all be dead.

10^{12} O_2 molecules are consumed everyday and H_2O_2 has a constant cellular concentration of 10^{-9}-10^{-7} M. It is inconceivable and counterintuitive that oxygen or its EMODs are deadly (toxic).

NO WANTON EMOD DESTRUCTION

As opposed to earlier concepts, **ROS** interaction with proteins does not invariably lead to irreversible oxidative damage. H_2O_2 is poorly reactive: it can act as a mild oxidizing or
as a mild reducing agent, but it does not oxidize most biological molecules readily, including lipids, **DNA** and proteins,
especially in the absence of transition metal ions. (Halliwell,
2000)

A lack of oxygenation leads to a lack of oxidation, which leads
to or "allows" for disease manifestation. Our body has "an
intrinsic continually functioning oxidative defensive system"
to keep cancer from growing, plaques from aggregating, her-
petic lesions from developing and cataracts from coalescing.

**The accumulated evidence on antioxidants
and EMODs can be maddeningly contradictory.
Antioxidant vitamins are no longer about science,
they are about marketing.**

I have 316 reports, on 15 million+ subjects, showing "disappointing" results with antioxidants in preventing or reversing disease. Thus, prudent questions are, "Why take them in
the absence of a deficiency state?" "Why waste money for
marginally effective or harmful pills?" "Are the antioxidant
companies driven by market forces and not by medical science and are profiteers duping an unsuspecting public?"

Elevated levels of cellular oxidative stress (**EMODs**) represent a specific vulnerability of malignant cells and exposure
to cytotoxic drugs is known to induce apoptotic death in
cancer cells.

Chemotherapy, radiation and photodynamic therapy are all strongly prooxidant processes and depend on a constant high oxygen level and steady blood flow within a tumor.

The chemotherapeutic agents doxorubicin, mitomycin C, etoposide and cisplatin are superoxide generating agents. The anti-estrogen tamoxifen has been shown to increase EMOD induced apoptosis within carcinoma cells in vitro. The taxanes, vinca alkaloids, antifolates, and nucleoside and nucleotide analogues, generate only low levels but all drugs generate some free radicals as they induce apoptosis in cancer cells. (Conklin, 2004).

> Many vitamins and supplements classified
> as antioxidants are actually redox agents,
> meaning they act as antioxidants in some instances
> and prooxidants in others.

EMOD (redox) CONTROL
(signaling, regulation)

EMODs regulate vital pathways i.e., energy metabolism, survival/stress responses, apoptosis, inflammatory response, oxygen sensing, redox homeostasis, fertilization, survival kinase activation, ion channel regulation, apoptosis signaling, preconditioning, necrosis, inflammatory system, regulation of vascular tone, the activity of HIF (hypoxia inducible factor) etc.

One must use extreme care when altering antioxidant defenses. The redox balance is a critical aspect of all aerobic life. EMODs are signaling mechanisms for a vast range of vital metabolic pathways and networks.

Synthesis of thyroid hormones requires hydrogen peroxide for oxidation and incorporation of iodine into thyroglobulin. **DUOX** provides the hydrogen peroxide for this reaction.

EMOD SOURCES

Exponential numbers of oxygen molecules are hurriedly scurrying throughout my body and brain, shuttling down my electron transport chain and transferred by **NOX** and cytochrome P_{450} enzymes, carrying on instantaneous corporeal and cellular cross talk, protecting me and generating energy-rich **ATP**, thus, allowing me to utilize the combined prooxidant je ne sais quoi to present this lecture to you.

EMODs are formed via important biological systems, including the electron transport chain, **NADPH** oxidase, xanthine oxidase, prostaglandin synthesis, reduced riboflavin, nitric oxide synthetase, reperfusion injury, the cytochrome P450s, activated neutrophils and phagocytic cells. Outside sources of EMODs include drugs, antibiotics that depend on quinoid groups or bound metals for their action such as nitrofurantoin, and anti-cancer drugs such as doxorubicin, cisplatin, bleomycin and methotrexate, and pesticides, transition metals, tobacco smoke, alcohol, environmental radiation and high temperature, radiation treatment, inhalation of inorganic particles such as asbestos and silica, and ozone inhalation and even fever. **NOX** enzymes are even involved in the respiratory burst that occurs during fertilization.

The biggest controllable **EMOD** source of all,
behind normal breathing,
is **EXERCISE**,
which medical science has repeatedly validated
is good for us.
Antioxidants block the good effects of exercise.

All three layers of the vascular wall [intima (i.e., endothelial cells), media (i.e., smooth muscle cells), and adventitia (i.e., fibroblasts and macrophages)] express **NOX** family members.

A Great Biochemical Wonder, EMODs

It ranks amongst the greatest biochemical wonders of evolution that the most crucial and widespread small molecular EMODs, which are purportedly the alleged "enemies within" are ever present and essential occupants of the most sensitive intracellular organelle control systems for obligate aerobes. In fact, EMODs, such as hydrogen peroxide, can not be excluded from cells or their intracellular compartments.

The naïve notion that the EMODs are inherently deleterious is counterintuitive to the very basis of the concept of the Darwinian process itself and to sound evolutionary biochemical principles.

Because of public health concerns and the widespread use of antioxidant vitamins, we can no longer ignore these warnings or attempt to explain them away as so-called "paradoxes."

There are no magical antioxidant cures.

Pamela Mason, of the Health Supplements Information Service, said: "Antioxidant vitamins, like any other vitamins were never intended for the prevention of chronic disease and mortality. They are not magic bullets."

Andrew Shao, vice president for scientific and regulatory affairs Council for Responsible Nutrition, said, "Supplements are just one tool that people need to incorporate into their lifestyle to stay healthy. We can't expect just to take supplements and that's going to prevent cancer. That simply isn't the way it works."

DEFICIENCY AND OVEREXPRESSION STUDIES

Acatalasemics basically live a normal life. Humans with genetic deficiency of catalase ("acatalesemia") suffer few ill effects and genetic deletion of the catalase gene in mice is not detrimental. (Eaton, 1995)

Glutathione peroxidase 1 (GPx1 knockout mice) are grossly phenotypically normal and have a normal lifespan (Muller, 2007). CuZnSOD and GPx1 knockout mice develop normally and show no marked pathologic changes under normal physiologic conditions and no effects on animal survival under hyperoxia. (Ho, 1998)

MnSOD may be the most biologically important of the big three (CAT, SOD and GPx) since mice lacking this gene die soon after birth. Mice lacking CuZnSOD have a shortned lifespan, while EC-SOD lacking mice have minimal defects (Halliwell, 1999).

Excess scavenging of superoxide by overexpressed SOD genes in transgenic mice leads to impaired long-term potentiation, a crucial step in the generation of memory. (Thiels, 2000)

Mice that were bred to overexpress GPx to up to three times above normal developed hyperglycemia, hyperinsulinemia and elevated plasma leptin, and became 36 percent heavier and twice as fat as did control mice. (Pre-type 2 diabetes)

Acute coronary syndrome patients with high plasma levels of GPx have a worse prognosis.

Infants are born with cancers, heart disease, fatty streaked arteries, diabetes and cataracts, diseases attributed by the FRT to accumulated EMOD products. Yet, these babies have not had time for the stochastic accumulation of ROS products.

QUANTIFYING ANTIOXIDANT ACTIVITY

There are numerous discrepant, conflicting and inconsistent ways of attempting to quantify antioxidant activity using a range of lab based assays, including the ferric reducing ability of plasma (FRAP), the oxygen radical absorbance capacity (ORAC) and the Trolox equivalent antioxidant capacity (TEAC). Test tube studies may have little relevancy to the living/breathing cell.

OXIDATIVE STRESS MEASUREMENT
(markers or biomarkers)

-   measurement of lipid oxidation products such as conjugated dienes, 4-hydroxynonenal (4-HNE) levels, malondialdehyde or thiobarbituric acid reactive substances (TBARS) in tissue, blood or urine;
-   modified DNA bases (8-hydroxydeoxyguanosine) 8-oxoguanine, 8-hydroxy-2'-deoxyguanosine and/or DNA adducts, DNA breaks in peripheral blood cells or urine;
-   oxidized proteins; increased GSSG;
-   vitamin E or vitamin C levels in blood fractions;
-   catalase, Gpx or superoxide dismutase levels in blood fractions;
-   volatile gases such as pentane or ethane in expired breath;

- total peroxyl radical trapping antioxidant power of serum (TRAP assay);
- auto-oxidative (non-cyclooxygenase-derived) eicosanoids and prostanoids, 8-epiprostaglandin F2a, increased 8-iso-prostane in plasma;
- and the in vitro oxidation of blood fractions such as LDL;
- transient enhancement of heme oxygenase 1;
- ascorbate free radical, salicylate, glutathione antioxidant system, advanced oxidation protein products, ubiquino/ubiquinone ratio;

There are over 40 methods for measurement of oxidant stress or antioxidant capacity; all subject to artifactual errors, having little concordance and remaining problematic.

Those who question the FRT:

Howes, Segal, Linnane, Gems, Hekimi, Braeckman, Buffenstein

The most basic reason for the failure of the FRT is that the theory is wrong. Prooxidants are not deleterious, in normal living systems. They are crucial, low toxicity agents that sustain all aerobic life forms, provide pathogen defense and neoplasia protection. It is that simple.

We are witnessing the largest global human experiment in the history of the planet with antioxidant overkill.

29

CHAPTER THREE

Here is "the scientific evidence-based truth" exposed by Howes

The Orthomolecular Medicine News folks are good people but they have possibly overlooked a few hundred scientific facts. Certainly, the antioxidant vitamins are included in the overall category of "vitamins." So, check this section out and decide for yourself.

HARMFUL EFFECTS OF ANTIOXIDANTS by category

MORTALITY (number of deaths) (14 study reports)

1. - CARET (beta carotene and retinol) study found **28% increase in lung cancer; 26% increase in CVD** (nonsignificant); **17% increase in total mortality** among treatment group. This study was stopped 21 months earlier than planned. (Omenn et al, 1996)

2. - When beta-carotene was combined with retinol, data from a single study showed that there was a statistically significant, **increased risk of lung cancer incidence and mortality** in people with risk factors for lung cancer who took both vitamins. (Caraballoso et al., 2003)

3. - In the Women's Angiographic Vitamin and Estrogen Study (WAVE), **postmenopausal women with coronary disease on hormone replacement therapy given vitamin E plus vitamin C had an unexpected significantly higher all-cause mortality rate**

and a trend for an increased cardiovascular mortality rate. (Waters et al, 2002)

4. - The Medical Research Council/British Heart Foundation **Heart Protection Study**, which is **the study cited by Dr. Gibbons, randomized 10,269 patients to 660 IU/day of vitamin E and 10,267 to placebo control. The vitamin E group was associated with about a 10% increase in mortality.** (Gibbons quote, 2002) (MRC/BHF, 2002)

5. - **A high vitamin C intake from supplements is associated with an increased risk of cardiovascular disease mortality in postmenopausal women with diabetes.** (Iowa Women's Health study) (Lee et al, 2004)

6. - **antioxidant supplements significantly increased mortality. Beta-carotene and vitamin A and beta-carotene and vitamin E significantly increased mortality. When the selenium trials were excluded, both analyses showed a statistically significant increase in mortality, which was particularly strong in patients taking beta carotene and vitamin A.** (Cochrane Database Syst Rev.) (Bjelakovic et al, 2004)

7. - **high doses of vitamin E increased mortality.** (Miller et al., 2004)

8. - Conservatively, **the supplements increase the likelihood of dying by about 5 percent. When looked at separately, they found that Vitamin A increased death risk by 16 per cent, beta carotene by 7 per cent and Vitamin E by 4 per cent.** (Bjelakovic et al, 2007)

9. - mortality was increased in vitamin E users who had a history of stroke, coronary bypass graft surgery, or myocardial infarction and, independently, in those taking nitrates, warfarin or diuretics. (Hayden et al, 2007)

10. - antioxidant supplements **significantly increased mortality in a fixed-effect model. Vitamin A, beta-carotene, and vitamin E may increase mortality.** (Bjelakovic, Nikolova, Gludd, Simonetti and Gludd, 2008 Apr)

11. - **Antioxidant supplements had no significant effect on mortality in a random-effects model meta-analysis** but **significantly increased mortality in a fixed-effect model meta-analysis.** CONCLUSIONS: **There was no evidence that the studied antioxidant supplements prevented gastrointestinal cancers. On the contrary, they seem to increase overall mortality.** (Bjelakovic, Nikolova, Simonette and Gludd, 2008 Sept)

12. - **Beta carotene supplementation was associated with an increase in the incidence of cancer among smokers and with a trend toward increased cancer mortality.** (Bardia et al, 2008)

13. - **esophageal cancer deaths increased 14% among those aged 55 years or older. Vitamin A and zinc supplementation was associated with increased total and stroke mortality. (Qiao et al, 2009)**

14. - **Among participants with a dietary vitamin C intake above the median of 90 mg/day, vitamin E increased mortality among those aged 50-62 years by 19%.** (Hemila and Kaprio, 2009 Apr)

Yes, antioxidants can kill you.

In the 1980s, an injectable form of vitamin E, known as E-Ferol, was responsible for the deaths of 38 babies. This was verified by court hearings. Please remember that research has tragically shown that surviving infants who received E-Ferol injections were at an increased lifetime risk for reproductive problems, cervical and vaginal cancer, and other health problems.

Tragic deaths of 38 infants by lethal IV antioxidant vitamin E

Also, please remember the **fatal syndrome characterized by progressive clinical deterioration with unexplained thrombocytopenia, renal dysfunction, cholestasis, and ascites developed in certain infants throughout the United States who had received E-Ferol, an intravenous vitamin E supplement. (THE TRAGIC CASE HISTORY OF INTRAVENOUS VITAMIN E (The New York Times) May 27, 1984 By PHILIP M. BOFFEY).**

Prolonged exposure to E-Ferol was associated with progressive intralobular cholestasis, inflammation of hepatic venules, and extensive sinusoidal veno-occlusion by fibrosis. E-Ferol, contained 25 units per milliliter of dl-alpha-tocopheryl acetate solubilized with 9% polysorbate 80 and 1% polysorbate 20. They proposed that vasculocentric hepatotoxicity is the basis for the observed clinical syndrome that represents the cumulative effect of one or more of the constituents of E-Ferol. (Bove et al, 1985)

All affected infants received E-Ferol; some affected infants received up to 1 ml or more daily. **Both outbreaks ceased shortly after use of E-Ferol was discontinued. Three were jailed for selling the drug (vitamin E) that killed 38 babies.**

The Center for Drug Evaluation and Research, FDA, Rockville, Maryland, concluded that the use of E-Ferol in these neonatal intensive care units was associated with increased morbidity and mortality among exposed infants. (Arrowsmith et al, 1989)

As I mentioned above, "research has shown infants who received E-Ferol injections are at an increased lifetime risk for reproductive problems, cervical and vaginal cancer, and other health problems." This is of grave importance.

Here is another basic point: the increase in disease risk and mortality seen with the antioxidant vitamins can not be expected to increase the life span. Obviously, they would be expected to decrease or shorten the life span.

Additionally, for those that say that there have been only a "few" negative studies with the antioxidant vitamins, there are now hundreds (over 316) which I have compiled and reported on. That is, indeed, shocking!

Also, please remember the work of Drisko et al with IV mega-doses of vitamin C which resulted in 2 deaths and other adverse effects. (Padayatty et al, 2010)

Then check out the Japanese deaths from the antioxidant, clioquinol.

Antioxidant vitamin pushers are always saying, "If antioxidant vitamins are so bad, where are the dead bodies?" So, I just listed some of the dead bodies for them!

Since antioxidants increase cancer, heart disease and strokes, it is likely that there are lots of dead bodies attributable to them, but the antioxidants are not identified as the cause of death.

Additionally, the above studies show that the antioxidants increase "overall mortality," which means they increase your chances of dying, which contributes to more dead bodies!

The harmful effects of antioxidant overloading are insidious and not immediate, as with cyanide. But, they are just as real.

CANCER (34 study reports)

1. - ATBC (alpha tocopherol and beta carotene) study found **50% increase in hemorrhagic stroke deaths among vitamin E group; 11% increase in ischemic heart disease deaths among β-carotene group; 18% increase in lung cancer among β-carotene group.** (Heinonen et al, 1994, ATBC)

2. - CARET (beta carotene and retinol) study found **28% increase in lung cancer; 26% increase in CVD** (nonsignificant); **17% increase in total mortality** among treatment group. This study was stopped 21 months earlier than planned. (Omenn et al, 1996, CARET)

3. - **Positive association between retinol and advanced prostate cancer.** (Anderssen et al. 1996)

4. - **beta-Carotene supplementation was associated with increased lung cancer risk. beta-Carotene supplementation at pharmacologic levels may modestly increase lung cancer incidence in cigarette smokers.** (Albanes et al, 1996)

5. - the **possibility of increased risk of breast cancer in women taking folic acid supplements throughout pregnancy. The risk of death from breast cancer was much higher in women who had received high doses of the supplement.** (Giovannucci et al, 1998)

6. - Multivitamin and combination use had minimal effect on cancer mortality overall, **although mortality from all cancers combined was increased among male current smokers who used multivitamins alone or in combination with vitamin A, C, or E.** (Watkins et al, 2000)

7. - **point estimates suggested a possible decrease in second head and neck cancer risk but a possible increase in lung cancer risk.** (Mayne et al, 2001)

8. - **Other subjects, with normal or higher selenium levels, did not benefit and may have an increased risk for prostate cancer.** Vitamin E supplements in higher doses (> or =100 IU) were also associated with a higher risk of aggressive or fatal prostate cancer in nonsmokers from a past prospective study. (Moyad et al. 2002)

9. - **Breast cancer–specific survival (i.e., patients censored only at death from breast cancer) and disease-free survival were shorter in the nutrient-supplemented group.** (Lesperance et al, 2002)

10. - When beta-carotene was combined with retinol, data from a single study showed that there was a statistically significant, **increased risk of lung cancer incidence and mortality** in people with risk factors for lung cancer who took both vitamins. (Caraballoso et al., 2003)

11. - **For participants who smoked cigarettes and also drank more than one alcoholic drink per day, beta carotene doubled the risk of adenoma recurrence.** (Baron et al, 2003)

12. - **selenium supplementation was associated with statistically significantly elevated risk of squamous cell carcinoma** and of total nonmelanoma skin cancer. (Duffield-Lillico, 2003)

13. - **Regular multivitamin use was associated with a small increase in prostate cancer death rates.** (Stevens et al, 2005)

14. - **patients receiving alpha-tocopherol supplements had a higher rate of second primary cancers during the supplementation period. alpha-Tocopherol supplementation produced unexpected adverse effects on the occurrence of second primary cancers and on cancer-free survival.** Patients taking an antioxidant were 1.65 times more likely to suffer a return of their original cancer during the three years they were on the supplement.

The risk was highest among those taking only vitamin E (1.86 times higher). (Bairati et al, 2005 Apr 6)

15. - supplementation with beta-carotene was discontinued because of ethical concerns. **Quality of life was not improved by the supplementation. The rate of local recurrence of the head and neck tumor tended to be higher in the supplement arm of the trial. The rate of local recurrence of the head and neck tumor tended to be higher in the supplement arm of the trial. This trial suggests that use of high doses of anti-oxidants as adjuvant therapy might compromise radiation treatment efficacy.** (Bairati et al, 2005 Aug 20)

16. - **Cigarette smoking, alcohol drinking and green tea consumption were significantly associated with an increased risk of esophageal cancer. The population attributable fractions of esophageal cancer incidence that was attributable to smoking, alcohol drinking and green tea consumption were 72.0%, 48.6%, and 22.1%, respectively.** (Ishikawa et al, 2006)

17. - **A trend for increased risk of oral pre-malignant lesions was observed with vitamin E, especially among current smokers and with vitamin E supplements. Beta-carotene also increased the risk among current smokers.** (Maserejian et al, 2007)

18. - **In women, the incidence of skin cancer (SC) was higher in the antioxidant group and the incidence of melanoma was also four-fold higher in the antioxidant group for women.** (Hercberg et al, 2007, VITAL)

19. - **esophageal cancer excess was found with long-term follow-up of older Chinese patients (the Linxian study by Blot et al.) treated with selenium, β-carotene, and vita-**

min E supplements (Blot et al, 1993) (NIH State-of-the Science Panel. 2007)

20. - **Supplemental vitamin E was associated with a small increased risk of lung cancer.** (Slatore et al, 2008)

21. - **Beta carotene supplementation was associated with an increase in the incidence of cancer among smokers and with a trend toward increased cancer mortality.** (Bardia et al, 2008)

22. - **Regular use of multivitamin-multimineral supplements may be associated with higher mean breast density among premenopausal women.** (Berube et al, 2008)

23. - **There were statistically nonsignificant increased risks of prostate cancer in the vitamin E group** but not in the selenium + vitamin E group. **The trial was stopped ahead of its original 12 year deadline because of a lack of any noticeable benefit.** (Lippman et al, 2009, SELECT)

24. - **Antioxidant supplement use is associated with melanoma risk in light of recently published data from the Supplementation in Vitamins and Mineral Antioxidants (SUVIMAX) study, which reported a 4-fold higher melanoma risk in women.** (Asgari et al, 2009) referred to data from the SUVIMAX study and not Asgari's study.

25. - **Men who used zinc for ten years or more, either in a multivitamin or as a supplement, had an approximately two-fold increased risk of prostate cancer.** (Zhang et al. 2009)

26. - **Prostate cancer over a 10-year period was 9.7% in the folic acid group and 3.3% in the placebo group.** (Figueiredo et al, 2009)

27. - **esophageal cancer deaths increased 14% among those aged 55 years or older. Vitamin A and zinc supplementation was associated with increased total and stroke mortality.** (Qiao et al, 2009)

28. - **Longer duration of use of individual beta-carotene, retinol, and lutein supplements** (but not total 10-year average dose) **was associated with statistically significantly elevated risk of total lung cancer and histologic cell types. (VITAL) study.** (Satia et al, 2009)

29. - **In the meta-analyses of the recent cohort studies, the highest green tea consumption was shown to significantly increase stomach cancer risk using the crude data, but no significant association between them was seen when using the adjusted data.** (Myung, Int J Cancer. et al, 2009)

30. - **there was limited to moderate evidence that the consumption of green tea reduced the risk of lung cancer, especially in men, and urinary bladder cancer or that it could even increase the risk of the latter.** (Boehm et al, 2009)

31. - **Multivitamin use was associated with a statistically significant increased risk of breast cancer.** Use of **multivitamins was linked to a statistically significant 19 per cent increased risk of breast cancer.** (Larsson et al, 2010)

32. - **quercetin and ferulic acid are able to aggravate, if not induce, nephrocarcinoma in mice.** (Chiu-Lan Hsieh et al, 2010)

33. - **antioxidant administrations after the tumor's appearance accelerated tumor growth and favored metastases.** Continuous administration of pequi oil inhibited the tumor's growth, while the same protocol with **vitamins E and C accelerated it (tumor growth),** favoring metastasis and increasing oxidative stress on erythrocytes. (Miranda-Vilela AL, et al, 2011)

34. - In vitro exposure to **genistein/quercetin induced higher numbers of MII rearrangements in bone marrow cells of Atm-ΔSRI mutant mice compared with wt mice.** Prenatal **exposure to flavonoids associated with higher frequencies of MII rearrangements and a slight increase in the incidence of malignancies in DNA repair-deficient mice.** These data suggest that **prenatal exposure to both genistein and quercetin supplements could increase the risk on MII rearrangements** especially in the presence of compromised DNA repair. (Vanhees et al, 2011)

HEART DISEASE (19 study reports)

1. - ATBC (alpha tocopherol and beta carotene) study found **50% increase in hemorrhagic stroke deaths** among vitamin E group; **11% increase in ischemic heart disease deaths among β-carotene group; 18% increase in lung cancer among β-carotene group.** (Heinonen et al, 1994)

2. - ATBC study found **significantly more deaths from fatal coronary heart disease in the beta-carotene and combined alpha-tocopherol and beta-carotene groups.** (Rapola et al, 1997)

3. - the **Atherosclerosis Risk In Communities (ARIC)** group found individuals with **the highest carotid IMT to have lower levels of plasma carotenoids but higher alpha-tocopherol and retinol levels compared to controls** (Iribarren et al, 1997)

4. - **vitamin E decreased platelet function.** (Calzada et al, 1997)

5. - **Probucol has been pulled off the market due harmful effects and the likelihood of cardiac arrhythmias.** (Tardif et al, 1997)

6. - **homozygous familial hypercholesterolemia, intima-to-media thickness (FH IMT study) increased with vitamin E supplements.** (Raal et al, 1999)

7. - **When used in combination with simvastatin/niacin, antioxidants negated the benefit of the latter on plasma lipid profile and stenosis progression.** (Brown et al, 2001)

8. - In the Women's Angiographic Vitamin and Estrogen Study (WAVE), **postmenopausal women with coronary disease on hormone replacement therapy given vitamin E plus vitamin C had an unexpected significantly higher all-cause mortality rate and a trend for an increased cardiovascular mortality rate.** (Waters et al, 2002)

9. - **Men with high plasma gamma-tocopherol levels had an increased risk of nonfatal and fatal myocardial infarction (MI).** (Hak et al. 2003)

10. - **A high vitamin C intake from supplements is associated with an increased risk of cardiovascular disease mortality in postmenopausal women with diabetes.** (Iowa Women's Health study) (Lee et al, 2004)

11. -**beta-Carotene seemed to increase the post-trial risk of first-ever non-fatal MI.** (Thornwall et al., 2004)

12. - **antioxidant therapy (1,000 mg/day of vitamin C + 800 IU/day of vitamin E) was associated with improvement of coronary atherosclerosis in diabetic women with two copies of the haptoglobin 1 gene but worsening of coronary atherosclerosis in those with two copies of the haptoglobin 2 gene.** (Levy et al, 2004)

13. - **Another subgroup finding in HOPE-TOO was a vitamin E–associated increased risk of heart failure incidence that**

appeared in a secondary end point analysis in the 4.5-year report and persisted in the 7-year extended follow-up, as did the risk of hospitalization for heart failure. Patients in the vitamin E group had a higher risk of heart failure and hospitalization for heart failure. (Lonn et al, 2005)

14. - the rate of low-birth-weight babies was higher and the rate for gestational hypertension was higher for women in the vitamin group. Women in the vitamin group had an increased risk of being hospitalized antenatally for hypertension and having to take antihypertensive medication. In addition, a subgroup of women in the vitamin group had a higher frequency of abnormal liver-function tests. (Rumbold et al, 2006)

15. - The mean increase in intima-media thickness over time in the vitamin E group was 0.0041 mm/year faster than placebo. (Magliano et al, 2006)

16. - Among patients without chronic hypertension, there was a slightly higher rate of severe preeclampsia in the study (vitamin C and E) group. (Spinnato et al. 2007)

17. - Treatment with either alpha- or mixed tocopherols significantly increased BP, pulse pressure and heart rate in individuals with type 2 diabetes (Ward et al, 2007)

18. - There was a small, though significant increase in postoperative blood loss after cardiac surgery among patients treated with NAC. (Naughton et al, 2008)

19. - elevated serum vitamin E concentrations are associated with abnormal markers of atherosclerosis and may increase the risk of cardiovascular complications in HIV-infected adults. (Falcone et al, 2010)

STROKE (5 study reports)

1. - ATBC (alpha tocopherol and beta carotene) study found **50% increase in hemorrhagic stroke deaths among vitamin E group; 11% increase in ischemic heart disease deaths among β-carotene group; 18% increase in lung cancer among β-carotene group.** (Heinonen et al, 1994)

2. - **Alpha tocopherol increased risk of subarachnoid hemorrhage 50% and beta carotene increased risk of intra-cerebral hemorrhage 62% and alpha tocopherol increased risk of fatal subarachnoid hemorrhage 181%** *(Leppala et al. 2000)*

3. - **Vitamin E was associated with an increased risk of hemorrhagic stroke** (Sesso et al, 2008)

4. - **esophageal cancer deaths increased 14% among those aged 55 years or older. Vitamin A and zinc supplementation was associated with increased total and stroke mortality.** (Qiao et al, 2009)

5. - **Vitamin E increased the risk for hemorrhagic stroke by 22%** and reduced the risk of ischemic stroke by 10%. (**indiscriminate widespread use of vitamin E should be cautioned against**) (Schurks et al. 2010)

LUNG FUNCTION (3 study reports)

1. - **Vitamin E positively associated with productive cough. Beta carotene associated with wheezing** (Grievink et al. 1998)

2. - **Vitamin E intake showed a positive association with productive cough. The intake of beta-carotene had a positive association with wheeze. (Grievink et al. 1998)**

3. - **Early vitamin supplementation (multivitamins) is associated with increased risk for asthma in black children and
food allergies in exclusively formula-fed children** (**Milner**
et al. 2004)

DIABETES RELATED (6 study reports)

1. - **Vitamin E associated with prediabetic changes in glucose
metabolism** (Skrha et al. 1997)

2. - **Vitamin C associated with increased severity of diabetic
retinopathy. For those taking insulin, beta carotene and vitamin E was associated with increased severity of diabetic retinopathy** (Mayer-Davis et al. 1998)

3. - **An increase over time in vitamin C intake from the first to
ninth deciles was associated with a risk for increased severity of diabetic retinopathy (DR). Increased intake of vitamin
E was associated with increased severity of DR among those
not taking insulin. Among those taking insulin, increased
intake of beta-carotene was associated with a risk for severity of diabetic retinopathy (DR).** (Mayer-Davis et al. 1998)

4. - **A high vitamin C intake from supplements is associated
with an increased risk of cardiovascular disease mortality in
postmenopausal women with diabetes.** (Iowa Women's Health
study) (Lee et al, 2004)

5. - **antioxidant therapy (1,000 mg/day of vitamin C + 800 IU/
day of vitamin E) was associated with improvement of coronary atherosclerosis in diabetic women with two copies of the
haptoglobin 1 gene but worsening of coronary atherosclerosis
in those with two copies of the haptoglobin 2 gene.** (Levy et al,
2004)

6. - **Treatment with either alpha- or mixed tocopherols significantly increased BP, pulse pressure and heart rate in individuals with type 2 diabetes** (Ward et al, 2007)

PRETERM BIRTH & Pregnancy
(7 study reports)

1. - **Women supplemented with vitamin C alone or combined with other supplements were at increased risk of giving birth preterm.** (Rumbold et al, Apr 18, 2005. CD004072)

2. - **the rate of low-birth-weight babies was higher and the rate for gestational hypertension was higher for women in the vitamin group. Women in the vitamin group had an increased risk of being hospitalized antenatally for hypertension and having to take antihypertensive medication.** In addition, **a subgroup of women in the vitamin group had a higher frequency of abnormal liver-function tests.** (Rumbold et al, 2006)

3. - Concomitant supplementation with vitamin C and vitamin E does not prevent pre-eclampsia in women at risk, but does **increase the rate of babies born with a low birth weight.** (Poston et al., 2006)

4. - **the rate of low-birth-weight babies was higher and the rate for gestational hypertension was higher for women in the vitamin group. Women in the vitamin group had an increased risk of being hospitalized antenatally for hypertension and having to take antihypertensive medication.** In addition, **a subgroup of women in the vitamin group had a higher frequency of abnormal liver-function tests.** (Rumbold et al, 2006)

5. - **Among patients without chronic hypertension, there was a slightly higher rate of severe preeclampsia in the study (vitamin C and E) group.** (Spinnato et al. 2007)

6. - **Preterm mothers had higher vitamin C concentrations.** (Joshi et al. 2008)

7. - **Concomitant supplementation of vitamins C and E during pregnancy has been reportedly associated with low birth weight, the premature rupture of membranes and fetal loss or perinatal death in women at risk for pre-eclampsia. These changes may probably lead to the impairment of placental function and suboptimal growth of the fetus** (Hung et al, 2010)

INFECTIONS (2 study reports)

1. - **Among participants who obtained 90 mg/d or more of vitamin C in foods, vitamin E supplementation increased tuberculosis risk by 72%. Our finding that vitamin E seemed to transiently increase the risk of tuberculosis in those who smoked heavily and had high dietary vitamin C intake should increase caution towards vitamin E supplementation for improving the immune system.** (Hemila and Kaprio, 2008 Oct)

2. - **They had found a 14% higher incidence of pneumonia with vitamin E supplementation in a subgroup of the Alpha-Tocopherol Beta-Carotene Cancer Prevention (ATBC) Study cohort: Vitamin E increased the risk of pneumonia in participants with body weight less than 60 kg, and in participants with body weight over 100 kg.** (Hemila and Kaprio, 2008 Nov)

EXERCISE (3 study reports)

1. - **The administration of vitamin C significantly (P=0.014) hampered endurance capacity.** (Gomez-Cabrera et al, 2008)

2. - **Antioxidant supplementation may actually interfere with the cellular signaling functions of ROS, thereby adversely affecting muscle performance. Since the potential for long-term harm does exist, the casual use of high doses of anti-oxidants by athletes and others should perhaps be curtailed.** (McGinley et al. 2009)

3. - **Exercise increased parameters of insulin sensitivity (GIR and plasma adiponectin) only in the absence of antioxidants in both previously untrained (P < 0.001) and pretrained (P < 0.001) individuals.** (Ristow et al, 2009)

EYE DISEASES (3 study reports)

1. - **antioxidant enzymes might be implicated in the etiology of cataract.** (Delcourt et al, 2003) (POLA study)

2. - **Higher beta-carotene intake was associated with an increased risk of age-related macular degeneration (AMD).** (Tan et al. 2008)

3. - **Among women aged 65 y and over, vitamin C supplement use increased the risk of cataract by 38% (95% CI: 12%, 69%). Vitamin C use among hormone replacement therapy users compared with that among nonusers of supplements or of hormone replacement therapy was associated with a 56% increased risk of cataract.** (Rautiainen et al, 2010)

SKELETAL (5 study reports)

1. - increasing retinol became negatively associated with skeletal health at intakes not far beyond the recommended daily allowance **(RDA)**, intakes reached predominately by supplement users. the **Rancho Bernardo Study.** (Promislow et al, 2002)

2. - women in the highest quintile of total vitamin **A** intake (≥3000 µg/d of retinol equivalents **[RE]**) had a significantly elevated relative risk **(RR)** of hip fracture. (Feskanich et al, 2002)

3. - **The risk of fracture was highest among men with the highest levels of serum retinol. The risk of fracture was further increased within the highest quintile for serum retinol. Men with retinol levels in the 99th percentile (>103.12 µg per deciliter [3.60 µmol per liter]) had an overall risk of fracture that exceeded the risk among men with lower levels by a factor of seven.** (Michealsson et al, 2003)

4. - **Patients with** contrast-induced nephropathy (CIN) **and pre-existing renal insufficiency had worse clinical outcomes with acetyl cysteine (NAC).** (Boccalandro et al. 2003)

5. - **a modest increase in total fracture risk with high vitamin A and retinol intakes was observed in the low vitamin D-intake group.** (Caire-Juvera et al. 2009) **(Women's Health Initiative Observational Study)**

MISCELLANEOUS (9 study reports)

1. - isotretinoin is **associated with significant adverse systemic effects. (adverse mucocutaneous effects and serum triglyceride elevations)** (Tangrea et al, 1992, 1993)

2. - **acetyl cysteine or ascorbate and alpha tocopherol together induced further DNA damage to human sperm.** (Hughes et al, 1998)

3. - **sperm motility was significantly decreased by antioxidants vitamin C and alpha-tocopherol.** (Donnelly et al, Fertil Steril. 1999)

4. - **both ascorbate and alpha-tocopherol in combination to sperm preparation medium actually induced DNA damage and intensified the damage induced by H_2O_2.** (Donnelly et al, Mutagenesis. 1999)

5. - **Patients with** contrast-induced nephropathy (CIN) **and pre-existing renal insufficiency had worse clinical outcomes with acetyl cysteine (NAC).** (Boccalandro et al. 2003)

6. - **Vitamin E increased the risk of epistaxis** (Lee et al. 2005)

7. - **the trial in Tanzania found evidence that vitamin A supplementation increased the risk of mother to child transfer (MTCT) of HIV.** (Wiysonge et al, 2005)

8. - **More Alzheimer's participants taking Vitamin E suffered a fall.** (Isaac et al, 2008)

9. - **This "disappointing" study** warns that **indiscriminate use of high-dose Vitamin E supplementation does more harm than good.** (Dotan et al, 2009)

In summary, some antioxidant study reports have selectively shown the following:

Antioxidants have increased mortality by as high as 19% and 17%.

Antioxidants have increased lung cancer by 28% and 18%, esophageal cancer deaths by 14% and 22.1%, breast cancer by 19%, hemorrhagic stroke deaths by 50%, ischemic heart disease by 11%, and cardiovascular disease by 18%.

Antioxidants have increased prostate cancer deaths, elevated the risk of squamous cell carcinoma, doubled the risk of adenoma recurrence, increased the rate of second primary cancers, and increased recurrence of head and neck tumors.

Antioxidants increased the incidence of melanoma in women by 4 fold. (Hercberg et al, 2007, VITAL)

Patients taking an antioxidant were 1.65 times more likely to suffer a return of their original cancer.

Three major studies were stopped early due to unexpected harmful outcomes and adverse effects for study participants: ATBC, CARET and SELECT. (Heinonen et al, 1994, ATBC) (Omenn et al, 1996, CARET) (Lippman et al, 2009, SELECT)

Antioxidants have increased ischemic heart disease, deaths from fatal coronary heart disease, increased risk of nonfatal and fatal myocardial infarction, decreased platelet function, increased intima-to-media thickness, negated statin effects, increased risk of hospitalization for heart failure and hypertension, altered liver function tests, increased blood pressure, and increased blood loss after cardiac surgery.

Antioxidants have increased hemorrhagic stroke deaths as much as 22% to 50%, increased risk of subarachnoid hemorrhage 50% and increased risk of intracerebral hemorrhage 62% and increased risk of fatal subarachnoid hemorrhage 181%.

Antioxidants have increased wheezing, productive coughs, and risk of asthma.

In diabetics, antioxidants have increased pre-diabetic changes in glucose metabolism, increased severity of diabetic retinopathy, and increased blood pressure.

Antioxidant have increased preterm deliveries, premature rupture of membranes and low birth weight, increased gestational hypertension, increased risk of hospitalization, altered liver function tests, increased risk for severe preeclampsia, and fetal loss or perinatal death in women at risk for preeclampsia.

Antioxidants have increased risk of tuberculosis by 72% and pneumonia by 14%, indicating an altered immune system.

Antioxidants have adversely affected muscle performance and hampered endurance capacity.

Antioxidants increased risk of cataracts by 38% and by 56% in hormone replacement users, and increased risk of age-related macular degeneration (AMD).

Antioxidants increased the risk of hip fractures, damaged sperm DNA, increased epistaxis and mother to child transfer of HIV.

Other than that....

Now, you can make up your own mind.

CHAPTER FOUR

MAJOR MEDICAL AND SCIENTIFIC ORGANIZATIONS WHICH DO NOT RECOMMEND THE USE OF ANTIOXIDANT VITAMINS

Many health-related committees and major health organizations do not recommend that individuals take these vitamin and antioxidant supplements, in the absence of a vitamin deficiency. (Howes R.M. 2009, Am J Cosm Surg)

THE FOLLOWING LIST PROVIDES THE MAJOR MEDICAL AND SCIENTIFIC ORGANIZATIONS WHICH DO NOT RECOMMEND THE USE OF ANTIOXIDANT VITAMINS

The following either do not recommend antioxidant vitamins or have found inconclusive evidence of their benefit:

- The U.S. Food and Drug Administration (FDA)
- The American Heart Association (AHA)
- The American Cancer Society (ACS)
- The National Cancer Institute (NCI)
- Institute of Medicine of the National Academies
- The American College of Cardiology
- The American College of Chest Physicians (ACCP)
- The American Diabetes Association
- The American Academy of Family Physicians

- Scientific Statement From the American Heart Association and the American Diabetes Association
- The American College of Cardiology/American Heart Association Task Force on Practice Guidelines
- United States Preventive Services Task Force (USPSTF)
- The American Cancer Society Guidelines on Nutrition and Physical Activity for Cancer Prevention
- The Nutrition Committee of the American Heart Association Council on Nutrition, Physical Activity, and Metabolism
- The AHA Scientific Position of the American Heart Association
- The Canadian Task Force on Preventive Health Care (CTFPHC)
- Food and Nutrition Board, Institute of Medicine
- The Food and Nutrition Board of the National Academy of Sciences
- National Academy of Sciences
- The 2006 AHA Diet and Lifestyle Recommendations
- The Medical Letter
- The Oregon Health and Science University
- Food Standards Agency/ the British Nutrition Foundation (BNF)
- Quackwatch
- American College of Cardiology Foundation Task Force on Clinical Expert Consensus Documents
- National Institutes of Health State-of-the-Science Conference
- The American Heart Association Atherosclerosis, Hypertension, and Obesity in Youth Committee, Council of Cardiovascular Disease in the Young, With the Council on Cardiovascular Nursing
- The Physicians Health Study
- The 2008 VITAmins and Lifestyle (VITAL) study
- The Physicians' Health Study II Randomized Controlled Trial

- **The Swedish Council of Technology Assessment**
- **National Heart Foundation of Australia's Nutrition and Metabolism Advisory Committee**

Although their conclusions are not iron clad, many prestigious scientific organizations have concluded that, "taking antioxidant vitamins - such as vitamins A, C and E - serves no purpose, and in some cases could likely be harmful."

Such a list is rather astounding because broadcast media presents a never-ending cycle of advertisements pushing the wonders of antioxidants and antioxidant vitamins. One would assume that such advertisements would have the backing of major medical and scientific organizations, but that is not the case.

The above 32 conclusions or recommendations are apparently some of the best kept secrets in America, since antioxidants are being fortified or added to a wide spectrum of commercial products including foods, cosmetics, dermatologics, pet products, beverages, energy drinks, energy bars, fruits drinks, fruit juices, chewing gum, shampoos, etc. **Genetic engineers are hurriedly creating "super foods," which will be "antioxidant-rich."** (Howes R.M. 2009, Am J Cosm Surg)

My books, *Death In Small Doses? Antioxidant Overkill, Dangers Of Excessive Antioxidant In Cancer Patients* and *Heart Disease and Antioxidant Failures*, address the problems of antioxidant overloading.

EMODs (electronically modified oxygen derivatives) are not as harmful as you might've imagined, not as destructive as you might've thought and not the "inner enemy" you might've feared. In fact, your life depends on them.
R. M. Howes, M.D., Ph.D.
5/28/10

I am becoming **"anti-antioxidant"** in my approach to disease prevention and cure, other than their serving as co-oxidants or preoxidants.

CHAPTER FIVE

Prologue

Just as with chemotherapy in elderly patients, antioxidants should be used only if they are supported by evidence from clinical trials or by clinical practice guidelines. Unfortunately, clinical trials have shown that antioxidants have been a failure in the predictable prevention or cure of disease.

Antioxidants have insufficient data to support their use or have been shown to be ineffective or have not been supported by the data or by a compelling rationale!

People who received these antioxidants were exposed to significant risk without proven benefits at an estimated cost of billions per year just for the antioxidant-containing supplements.

People with conditions that do no respond to standard therapies should either be looking for clinical trials, where there is a chance for antioxidant benefit, or should have been thinking about shifting toward a more nutritious diet or increasing exercise as tolerated.

People should not face the risks, discomforts and costs of antioxidants, which can be quite harmful and have not provided a benefit in previous studies.

Antioxidants been subjected to the rigorous scrutiny given to research published in established medical journals.

We must continue to raise the question, "How damaging are EMODs in the living/breathing cell? To me, the answer is obvious. EMODs are omnipresent and ubiquitous and are therefore of low toxicity, if not a necessity, in the respiring cell.

Also, we must continue to ask, "How dangerous are antioxidants in the respiring cell?" That answer is also rather obvious. I have accumulated 90 study reports showing their harmful potential. I have repeatedly pointed out the dangers of excessive antioxidants, such as cholesterol, uric acid, estrogen, testosterone, lactic acid, and hypervitaminosis.

What else do you need to know?

Cancer survivors

There are nearly 12 million cancer survivors in the United States and an additional 4,000 Americans join those ranks every day and 28 million people around the world are living with cancer and as cancer survivors today.

Tragically, a large percentage of these survivors are injudiciously and carelessly taking antioxidant-containing supplements on a daily basis. They are inadvertently and unknowingly shielding their cancer cells from being killed oxidatively by chemotherapy, photodynamic therapy and/or radiation therapy. They are being misled to block the specific intent of their anti-cancer therapy. This is wrong!

They must be informed of the potentially disastrous consequences of antioxidant ingestion in premalignant cancerous conditions or by current cancer victims and cancer survivors. Failure to do so is totally unacceptable and negligent.

The current scientific evidence clearly shows the dangers of excessive antioxidants in cancer patients. In the end, we must acknowledge and heed the scientific evidence, as opposed to following the misleading marketing campaigns of those profiting from antioxidant-containing supplement sales!

Tell the cancer survivors the scientific truth. Then, and only then can they make intelligent decisions regarding ingestion of the potentially dangerous antioxidant-containing supplements.

Any less goes against the very bedrock, Hippocratic precept of the practice of medicine: Physician, "first, do no harm!"

Human cancer cell types (27 human & 9 murine) shielded by antioxidants

If you are worried about cancer, just take a look at the human cell types that are protected by antioxidants from cell death in lab experiments.

Unbelievably, there are twenty seven (27) types of human cancer cell types and nine (9) murine cancer cell types that can be killed by EMODs and in which the killing can be blocked by antioxidants, thereby providing antioxidant protection and shielding of the cancer cells. Published data has shown that antioxidants blocked the killing of the following human and murine (rodent) cancer cell types by EMODs:

- **human breast cancer** (J. Nutr. 134, 2004) (Gundimeda et al, 1996) (Peralta et al, 2006) (Aykin-Burns et al, 2009) (Xiao et al, Mol Cancer Ther. 2006)
- **human prostate carcinoma** (Xiao et al, 2006) (Wu et al, 2005) (Singh et al, 2005) (Cho et al, 2005) (Milanesa et al, 2000)
- **human non-small cell lung cancer** (Ling et al, 2003) (Wu et al, 2006)
- **human colon adenocarcinoma** (Wenzel et al, 2005)

- **human colon cancer** (Wenzel et al, 2004) (Aykin-Burns et al, 2009)
- **human colorectal carcinoma** (Chen et al, 2004) (Gali-Muhtasib et al, 2008)
- **human ovarian cancer cells** (Pak et al, 2011)
- **human melanoma** (Marcin et al, 2005) (Okroj et al, 2006) (Nishikawa et al, 2004) (Grimm et al, 2011)
- **human metastatic melanoma** (Kirshner et al, 2008)
- **human head and neck cancer** (Mattson et al, 2009)
- **human lymphoma** (J. Nutr. 134, 2004) (Mansat-De Mas et al, 1999)
- **human leukemia** (Hileman et al, 2004) (McKallip et al, 2006) (Hou et al, 2005) (Feng et al, 2007) (Yedjou et al, 2008) (Hiraoka et al, 1998)
- **human hepatoma** (Wu et al, 2004) (Wu, Ng, Lin, 2004)
- **human hepatocellular liver carcinoma** (Shimoda et al, 2003)
- **human pancreatic cancer** (Maehara et al, 2004)
- **human multiple myeloma** (Grad et al, 2001) (Ahmad et al, 1997) (Gupta et al, 2000) (Nakazato et al, 2005) (Isham et al, 2007)
- **Burkitt's lymphoma** (Ahmad et al, 1997) (Gupta et al, 2000) (Nakazato et al, 2005) (Ahmad et al, 1997)
- **human chronic lymphocytic leukemia** (Kay, 2006) (Chandra et al, 2003) (Shanafelt et al, 2005) (Mow et al, 2002) (Biswas S, et al, 2010)
- **human acute myeloid leukemia** (Kay, 2006) (Chandra et al, 2003) (Shanafelt et al, 2005) (Mow et al, 2002)
- **human promyelocytic leukemia** (Hou et al, 2005)
- **human erythromyeloid leukemia** (Wagner et al, 2000)
- **human epithelial cancer cells** (breast and colon) (Aykin-Burns et al, 2009)
- **human endometrial cancer** (Llobet et al, 2008)
- **human bladder cancer cells** (Miyajima et al, 1999)
- **human invasive bladder cancer** (Miyajima et al, 1999 - human bladder cancer KU-1 cell line)
- **human glioblastoma cells** (Lee et al, 2004)
- **human osteosarcoma** (Ahmad et al, 2005)
- **murine pheochromocytoma** (Jang, Surh, 2001)

- **murine retinoblastoma** (Salganik et al, 2000)
- **murine thymoma** (Tome et al, 2001)
- **murine lymphoma- six cell types** (Nathan et al, vol 153, 1981)
- **murine leukemia** (Wagner et al, 1996)
- **murine fibrosarcoma** (Teicher et al, 1994)
- **murine neuroblastoma** (Prasad et al, PNAS. 1979)
- **murine mammary cancer** (Bracke et al, 1999)
- **murine brain cancer** (Zeisel (2), 2004)

Part of this list was compiled in 2009 and has been updated for this book. (Howes, Philica. Feb 7, 2009). **It is evident to many investigators that the in vitro apoptogenic agents function as pro-oxidants.** (Hail et al, 2008). Note: References for this list are available in my book, *Dangers of Excessive Antioxidants In Cancer Patients.*

"Antioxidant" is the buzz word

The buzz word in supplement and neutraceutical sales has been "antioxidants" for the past two decades and antioxidants has been promoted as a virtual "fountain of youth and adjunct to healthy living."

Over half of the American population, cancer survivors and the medical profession have hitched a ride on the antioxidant bandwagon.

Foreceful marketing has placed antioxidants in products ranging from vitamin water to energy drinks, pizza dough to cake mix, from Rice Crispies to V8 juice, from doughnuts to dog food and from bread to bubble gum.

By simply popping an antioxidant pill, one could allegeldy protect themselves from cancer, heart disease, strokes, diabetes, arthtitis, cataracts, macular degeneration, Alzheimer's disease, Parkinson's disease, COPD, hypertension, eclampsia, low birth weight, and on and on and on.

Miraculous qualities were associated with vitamin A, vitamin C, vitamin E, beta carotene, selenium, CoQ10, pomegranate juice, grape seed extract, pycnogenol, EGCG, ginko biloba, or anything else which could be misconstrued as being an antioxidant or having antioxidant properties.

Biased scientific studies were the rule of the day and observational and epidemiological studies were fanning the antioxidant-hyped fires of confusing and inconsistent data. The antioxidant bandwagon had been transformed into a run away train of nonsense/ nonscience.

However, the arrival of randomized controlled trials (RCTs) in the late 1990s was destined to derail that run away antioxidant train. It is now apparent that antioxidant proponents jumped on the wrong train and were carried into the realm of harmful antioxidant adverse effects. Real science has put a damper on the unjustified excitement aurrounding antioxidants and called into question widespread unsupported advertising and forceful marketing.

In October 2007, an online survey was administered to 900 physicians and 277 nurses by Ipsos Public Affairs for the Council for Responsible Nutrition (CRN), a trade association representing the dietary supplement industry. The "Life...supplemented" Healthcare Professionals Impact Study (HCP Impact Study) found that **72% of physicians and 89% of nurses in this sample used dietary supplements regularly, occasionally, or seasonally. Regular use of dietary supplements was reported by 51% of physicians and 59% of nurses. The most common reason given for using dietary supplements was for overall health and wellness (40% of physicians and 48% of nurses),** but more than two-thirds cited more than one reason for using the products. **When asked whether they "ever recommend dietary supplements" to their patients, 79% of physicians and 82% of nurses said they did.** (Dickinson et al, 2009).

CHAPTER SIX

"DISAPPOINTING" RESULTS

During my lifetime as a scientist and medical professional, I have never seen a more prejudicial era for conducting so-called scientific investigations. The outcome of hundreds of studies were reported as being "disappointing," when antioxidants failed to prevent, reverse or cure the myriad of diseases that oxygen free radicals were supposed to have caused.

Let me give you just a few examples of so-called "disappointing results."

- In spite of this, the results of interventional (antioxidant) supplementation studies in established disease have been **disappointing.** (Devereux, Seaton, 2005)

- Overall, early, promising results have grown increasingly inconsistent over time. The promise of early epidemiological studies on carotenoids and cardiovascular disease paved the way to largely **disappointing** results from several large prevention trials of beta-carotene. (Sesso, 2006)

- Recent meta-analyses of the clinical studies with vitamin E were rather **disappointing**. (Munteanu, Zingg, 2007)

- The outcome of recent large, prospective, randomized and placebo-controlled clinical studies does not encourage the use of vitamin E supplements. These **overall results have been disappointing.** (Stocker, 2007)

- To date, however, clinical trials of antioxidants for COPD have yielded **disappointing** results. (Foronjy et al, 2008)

- **Although antioxidants provide short-term improvement of endothelial function in humans, all studies of the effectiveness of preventive antioxidant therapy have been disappointing.** (Hadi, Suwaidi, 2007)

- **Recent evidence from intervention studies (CVD) with large doses of the antioxidant vitamins and other antioxidants in foods has been very disappointing.** (Bruckdorfer, 2008)

- RCTs have not yet supported a role for vitamins in primary or secondary prevention of CVD and have in some cases even indicated increased mortality in those with pre-existing late-stage atherosclerosis. **The trials that used a combination of vitamins that include beta-carotene have been disappointing.** (Honarbakhsh, Schachter, 2009)

- Despite biological plausibility, for the most part, **data derived from nutrient supplement trials using moderate to high doses of single nutrients or nutrient combinations (exceeding amounts to avoid nutrient deficiency) have been disappointing.** There is some evidence that **use of nutrient supplements intended to decrease CVD risk has resulted in unanticipated adverse consequences.** (Lichtenstein, 2009)

- Neuroprotective trials (in Parkinson's disease) using a single agent, such as an **antioxidant, have thus far generated disappointing results.** (Zhou et al, 2008)

- Although animal studies suggested that there may be a role for antioxidants (especially alpha-tocopherol) as therapy for heart failure of diabetic cardiomyopathy (HF), **the results obtained from human trials are disappointing.** (Celik et al, 2010)

- Various reasons, including incomplete knowledge of the mechanisms of action of these agents, lack of target specificity, and potential interindividual differences in therapeutic efficacy **preclude us from recommending any specific natural antioxidant for antihypertensive therapy at this time.** (Kizhakekuttu, Widlansky, 2010)

- **While a significant body of epidemiological and clinical data suggests that antioxidant-rich diets reduce blood pressure and cardiovascular risk, randomized trials and population studies using natural antioxidants have yielded disappointing** results. (Kizhakekuttu, Widlansky, 2010)

- **Despite this sound biological rationale and a number of pre-clinical and clinical lines of evidence, studies testing the effects of classical antioxidants such as vitamin C, vitamin E, or folic acid in combination with vitamin E have been disappointing** in cardiovascular disease. (Munzel et al, 2010)

- **Clinical trials using antioxidant therapies have been largely disappointing. Neither oxidant scavengers like N-acetylcysteine and vitamins E and C, nor xanthine oxidase inhibitor allopurinol have provided indisputable evidence of a clinical benefit despite numerous favourable studies in animal models** in myocardial reperfusion injury. (Braunersreuther, Jaquet, 2011)

Seemingly, none of these authors can bring themselves to say that, for the most part, the reliable scientific antioxidant studies have been outright **FAILURES!**

Their "disappointment" clearly shows their bias in terms of the expected outcome of their studies. It seems to me that, if the studies were conducted accurately, the investigtors should accept the end result as it is. There should be no disappointment or joyful celebration but we should work with the outcome to find the scientific truth and not keep trying to force the results to coincide with the failed predictions of the free radical theory with the failed or "disappointing" studies.

I must mention another important point, namely, none of these authors seem to question the theory upon which the studies were based, i.e., the free radical theory (FRT). The studies have not had predictability because the free radical theory is WRONG! Their studies were based on a nullified theory.

The free radical theory has been clearly invalidated by the 316 study reports in this book, demonstrating antioxidant unpredictability in preventing or curing various diseases, allegedly caused by oxidative free radicals.

If the FRT had been correct, the results would have proved its validity. Yet, it has been invalidated hundreds and hundreds of times in this book, in which I present over 316 such examples. When is enough... enough?

Worse still, antioxidants have been shown in over 90 study reports to increase the risk of common diseases, such as cancer, stroke and heart disease and to increase the risk of dying. Folks, please wake up.

In light of the data, which I present in this book, it is ludicrous to continue to try to force studies to conform to the erroneous predictions of the outdated free radical theory. We must adopt a new, advanced, improved theory (one which does have predictability and reproducibility) and which even makes good common sense.

For that reason, I have previously proposed my unified theory and the ROSI syndrome, whereby an EMOD insufficiency is the basis of disease allowance, causation and coexistence. Please refer to my other books to get detailed explanations for all of this.

Consensus statements on the fall of the free radical theory

Some authors argue that the hypothesis that antioxidants could prevent chronic diseases has now been disproved and that the idea was misguided from the beginning. (Hail et al, 2008)

The hype about antioxidants creates a false sense of security. There is no such thing as a super-food. (Jeffrey Blumberg, 2005)

The scientific understanding of radicals has not yet led to any therapeutic application. (Wingler et al, 2009)

For many years, scavenging already formed radicals with antioxidants was considered to be the most promising therapeutic approach, but clinical trials based on this principle have yielded mostly negative results. (Wingler et al, 2009)

The current lack of sufficient data does not permit the systematic recommendation of anti-oxidants. (Bonnefoy et al, 2002)

The results of the large, randomized controlled studies, has been disappointing and arguably provides the strongest evidence against the oxidative modification hypothesis of atherosclerosis. (Stocker, Keaney, 2004)

Supplementation with antioxidant vitamins does not lower serum lipid and lipoproteins or blood pressure. (Hodis et al, 2002).

Collectively, for the most part, clinical trials have failed to demonstrate a beneficial effect of antioxidant supplements on CVD morbidity and mortality. With regard to the meta-analysis, the lack of efficacy was demonstrated consistently

for different doses of various antioxidants in diverse population groups. (Kris-Etherton et al, 2004)

The relevance of the antioxidant hypothesis for the treatment of patients with atherosclerosis has not been definitively proven. Results of randomized trials with 'antioxidant' vitamins have been disappointing. (Tardif, 2006)

The obtained results suggest that supplementation of antioxidants cannot be recommended for the normal population. (Siekmeier et al, 2006)

Overall results of clinical studies investigating antioxidant effects have been disappointing given the consistent and promising findings from experimental investigations, clinical observations, and epidemiological data. (Parvicini, Touyz, 2008)

Much of the U.S. public has been misled by manufacturers and have an unhealthy faith in the benefits of nutritional supplements (Blendon et al, 2001)

Ironically, clinical trials of antioxidant supplements have shown that the oxidative theory lacks predictability but antioxidants continue to be widely used, even by physicians (Muntwyler et al, 2002)

There is **no convincing evidence that taking supplements of vitamin C prevents any disease. No one should take beta carotene supplements.** (Vitamin supplements. The Medical Letter on Drugs and Therapeutics, 1998)

Findings add to the growing body of evidence that certain supplemental antioxidant regimes have limited benefit in patients with cardiovascular disease. (Freedman, 2001)

The absence of efficacy and safety data from randomized trials precludes the establishment of population-wide recommendations regarding vitamin E supplementation. (Tribble et al, 1999)

In 2003, the U.S. Preventive Services Task Force (USPSTF) concluded there is insufficient scientific evidence to recommend vitamin supplements as a way to prevent cancer or heart disease. (U.S. Preventive Services Task Force, New Topic, 2003)

In 2004, after reviewing their results, the American Heart Association Council on Nutrition, Physical Activity, and Metabolism concluded that antioxidants have little or no proven value for preventing or treating cardiovascular disease.

In 2007, Cochrane Collaboration team concluded that commonly taken antioxidant supplements may do more harm than good. (Bjelakovic et al, 2007)

With respect to antioxidants and other phytochemicals, the key question is whether supplementation has been proven to do more good than harm. So far, the answer is no, which is why the FDA will not permit any of these substances to be labeled or marketed with claims that they can prevent disease. (Barrett, Quackwatch report, 2003)

Intervention studies have failed to show a consistent beneficial effect of high doses of antioxidant supplementation against chronic diseases, including cardiovascular disease. (Barrett, Quackwatch report, 2003)

The initial promising reports on beneficial effects with antioxidant therapies against atherosclerosis, derived from observational studies, were followed by generally negative

results reported from large randomized controlled trials. (Riccioni et al, 2007)

Clinical trials with classic vitamin antioxidants have generally failed to demonstrate any benefit in cardiovascular outcomes. (Sachidanandam et al, 2005)

Despite the wealth of data supporting a role of ROS in hypertension and other cardiovascular diseases, treatment with commonly employed antioxidants have failed, and in some cases have proven harmful, prompting a reconsideration of the concept of oxidative stress. (Harrison et al, 2007)

Even if antioxidant supplementation is receiving growing attention and is increasingly adopted in Western countries, supporting evidence is still scarce and equivocal. (Fusco et al, 2007)

Given the lack of efficacy of antioxidants in clinical trials to date, antioxidant vitamin combinations above the recommended dietary allowances should not be recommended for prevention or treatment of cardiovascular disease. (Kuller, 2001)

The existing scientific database does not justify routine use of antioxidant supplements for the prevention and treatment of CVD. (Kris-Etherton et al, 2004)

There is enough evidence from large, well-designed studies to discourage the use of vitamin E in Parkinson's disease, cataract, and Alzheimer's disease. (Pham and Plakogiannis, 2005)

It has now been proven that the "biological rationale rendered by the oxidative modification hypothesis" is wrong. (Steinberg et al, 1989)

The overwhelming majority of large, randomized and prospective trials of antioxidant supplements in CVD have yielded disappointing results. (Dagenais et al, 2000) (ATBC Study, 1994) (GISSI-Prevenzione Investigators;1999) (Yusuf et al. 2000 HOPE) (Omenn et al, 1996) (Hennekens et al, 1996) (Wilson et al, 1973)

A number of prospective, randomized, placebo-controlled, 3- to 6-year clinical trials (HOPE, GISSI, ATBC, Hennekens study, Omenn's study, Brown's study, MRC/BHF, Vivekananthan's meta-study, Miller's meta-study) have been published, testing the effect of vitamin E and other antioxidant vitamins or their combinations on clinical manifestations of cardiovascular disease and cancer. (Yusuf et al. 2000 HOPE) (GISSI-Prevenzione Investigators;1999) (Virtamo et al, 1998) ATBC Study; (Hennekens et al, 1996) (Omenn et al, 1996) (Brown et al, 2001) (MRC/BHF, 2002) (Vivekananthan et al, 2003) (Miller et al, 2005). These trials have surprisingly yet consistently shown that commonly used antioxidant vitamin regimens (vitamins E, C, beta carotene, or a combination thereof) do not significantly reduce overall cardiovascular events or cancer.

By extending HOPE and adding to the growing list of neutral prospective vitamin E trials (HOPE, GISSI-IV, ATBC, HPS, HATS), this report effectively closes the door on the prospect of a major protective effect of long-term exposure to vitamin E, taken in moderately high dosage, against complications of atherosclerosis and overall cancer incidence. (Brown et al, 2001)

The public health viewpoint would have to be that there's really nothing to support widespread use of these vitamins. (Dr. Ian Graham quote, a professor of cardiology, 2005)

Antioxidants – compounds in foods and supplements that prevent cell damage – may actually increase the chances of getting diabetes, at least in the early stages, antioxidants may contribute to early development of insulin resistance. (Tiganis quote, 2009)

Although some levels of antioxidant vitamins and minerals in the diet are required for good health, there is considerable doubt as to whether these antioxidant supplements are beneficial or harmful, and if they are actually beneficial, which antioxidant(s) are needed and in what amounts. (Stanner et al, 2004)

Some but not all studies have reported a direct association between uric acid (an antioxidant) concentrations and atherosclerosis, hypertension, and cardiovascular mortality. (Rao et al, 1991)

A 2009 review of experiments in mice concluded that almost all manipulations of antioxidant systems had no effect on aging. (Pérez et al, 2009)

Antioxidant vitamin supplementation has no detectable effect on the aging process, so the effects of fruit and vegetables may be unrelated to their antioxidant contents. (Thomas quote, 2004)

Antioxidant supplements do not appear to increase life expectancy in humans. (Green, 2008)

Antioxidant supplements have no clear effect on the risk of chronic diseases such as cancer and heart disease. (Stanner, 2004)

Nutrition expert, Marian L. Neuhouser, said, **"Consumers spend money on dietary supplements with the thought that they are going to improve their health, but there's no evidence for this."**

In 2006, Dr. David Gems said in Science Daily. **"But there is no clear evidence that dietary antioxidants can slow or prevent aging." "Oxidative damage is clearly not a universal, major driver of the aging process." "The free radical theory of aging has filled a knowledge vacuum for over 50 years now, but it just doesn't stand up to the evidence."**

Hekimi believes that the findings of Van Raamsdonk (i.e., removing SODs from C. elegans increases their lifespan) throws a wrench in the entire free radical theory of aging. Bart Braeckman of Ghent University also does not think that the free radical theory is the only answer. Braeckman states, "The final conclusion was similar in all these papers: there is a problem with the free radical theory."

References for this section on the FRT can be found in my book entitled, *Heart Disease and Antioxidant Failures.*

CHAPTER SEVEN

The epiphany of 4-4-11

For years, I had questioned the role of antioxidants in foods and this epiphany resolved this issue for me.

I believe that the antioxidant overkill is doing the following: the over supply of electrons by the electron donating antioxidants are allowing the 4 electron reduction of ground state oxygen (O_2) to proceed to the final step, i.e., the formation of water. In doing so, this in effect "removes" the beneficially acting EMOD intermediates of superoxide, hydrogen peroxide, the hydroxyl radical, nitric oxide and peroxynitrite and "supplies" the non-reacting (non-protective) water to or within the milieu (intra- or extracellular). This, in effect, "quenches" the EMODs or ROS and it also prevents the protective action (messaging) of the reactive EMODs against pathogens and cancer.

This is an entirely new paradigm. This is how it is happening and explains the electron donating nature of the antioxidants and the electron accepting nature of oxygen. This explains "how" the antioxidant overkill causes a harmful effect!!!!!

The antioxidant overload negates the beneficent actions of the intermediate EMODs. However, when the electron donating antioxidants are present in limited amounts, they serve as pre-oxidants or co-oxidants and form the crucial active intermediates and reactive EMODs, after interacting with triplet ground state oxygen.

Finally, this explains the need for average amounts of antioxidants contained within a balanced nutritious diet, containing fresh fruits and vegetables. It also explains the failures of the antioxidant supplement studies, which were supplying an antioxidant overkill, thereby negating the protective EMOD effects against cancer and pathogens.

It has always puzzled me that antioxidants are "necessary" to form EMODs. In fact, this is why I referred to them as pre-oxidants or co-oxidants. The query for me was, "Then, how could the antioxidants be harmful?" The answer is that without them, ground state oxygen does not have an adequate source of available electrons to form the intermediate EMODs, i.e., the 1e (e, electron), 2e and 3e reduction products. But, with a sufficient amount of antioxidant electrons, oxygen can accept the electrons to form superoxide, hydrogen peroxide, the hydroxyl radical, nitric oxide and peroxynitrite, the active EMOD forms.

The basic redox chemistry is correct in that oxygen is the ultimate electron acceptor and antioxidants are the agents donating the electrons. My approach does not violate the rules of thermodynamics governing such reactions. This also explains the reason that transition metals, such as iron or copper, can also donate electrons to ground state oxygen and act as prooxidants. It is via their contributions as electron donors that antioxidants are acting to form the protective prooxidant EMOD intermediates.

Part of the problem is that our diet and our environment supply us with adequate (and sometimes over supply us) with antioxidants. When we endeavor to "supplement" these sources, we risk providing an over load of the electron supply to oxygen and thus, it undergoes a 4e electron reduction to form water. Wow!!!! This is way cool.

**The secret to longevity
is to keep breathing oxygen.**
R. M. Howes, M.D., Ph.D.
6/8/11

**The only oxygen we need is
one pint per breath at the rate of
only a minimum of 21,600 times per day.
Other than that....**
R. M. Howes, M.D., Ph.D.
9/12/09

Five common antioxidant killers

Five common antioxidant killers, when in excess: uric acid, cholesterol, bilirubin, estrogen, testosterone.

TEN common harmful antioxidants

TEN common harmful antioxidants, when in excess: 1) cholesterol, 2) bilirubin, 3) uric acid, 4) estrogen, 5) testosterone, 6) vitamin E, 7) vitamin A (its precursor is beta carotene), 8) multivitamins, 9) selenium, 10) **NAC, N**-acetylcysteine.

13 Ineffective antioxidants

At this point, my studies have shown the following 13 antioxidants can be ineffective, many of which have shown considerable harmful potential in those without known vitamin deficiencies:
- Vitamin A, beta carotene
- Vitamin E, alpha tocopherol
- Vitamin C, ascorbic acid
- Selenium
- Ferulic acid
- Quercetin, a flavonol
- Lycopene, a carotenoid

- **N-acetylcysteine, NAC**
- **Pine bark extract, antioxidant oligomeric proanthocyanidin complexes**
- **Grape seed extract or EGCG, polyphenols decrease iron absorption**
- **Pomegranate juice**
- **Multivitamins (contain multiple antioxidants)**
- **Prenatal vitamins**

8 potentially harmful endogenous antioxidants

Also, high levels of the following 8 endogenous antioxidants can be associated with harm:
- **Cholesterol (atherosclerosis and heart disease)**
- **Uric acid (gout and heart disease)**
- **Bilirubin (brain damage)**
- **Estrogen (breast cancer, oral cancer)**
- **Testosterone (prostate cancer)**
- **Glutathione (heart failure)**
- **Catalase (CAT)**
- **Glutathione peroxidase 1 (GPx1)**

CHAPTER EIGHT

Sale of children's vitamins on the rise

It appears that we are continuing to overload on antioxidants. In July of 2010, Carlotta Mast with the Nutrition Business Journal said sales of children's formulas were on the rise. In 2009, **"Sales of children's vitamins and mineral products grew 25 percent at grocery stores, drug stores, and other mass market stores, and grew 10 percent in natural health food stores and vitamin specialty stores."**

The American Academy of Pediatrics, AAP, spokesperson Dr. Michael Cabana said vitamins and other supplements are only necessary when a child isn't eating a healthy, well-balanced diet. and even then, "Vitamins, by definition, are only needed in small amounts. So even the pickiest of eaters probably have enough of all the vitamins that they need," explains Dr. Cabana.

"Consider vitamins and supplements in the same category as medicines, because there can be interactions," explains Dr. Cabana.

Regularly, there's a new product that promises to provide you with extra nutritional support, a boost of immunity or a host of antioxidants to fight off diseases.

Take a walk down the pharmacy aisle and you'll see so many vitamin bottles, it can leave you breathless and confused: Which ones are necessary? Will taking them actually do anything helpful? Will they be harmful?

"Here's the point about vitamins," said Dr. Marc Siegel, an internist and FOX News Channel contributor. "**Most people who eat a well-balanced diet in an upscale socio-economic environment do not need vitamin supplements**, with the exception of calcium and vitamin D. Those are usually deficient in our diets."

Siegel said **vitamin A and E supplements are not necessary**. Instead, people should seek natural sources of these antioxidants.

"They are overused," Siegel said. **"These are antioxidants that were trendy for many years, and it was thought they protected you from heart disease and cancers, but it turns out they don't. So, I don't recommend those."**

A report, from the Children's National Medical Center in Washington, on more than **8,000 infants and found a possible link between the use of multivitamin supplements and the risk of asthma and food allergies.** It found **"an association between early infant multivitamin intake and asthma** among black infants and **an association between early infant multivitamin intake and food allergies** in formula-fed infants." **More than 50% of all toddlers in the United States are taking multivitamins.** (Journal of the American Academy of Pediatrics, 2004)

According to Orthomolecular News Service, over half of the U.S. population takes daily nutritional supplements. Even if each of those people took only one single tablet daily, that makes 155,000,000 individual doses per day, for a total of nearly 57 billion doses annually, in just the United States of America.

Use of Dietary Supplements Keeps Climbing: CDC

More than half of U.S. adults take dietary supplements, such as multi-vitamins and calcium, and their use jumped dramatically over a recent 20-year period, according to a new government report.

Between 1994 and 2006, the proportion of Americans using at least one dietary supplement jumped from 42 percent of adults to 53 percent, according to the U.S. Centers for Disease Control and Prevention.

"The increase in supplement use may be due to increased awareness and education about dietary supplement use," said lead report author Jaime Gahche, an associate service fellow in the CDC's National Health and Nutrition Examination Survey/Planning Branch at the National Center for Health Statistics.

Media attention on vitamin D likely boosted intake of that supplement, she said, and **massive advertising by the supplement industry may have influenced use of multivitamins**. But some experts say multivitamins may not be necessary.

However, since 2006, the growth in supplement usage has leveled off, she said. "We have reason to believe it should stay relatively stable," she added.

The data for the report were gathered from the National Health and Nutrition Examination Survey (NHANES) and included three time periods: 1988 to 1994, 1999 to 2002 and 2003 to 2006.

Among the report's other findings:

* More women (59 percent) than men (49 percent) use supplements.

- Use of **multivitamins**, the most commonly consumed supplement, rose from 30 percent in the earlier period to **39 percent** by 2003-2006.
- Calcium use by women 60 and older increased from 28 percent to 61 percent across the three time periods, but varied by race and ethnicity. More than two-thirds of white women 60 and older (65.7 percent) were taking calcium-containing supplements in 2003-2006, up from 58.9 percent in 1999-2002. Among Mexican American women, 52.3 percent were using calcium supplements in 2003-2006, an increase from 39.5 percent. Increased use of calcium among black women was not statistically significant, the researchers noted.
- Consumption of folic acid supplements, recommended for women of childbearing age, also varied by race and ethnicity, but stayed at about 34 percent from 1988 to 2006, Gahche's group found. White women aged 20 to 39 are twice as likely to take a supplement containing folic acid as blacks or Mexican Americans. Folic acid helps prevent neural tube defects in babies. Folate is found in green leafy vegetables and beans.
- Twenty-four percent of men and 30 percent of women took **vitamin D** in 1988-1994, a figure that stayed stable for younger adults but increased for men and women ages 40 to 59. Use of vitamin D, which is difficult to obtain in foods, also increased among women 60 and older, jumping from 49.7 percent to **56.3 percent** from 1999-2002 to 2003-2006.

Samantha Heller, a dietitian, nutritionist, exercise physiologist and clinical nutrition coordinator at the Center for Cancer Care at Griffin Hospital in Derby, Conn., said **supplement makers promote the idea that dietary supplements will provide people with all the nutrients they need. But, this is wrong.**

While some studies suggest a benefit in taking multivitamins, the most commonly used dietary supplement, others do not, Heller said. **Increasingly, multivitamins are being found to**

be ineffective in curbing or curing diseases, other than vitamin deficiency states.

"The general recommendation from health professionals is that we get the bulk of our nutrients, vitamins and minerals from foods, and that if we are eating a healthy diet we may not need most supplements," she said.

Both over- and under-supplementation can be a problem, she noted. "Supplement use is complicated in part because **many of the foods we eat are fortified with some vitamins and minerals** and not others," she said. I believe that we are being overloaded with anti-oxidants from many sources as described in my 2011 book entitled, *Antioxidant Overkill.*

Also, with up to 50 percent of Americans reportedly dieting at any given time, many miss out on essential vitamins and minerals, Heller said. "Combine this trend with the popularity of fast and junk food, which are nutrient-poor and calorie-rich, and it is difficult to make broad dietary supplement recommendations that apply to everyone," she said.

"I generally advise my patients to take a multivitamin, vitamin D, omega-3 fatty acids and for women, calcium, in addition to a healthy diet and lifestyle," Heller said. However, as you will see in my chapter in this book on omega-3 and fish oil, there is considerable controversy about its efficacy and safety.

"Growing children, athletes, the elderly and people with certain illnesses may need additional or different supplementation," she added. **Still, RCTs have shown very little positive effects from the various antioxidant supplements (and significant potential harm).**

Marketing prowess, not science based evidence, are responsible for the widespread use of the supplements, especially the antioxidant vitamins. Contrary to the impression that the antioxidant vitamins will decrease the risks for common diseases, they actually can increase the risk of cancer, heart disease, strokes and overall mortality.

I have researched this area extensively and have found over 316 scientific study reports showing the ineffectiveness of the antioxidants, including vitamins A, C and E. Of these, 90 studies show their harmful potential. They are especially dangerous for patients with pre-malignant conditions or for cancer survivors, in that I have found that they will shield 27 types of human cancer in cell cultures from being killed by electronically modified oxygen derivatives (EMODs), which are described in my comprehensive books, *Dangers of Excessive Antioxidants In Cancer Patients* and *Heart Disease and Antioxidant Failures*.

You owe it to yourself to read my other two recent books entitled, *Death In Small Doses?* and *Antioxidant Overkill*. Then, choose wisely.

Even though the heart has one of the highest requirements for continual oxygen consumption and consequent EMOD generation, the heart is relatively deficient in those enzymes responsible for clearing free radicals i.e., superoxide dismutase, catalase, and glutathione peroxidase. (Doroshow et al, 1980). Why?

> **Don't spend all your time, doing time,**
> **by being locked inside the confines of a false theory.**
> **Break down the doors and**
> **live on a perpetual intellectual high.**
> **Get drunk on intellectual freedom.**
> **Free to think as you see fit and**
> **as far as your dreams will reach.**
> **Hallelujah!**
> **I have escaped the prison of**
> **the free radi-crap theory.**
> R.M. Howes M.D., Ph.D.
> 5/15/11

In 2002, Asplund completed a review of antioxidants and their relation to cardiovascular disease. My conclusions were as follows:

My conclusions based on Asplund's paper (23 factoids):

- **a high dose of ascorbic acid (4.5 g day^{-1}) during 12 weeks
  did not effect lipoprotein(a) levels** (Bostom AG)
- There seems to be **no relationship between plasma levels of
  tocopherol and blood pressure** (Moran JP)
- **individuals with a genetically determined tocopherol
  deficiency do not have increased LDL oxidation** (Thomas
  SR)
- **α-Tocopherol given as food supplements in a dose of
  800 U day^{-1} during 3 months to hypercholesterolemic
  individuals did not affect the levels of total cholesterol,
  LDL cholesterol, HDL cholesterol or triglycerides despite
  a reduction in TBARS, a marker of oxidative stress**
  (Bierenbaum ML)
- **most of the scientific evidence indicates that there is no
  relationship between intake of ascorbic acid or tocopherol and various components of the fibrinolytic system**
  (Simon JA) (Koh KK)
- **Reduced platelet adhesiveness seems not to be associated with the antioxidative properties of tocopherol but
  to other mechanisms** (inhibition of protein kinases and membrane stabilization) (Steiner M)
- **people with high plasma levels of tocopherol tended
  to have an increased risk of death from heart disease**
  (Sahyoun NR)
- In the Seven Countries Study, mortality from coronary heart disease was followed for 25 years in 16 cohorts in seven countries.
  The **25-year trends in coronary mortality did not correlate with trends in dietary intake of β-carotene, tocopherol or ascorbic acid** (Kromhout D)
- **Serious adverse cardiovascular effects were noted in one
  of the studies in which there was a significantly increased**

risk for fatal or nonfatal intracerebral and subarachnoid haemorrhage in subjects taking α-tocopherol (The Alpha-Tocopherol, Beta-Carotene Cancer Prevention Study Group) (Leppala JM)

- there was also a significant increase in the risk for intracerebral haemorrhage in the group taking β-carotene, but no obvious mechanism to explain this is at hand (Leppala JM)

- In two of the trials, intake of carotene supplements was associated with an increased risk of lung cancer in cigarette smokers (The Alpha-Tocopherol, Beta-Carotene Cancer Prevention Study Group) (Omenn GS)

- In an RCT, pretreatment with α-tocopherol and ascorbic acid did not affect various indicators of myocardial damage (Kugiyama K)

- Investigators do not conclusively confirm reduced symptoms for intermittent claudications in the intervention group. No beneficial effects of tocopherol treatment was observed in a more stringent but still small, randomized trial not included in the Cochrane review (Leng GC) (Kleijnen J)

- the RCTs have all failed to confirm an effect of any of the antioxidant vitamins on the risk of cardiovascular events. Together, the eight RCTs of antioxidant vitamins as dietary supplements have covered close to 700 000 observation years, and more than 4000 fatal and nonfatal myocardial infarctions and strokes have been accumulated. This would give a sufficient statistical power to detect even a small effect (< 10% risk reduction) of antioxidant vitamins in the doses that have been used, if there were such an effect (Asplund)

- There was no tendency for ascorbic acid supplementation to be of benefit (relative risk for all deaths was 1.08; 95% CI 0.93–1.23) (Ness A, Role of antioxidant vitamins in prevention of diseases. Meta-analysis seems to exclude benefit of vitamin C supplementation)

- Carotene and ascorbic acid have not only antioxidant but also prooxidant properties that are particularly evident at high concentrations of the vitamins (Schwartz JL)
- **Ascorbic acid also enhances the absorption of dietary iron, a powerful prooxidant** (Gerster H)
- **it is very hard to reach the same intake by regular food as by taking vitamin supplements – the content of tocopherol in multivitamin tablets is 3–4 times higher than that reached by normal food intake and the content of tocopherol may be 40–50 times higher** (Byers T)
- **High intake of antioxidant vitamins is a component of a cluster of healthy behavior**
- **In RCTs, antioxidant vitamins as food supplements have convincingly been shown to have no beneficial effects on the risk for myocardial infarction or stroke**
- **In randomized controlled trials, however, antioxidant vitamins as food supplements have no beneficial effects in the primary prevention of myocardial infarction and stroke. Serious adverse events have been reported**
- **A report on increased risk of intracerebral and subarachnoid hemorrhage during treatment with carotene and tocopherol may caution against the use of antioxidant vitamins in healthy people**

Antioxidant vitamins as food supplements cannot be recommended in the primary or secondary prevention against cardiovascular disease (Asplund, 2002) (Asplund K. Antioxidant vitamins in the prevention of cardiovascular disease: a systematic review. J Intern Med. 2002 May;251(5):372-92). References in these conclusions can be found in R. M. Howes. *Reactive Oxygen Species Insufficiency (ROSI) as the Basis for Disease Allowance and Coexistence: Extraordinary Support for an Extraordinary Theory* Vol I, II & III. © 2008; 1564 pages.

Variables and covariants

Variables plague antioxidant studies and must be taken into account. They include studies occurring at differing time intervals, lengths of study, concentrations of vitamins, combinations of vitamins, source of vitamins, types of participants, varying states of health of participants, varying diets of participants, varying methods of statistical analysis, countries of origin, gender of participants, smoking and drinking habits of participants, varying body weights and BMIs, levels of daily exercise, varying levels of stress or depression of participants, types of questionaires used, varying study end points, genetic sway, environmental conditions, etc.

THIRTY VARIABLES IN ANTIOXIDANT & EMOD (REDOX) STUDIES:

Variables and co-variants

1) multitude of dietary variances (*total energy intake and fat, carbohydrates, protein, fiber, electrolyte intake*), 2) varying environmental factors (*sunshine exposure, excessive cold, heat, radiation exposure or treatment, second hand smoke, inhalation of inorganic particles such as asbestos and silica, ozone inhalation, socioeconomic factors, etc.*), 3) exercise level per se or lack thereof, 4) overall physical activity (*number of hours of television watched per week*), 5) degrees of obesity, 6) race, 7) sex, 8) education, 9) presence of fever, 10) smoking, 11) alcohol use, 12) oral contraceptive use, 13) synthetic hormone intake, 14) pregnancy history, 15) dietary antioxidant use, 16) herbal supplement use, 17) use of medications, 18) anthropometry (*height, weight, BMI, etc.*), 19) varying antioxidant dosage levels, 20) varying antioxidant combinations (*nutrient synergy*), 21) synthetic or natural vitamin sources (*L vs. D forms, alpha vs. gamma*

forms), 22) **improperly combined study groups, 23) use of improper exclusion criteria, 24) flawed statistical methods, 25) over generalization of findings, 26) accuracy of dietary records and questionnaires etc. 27) questionable species of animal studied** *(primates, invertebrates, exposure to insecticides or herbicides, etc.)*, **28) durations of treatments and 29) studies targeted at different diseases/conditions, 30) bias of investigators.**

Because it is not routinely possible to directly measure free radicals *in vivo*, scientists have approached this question by measuring the by-products that result from free radical reactions. However, there is limited concordance amongst the many bio-markers being currently used for measuring oxidant levels or oxidative activity.

Antioxidant supplementation has no effect on biomarkers of oxidant damage

The link between high fruit/vegetable intake and reduced chronic disease may be partly explained by antioxidant protection. To determine the effect of moderate antioxidant intake on biomarkers of oxidant damage, Jacob et al. assessed in vivo lipid and protein oxidation in 77 healthy men whose typical diet contained few fruits and vegetables (mean of 2.6 servings/d). The 39 nonsmokers and 38 smokers, age 20–51 y, were given a daily supplement (272 mg vitamin C, 31 mg all-rac-a-tocopherol, and 400 μg folic acid), or placebo, for 90 d with their usual diet. Blood and urine were taken at baseline and the end of the study for determination of lipid peroxidation products, including F_2-total and 8-isoprostanes, and protein carbonyls. Urine thiobarbituric acid reactive substances (TBARS) was the only oxidant damage marker that was significantly higher in smokers compared to nonsmokers ($P < 0.05$). **Supplementation increased plasma ascorbate and tocopherol, but had no effect on the oxidant biomarkers.** In healthy young men, the endogenous antioxidant defense system and a modest intake of dietary antioxidants are adequate to minimize levels of in vivo oxidant damage such that they cannot be differentiated by current methods. (Jacob and Ames, 2003). **I believe that this is a landmark paper, which also effectively serves to invalidate the free radical theory.**

Safety and cautionary statements

Assuredly, vitamin supplements have benefited those with vitamin deficiencies, but before recommending their use in healthy individuals or in attempts to block aging, one should review the research data as regards fact versus speculation. It is dominated by confusion and inconsistency.

Most evidence about the salutary benefits of antioxidant vitamin supplements is epidemiological, including observations and uncontrolled studies, and has been fraught with misinterpretation and conclusions with unsubstantiated claims. Experts urge caution with dietary supplements because anticipated benefits of antioxidant vitamin use, once widely thought to prevent cancer, heart disease and strokes, have failed to live up to their predicted expectations when put to rigorous testing.

This issue is discussed in great detail and is available in "The Howes Selective World Library of Oxygen Metabolism" at the www.iwillfind-thecure.org.

CHAPTER NINE

Cocoa, chocolate and flavonoids

According the Dr. Hyla Cass, the essence of all such products is the **cocoa** solids that give chocolate both its flavor and its flavonoids. The higher the cocoa content, the higher the flavonoid content, which is why dark chocolate is much more healthful than milk chocolate. (Don't even think about "white chocolate," which is not really chocolate at all, because it has no cocoa solids, just cocoa butter, a sweet-tasting fat.). Flavonoids are not called that because they taste good (they don't). The word is derived from the Latin *flāvus,* yellow, because many of these brightly colored plant pigments are yellow; others are red, orange, blue, or purple. Flavonoids (aka bioflavonoids, a synonym—all flavonoids are biological in origin) give most fruits, vegetables, and flowers their characteristic colors. All flavonoids belong to the larger class of compounds called *polyphenols,* **which are known for their strong antioxidant properties.**

Flavanols and *proanthocyanidins* (the correct term in this context) are the types of flavonoid that predominate in cocoa, as well as in tea, apples, and apricots. As a rule, the more colorful a fruit or vegetable, the more likely it is to be high in beneficial flavonoids. Berries are particularly rich in these compounds, and **blueberries** are the berry best of all, being loaded with potent anthocyanins and proanthocyanidins, two types of flavonoids. (On the other hand, blueberries have virtually no flavanols, a kind of flavonoid found abundantly in cocoa and tea.).

Based on a great body of epidemiological research, medical scientists believe that a diet rich in fruits and veggies is vital for good health and antiaging, owing in part to their diverse content of flavonoids and other polyphenols—and the greater the variety of these compounds, the better. (Lau, 2005)

There is debate that they can help prevent some of the major diseases of aging, such as cancer, heart disease, and certain neurodegenerative diseases.

How they accomplish this is an intriguing question, however. It was long believed that the biological activity of polyphenols (including flavonoids) occurred via their antioxidant properties, because experiments with cell cultures have consistently shown that they are, in fact, strong antioxidants—*in cell cultures*. There is the catch.

Much evidence obtained over the past decade has suggested that these laboratory results are red herrings. **In March 2007, scientists at the Linus Pauling Institute announced that flavonoids actually have little or no value as antioxidants and that their health benefits are likely the result of entirely different biochemical mechanisms** (Anon. Studies force new view on biology, nutritional action of flavonoids. Linus Pauling Institute, media release, March 5, 2007).

It has been proved, they said, that **flavonoids are poorly absorbed** (usually less than 5%) and that **what small amounts do get through to the circulation are rapidly metabolized to derivative compounds and excreted**.

Dr. Frei discounts the antioxidant properties of flavonoids

In a media release, Dr. Balz Frei, professor of biochemistry and biophysics and director of the Linus Pauling Institute, was quoted as saying,

"What we now know is that **flavonoids are highly metabolized, which alters their chemical structure and diminishes their ability to function as an antioxidant**.... If you measure the activity of flavonoids in a test tube, they are indeed strong antioxidants. But, with flavonoids in particular, **what goes on in a test tube is not what's happening in the human body**....We can now follow the activity of **flavonoids** in the body, and one thing that is clear is that **the body sees them as foreign compounds and is trying to get rid of them.** But this process of gearing up to get rid of unwanted compounds is inducing so-called Phase II enzymes that also help eliminate mutagens and carcinogens, and therefore may be of value in cancer prevention. Flavonoids could also induce mechanisms that help kill cancer cells and inhibit tumor invasion."

Dr. Frei also stated that "the flavonoids (or their bioactive metabolites) strongly influence cell signaling pathways and gene expression, with relevance to cancer, heart disease, and neurodegenerative disease." He believes that **the large increase in total antioxidant capacity of blood observed after the consumption of flavonoid-rich foods is not caused by the flavonoids themselves, but is most likely the result of increased uric acid levels.**

Please remember that **hyperuricemia is strongly associated with cardiovascular disease and gout.**

The above is a classic example of the way science works: an attractive theory, no matter how long in existence, no matter how widely believed or even cherished by its partisans, must be modified or discarded if new and convincing evidence refutes it. **There are dogmas in science, but they are written in sand, not stone—they are always subject to change.**

Commercial cocoa powders are largely fat-free, having only about 10–22% residual cocoa butter. In any case, **cocoa must be sweetened to make it palatable**, and here it's important to look for nonsugar sweeteners, preferably natural ones. An important thing to know about commercial cocoa products is that many are **"Dutched," meaning that they've been alkali-treated to mellow the inherent bitterness of cocoa.** The trouble with **this process is that it destroys most of the beneficial flavonoids.** Thus, consumers who want the health benefits of cocoa should look for products that have *not* been Dutched. Either way, the initial flavonoid content of cocoa depends on the variety of cacao tree *(Theobroma cacao)* from which it came, and the final content depends on a variety of factors in the production methods used.

Omega-3 fish oil

For sometime, I have pondered the activity of the omega-3 fatty acids, since it has a double bond and some claim that it is an antioxidant. Finally, I may have found the answer, which is the basis for their reported salutary effects in fighting disease. Omega-3s generate EMODs. Supplementation (polyunsaturated fatty acids (PUFAs) with **n-3 long-chain-PUFA (n-3 LCPUFAs) significantly increased oxidative stress** at rest and after a judo-training session (Flaire et al, 2010). In short, omega-3s can help maintain sufficient EMOD levels for pathogen and cancer protection because they are prooxidants.

Fish Oil Does Not Prevent Alzheimer's Disease

Omega-3 fish oil supplement producers have claimed curative powers for diseases such as cancer, heart disease, dementia, ADD, ADHD, depression, bipolar disorder, dyslexia, dyspraxia, obsessive compulsive disorder, headaches and migraines.

Also, it was claimed to decrease aggressive behavior, prevent learning disabilities and make kids smarter. Unsupported aggressive advertising has led to "soaring sales for fish oil."

However, disappointing studies are now coming forth and supplements do not have the same benefits as do natural foods containing these same ingredients. Even though claims of being "heart healthy" persist, **a large German study gave fish oil or dummy capsules to more than 3,800 people who had suffered a recent heart attack and found that after a year, there was no difference between fish oil pills or placebo**.

But, there is more bad news for fish oil. Even though data from a trial of over 800 older people initially showed that those who eat plenty of oily fish seem to have better cognitive function, a new study has found that omega-3 pills, promoted as boosting memory, did not slow mental and physical decline in older patients with Alzheimer's disease.

This $10 million project studied nearly 300 men and women, aged 76 on average, with mild to moderate Alzheimer's disease. They were randomly assigned to take either the omega-3 fish oil pill (DHA) or dummy pills daily for 18 months. Results were similar in both groups, in that **DHA provided no benefits in slowing Alzheimer's symptoms** nor did the pills work in a subgroup of participants with the mildest Alzheimer's symptoms.

The researchers concluded, "There is no basis for recommending DHA supplementation for patients with Alzheimer disease." Laurie Ryan, program director of Alzheimer's studies at the Institute on Aging, called the results **discouraging**. Thus, Alzheimer's disease remains basically untreatable and the "fish oil gold rush" may be slowing.

We realize that generally dietary supplements do not work, except in cases of known deficiencies or malabsorption syndromes. Still, sales of dietary supplements bring in about $27 billion annually.

We know that eating more heart-healthy omega-3 fats provided no additional benefit in a study of heart attack survivors who were already getting good care. There is little harmful side effects of fish oil when the ratio of omega-6 to omega-3 is properly balanced.

Please remember the words of expert, Dr. Alice Lichtenstein, "We need to be a little more cautious about the prediction of individual benefit of any nutritional supplements. People are so willing to embrace the simple answer, as if it's possible to crack a capsule over a hot fudge sundae and undo the harm of harmful diets and lack of exercise."

But, give me a chocolate sundae. I'll take my chances.

Fish Oils: Is There A Downside?

Please remember, even though there is a current fish oil craze, you can never be too careful about what you put into your body. As I mentioned, fish oil (especially omega-3) proponents have made wild claims for their curative powers for diseases such as cancer, heart disease, dementia, ADD, ADHD, depression, bipolar disorder, dyslexia, dyspraxia, obsessive compulsive disorder, headaches and migraines.

Such unsupported aggressive advertising has led to "a fish oil gold rush." Unfortunately, most supplements do not hold up under scientific testing and pill-forms of agents such as omega-3, vitamin E, beta carotene and vitamin C do not offer the same benefits as do natural foods containing these same ingredients.

When it comes to improving brain function, data from a trial of over 800 older people initially showed that those who eat plenty of oily fish seem to have better cognitive function. But factors such as education and mood explained most of the difference and a UK study has cast doubt on claims that eating oily fish can protect against dementia in old age.

The American Heart Association (AHA) recommends adults eat fish at least twice a week and for people with heart disease, **they advise 1 gram of omega-3 a day**.

Fish oil capsules are not for children or pregnant women, because the pills pose a bleeding risk and capsules should be stopped a week before any surgery.

We must continue to emphasize the basics of eating a balanced diet, exercising more, avoiding stress, not smoking and not being misled by unscrupulous advertisers. Every time we turn on the TV, we are being oversold on the latest miracle meds or supplements, like they were pitching OxyClean or Sham Wow. So, be on guard and protect your wallet and your health.

Omega-3: Up to 1,000 mg daily if you don't reliably eat fish such as salmon 2 or 3 times a week.

The Fish Oil Gold Rush

Slowly, Americans are becoming more aware that they are constantly being bombarded with unrestricted dietary supplement advertising and unsupported advertising of legal medical drugs but that does not stop their onslaught.

Currently, we are in the midst of "a fish oil gold rush" and the latest mantra is that fish oil (especially omega-3) can cure all varities of diseases. If only all of this were true, it would be a miracle but it is not all true.

Just as you should not rely on synthetic vitamin pills to replace fresh fruit and vegetables, you should not rely on omega-3 pills to replace the ingestion of fish with a high fat content.

These results do not mean that getting more of the essential nutrient has no value because studies have shown fish oil fats may reduce heart disease. Larger randomized studies are needed to clarify the benefits or lack of benefits with omega-3 supplements and results have been inconsistent.

As with marketing techniques for other supplements (such as anti-oxidants), in recent years, omega-3 has been added to or "fortified" in some foods such as margarine and eggs, or labeled to highlight the omega-3 content of foods like tuna fish in efforts to **capture the current fish fat craze.**

We must concentrate on eating a balanced diet, exercising more, avoiding stress, not smoking and not being misled by unscrupulous advertisers. People seem to ignore this advice even when their life depends on it. Get your medical advice from a physician and not from someone who is "pushing" supplements.

Fish oil does not accelerate weight loss

According to an online December 15, 2010 article, in the American Journal of Clinical Nutrition, fish oil capsules won't help boost weight loss if you're already dieting and exercising.

Among a group of overweight and obese adults enrolled in a diet and exercise program, **those who took omega-3 fatty acids didn't lose any more weight than those given placebo** capsules, Dr. Laura F. DeFina of The Cooper Institute in Dallas and her colleagues found.

There is prior evidence from animal studies that omega-3 fatty acids promote weight loss, DeFina and her team note, while studies in people have had mixed results. Because fish oil has many other potential health benefits, including cutting cholesterol, improving insulin sensitivity, and reducing blood pressure, "weight-loss programs associated with the use of omega-3 polyunsaturated fatty acids seemed appropriate to the investigators."

To investigate whether fish oil enhanced the results of a diet and exercise regime, the researchers randomly assigned 128 sedentary overweight or obese men and women to take five fish oil capsules (providing a total of three grams of omega-3 fatty acids) or five placebo capsules every day for 24 weeks.

Participants were also instructed to do 150 minutes a week of aerobic exercise and 20 to 30 minutes of strength exercises at least twice a week.

The people in the omega-3 group lost 5.2 kilograms, or about 11.5 pounds, on average, compared to 5.8 kilograms (nearly 13 pounds) for the placebo group, not a statistically significant difference. People in both groups lost more than five percent of their body weight, enough to produce health benefits.

At the end of the study, there was no difference between the groups in measures of heart disease risk, such as blood pressure and cholesterol levels. However, researchers found that omega-3 blood levels in the fish oil group increased to a level "previously found to have a positive cardiovascular effect."

Not everyone actually completed the 24-week program; there were only 81 participants at the end. But the results were similar when the researchers looked only at those 81 people.

Investigators said, "Whereas one may not enhance weight loss by taking supplements with this level of omega-3 fatty acids, the protective cardiovascular effect should still be realized because of the sheer increase in blood concentrations of the fatty acids."

One online jokester said, "The BP brand of fish oil is going on sale in November." That is not funny for those of us living in Louisiana.

Canola oil and omega-3s (fish oil)

Some of the following was excerpted from THE GREAT CAN-OLA SCAM Part 2 By Sally Fallon and Mary G. Enig

Animal studies on **Low Erucic Acid Rapeseed (LEAR)** oil were performed when the oil was first developed and have continued to the present. The results challenge not only the health claims made for canola oil, but also the theoretical underpinnings of the diet-heart hypothesis.

The first published studies on the new oil were performed in 1978 at the Unilever research facility in the Netherlands. (Viles et al, 1978)

The industry was naturally interested to know whether the new LEAR oil caused heart lesions in test animals. In earlier studies, **animals fed high-erucic-acid rape seed oil showed growth retardation and undesirable changes in various organs, especially the heart, a discovery that touched off the so-called "erucic acid crisis"** and spurred plant geneticists to develop new versions of the seed. (Viles et al, 1978)

The results of the LEAR study were mixed. **Rats genetically selected to be prone to heart lesions developed more lesions on the LEAR oil and the flax oil, than those on olive oil or sunflower oil, leading researchers to speculate that the omega-3 fatty acids (not erucic acid) in LEAR and flax oil might be the culprit.** But rats genetically selected to be resistant to heart lesions showed no significant difference between the four oils tested and LEAR oil did not cause heart problems in mice, in contrast to high-erucic oil which induced severe cardiac necrosis.

In 1979, researchers at the Canadian Institute for Food Science and Technology pooled the results of 23 experiments involving rats at four independent laboratories. All looked at the effects of LEAR and other oils on the incidence of heart lesions. They found that **saturated fats (palmitic and stearic acids) were protective against heart lesions but that high levels of omega-3 fatty acids correlated with high levels of lesions.** They found a lesser correlation with heart lesions and erucic acid. (Trenholm et al, 1979)

Studies carried out at the Health Research and Toxicology Research Divisions in Ottawa, Canada discovered that **rats bred to have high blood pressure and proneness to stroke had shortened life-spans when fed canola oil as the sole source of fat**. The results of a later study suggested that **the culprit was the sterol compounds** in the oil, which "make the cell membrane more rigid" and contribute to the shortened life-span of the animals. (Wallsundera et al, 2000)

Finally, studies carried out at the Health Research and Toxicology Research Divisions in Ottawa, Canada discovered that **rats bred to have high blood pressure and proneness to stroke had shortened life-spans when fed canola oil as the sole source of fat.** (Ratnayake et al, 2000)

Such diets have been **presented with great marketing prowess**, but in actuality they appear to be "payola for the food companies and con-ola for the public."

Omega-3 fatty acids fail glowing predictions

Even though the omega-3s have been touted as "miracle pills," the research on **omega-3 fatty acids** is not conclusive. While some studies indicate that omega-3 fatty acids may be helpful, others showed no effect. One explanation for this may be found in the fact that saturated fats help the body store and use omega-3 fatty acids more effectively.

In fact, **there is evidence that over consumption of omega-3 fatty acids in a diet lacking in saturated fats may actually be bad for the heart.**

In test animals, diets high in canola oil, which is **relatively high in omega-3 fatty acids but low in saturated fats, caused fibrotic heart lesions, vitamin E deficiencies and abnormal changes**

to the blood platelets. (Sauer et al, 1997) (Kramer et al, 1982) (Trenholm et al, 1979)

These studies all point in the same direction -- that **canola oil is definitely not healthy for the cardiovascular system.** Like rapeseed oil, its predecessor, **canola oil is associated with fibrotic lesions of the heart.** It also causes vitamin E deficiency, undesirable changes in the blood platelets and shortened life-span in stroke-prone rats when it was the only oil in the animals' diet. Furthermore, it seems to retard growth, which is why **the FDA does not allow the use of canola oil in infant formula.** (Federal Register, 1985)

When saturated fats are added to the diet, the undesirable effects of canola oil are mitigated. Most interesting of all is the fact that **many studies show that the problems with canola oil are not related to the content of erucic acid, but more with the high levels of omega-3 fatty acids and low levels of saturated fats.** (http://www.dcnutrition.com/news/Detail.CFM?RecordNumber=639. Accessed 6-4-11)

Although research on trans fatty acids found in hydrogenated fats has not received much publicity, it adds up to a strong case for the theory that these manufactured fats contribute to heart disease. The tragedy is that those who are trying to avoid saturated fats and cholesterol will probably eat more trans fatty acids, because these are used in foods promoted as low in saturated fat and cholesterol. Additionally, soy products may lead to arrhythmias and it is said that **areas of the world where coconut is consumed have low levels of heart disease.**

Many recent studies show that **fish oil has failed its marketing claims.** Specifically, **it is no help for heart patients, does not forestall Alzheimer's disease, does not prevent depression, and does not make babies smarter.**

Back in 2005, **a Journal of the American Medical Association report showed that fish oil may actually increase the risk of cardiac arrhythmias in some patients.** (Riatt et al, 2005)

Also, in 2005, JAMA reported that **fish oil does not prevent cancer.** (MacLean et al, 2005)

The following year in 2006, the "British Medical Journal reported that **omega-3 fatty acids have no heart-health benefit. Among nearly 4,000 heart attack patients, no difference was seen between those who consumed omega-3 supplements and those who took placebo pills.** (Hooper et al, 2006)

In 2009, other investigators agreed when they found that **consuming fish does not reduce the risk of heart failure heart failure.** (Dijkstra et al, 2009, the Rotterdam Study)

Then in 2010, the New England Journal of Medicine reported **similarly dismal results with heart patients given omega-3 fatty acids in addition to standard drug therapy.**They had **no reduction in cardiovascular events.** (Krombout et al, 2010)

In the **Alpha Omega Trial**, as reported by Kromhout et al supplementation with a combination of n–3 eicosapentaenoic acid (EPA) and n–3 docosahexaenoic acid (DHA), n–3 alpha-linolenic acid (ALA), or both EPA–DHA and ALA **did not significantly reduce the rate of major cardiovascular events among patients who had had a myocardial infarction** and who were receiving conventional state-of-the-art therapy. In contrast to the Alpha Omega trial, **the Lyon Diet Heart Study (LDHS), which also involved patients who had had a myocardial infarction, was stopped early because a clinical benefit from a margarine intervention had been demonstrated.** In the LDHS, a canola-based margarine that was high in n–3 fatty acids, low in n–6 fatty acids, and high in n–9 fatty acids was the only study-supplied intervention. (Krombout et al, 2010) (de Lorgeril et al, 1994)

The margarine in the Alpha Omega trial had 2.3 times as much n–6 linoleate as the margarine in the LDHS, which probably further weakened the power to demonstrate benefit. (Ramsden et al, 2010)

Paraphrasing from a previous editorial: **"only omega-3 trials that also reduced n–6 polyunsaturates found reductions in cardiovascular and all-cause mortality."** (Leaf, 1999, the Lyon Diet Heart Study)

Surprisingly, in 2009, Harvard linked fish and omega-3 fats to type 2 diabetes. Following **195,204 adults** for 14 to 18 years, researchers reported in 2009 that they had found that **the more fish or long-chain omega-3 fatty acids participants consumed, the higher their risk of developing diabetes.** (Kaushik et al, 2009)

Meanwhile, fish oil manufacturers pinned their hopes on brain function. Maybe fish oil could make you smarter. But in 2010, researchers dashed those hopes also. A group of **867 elderly people** were randomly assigned to either **a fish-oil supplement** or placebo. After two years of supplementation, **elderly adults showed no benefit at all in tests for reaction time, spatial memory, and processing speed measurements.** (Dangour et al, 2010)

A later 2010 JAMA report showed that **omega-3 supplements do not slow mental decline in Alzheimer's patients.** (Quinn et al, 2010)

And at the other end of the age spectrum, babies get no benefit either from omega-3s. A JAMA report showed that **consumption of fish oil during pregnancy does not benefit babies' cognitive development.** (Makrides et al, 2010)

In these reports, **fish oil is sounding more like snake oil.** The new findings linking higher DHA levels to cancer add yet another reason to use caution with fish oil supplements.

Omega-3 (fish oil) Summary

- animals fed high-erucic-acid rape seed oil showed growth retardation and undesirable changes in various organs, especially the heart, a discovery that touched off the so-called "erucic acid crisis." (Viles et al, 1978)

- **Rats genetically selected to be prone to heart lesions developed more lesions on the** Low Erucic Acid Rapeseed (LEAR) **oil and the flax oil, than those on olive oil or sunflower oil, leading researchers to speculate that the omega-3 fatty acids (not erucic acid) in LEAR and flax oil might be the culprit.** But rats genetically selected to be resistant to heart lesions showed no significant difference between the four oils tested and LEAR oil did not cause heart problems in mice, in contrast to high-erucic oil which induced severe cardiac necrosis.

- In 1979, researchers at the Canadian Institute for Food Science and Technology pooled the results of 23 experiments involving rats at four independent laboratories and found that **saturated fats (palmitic and stearic acids) were protective against heart lesions but that high levels of omega-3 fatty acids correlated with high levels of lesions.** (Trenholm et al, 1979)

- the Health Research and Toxicology Research Divisions in Ottawa, Canada discovered that **rats bred to have high blood pressure and proneness to stroke had shortened life-spans when fed canola oil as the sole source of fat**. The results of a later study suggested that **the culprit was the sterol compounds** in the oil, which "make the cell membrane more rigid" and contribute to the shortened life-span of the animals. (Wallsundera et al, 2000)

- **rats bred to have high blood pressure and proneness to stroke had shortened life-spans when fed canola oil as the sole source of fat.** (Ratnayake et al, 2000)

- there is evidence that over consumption of omega-3 fatty acids in a diet lacking in saturated fats may actually be bad for the heart. In test animals, diets high in canola oil, which is **relatively high in omega-3 fatty acids but low in saturated fats, caused fibrotic heart lesions, vitamin E deficiencies and abnormal changes to the blood platelets.** (Sauer et al, 1997) (Kramer et al, 1982) (Trenholm et al, 1979)

- **canola oil is definitely not healthy for the cardiovascular system.** Like rapeseed oil, its predecessor, **canola oil is associated with fibrotic lesions of the heart.** It also causes vitamin E deficiency, undesirable changes in the blood platelets and shortened life-span in stroke-prone rats when it was the only oil in the animals' diet. Furthermore, it seems to retard growth, which is why **the FDA does not allow the use of canola oil in infant formula.** (Federal Register, 1985)

- **many studies show that the problems with canola oil are not related to the content of erucic acid, but more with the high levels of omega-3 fatty acids and low levels of saturated fats**. (http://www.dcnutrition.com/news/Detail. CFM?RecordNumber=639. Accessed 6-4-11)

- soy products may lead to arrhythmias and it is said that **areas of the world where coconut is consumed have low levels of heart disease.**

- **fish oil has failed its marketing claims.** Specifically, **it is no help for heart patients, does not forestall Alzheimer's disease, does not prevent depression, and does not make babies smarter.**

- in 2005, **a Journal of the American Medical Association report showed that fish oil may actually increase the risk of cardiac arrhythmias in some patients.** (Riatt et al, 2005)

- in 2005, JAMA reported that **fish oil does not prevent cancer.** (MacLean et al, 2005)

- in 2006, the "British Medical Journal reported that **omega-3 fatty acids have no heart-health benefit. Among nearly 4,000 heart attack patients, no difference was seen between those who consumed omega-3 supplements and those who took placebo pills.** (Hooper et al, 2006)

- In 2009, other investigators agreed when they found that **consuming fish does not reduce the risk of heart failure heart failure.** (Dijkstra et al, 2009, the Rotterdam Study)

- in 2010, the New England Journal of Medicine reported **similarly dismal results with heart patients given omega-3 fatty acids in addition to standard drug therapy.** They had **no reduction in cardiovascular events.** (Krombout et al, 2010)

- In the **Alpha Omega Trial**, supplementation with EPA, DHA and ALA **not significantly reduce the rate of major cardiovascular events among patients who had had a myocardial infarction. However, in** contrast to the Alpha Omega trial, **the Lyon Diet Heart Study (LDHS), which also involved patients who had had a myocardial infarction, was stopped early because a clinical benefit from a margarine intervention had been demonstrated.** (Krombout et al, 2010) (de Lorgeril et al, 1994)

- "**only omega-3 trials that also reduced n–6 polyunsaturates found reductions in cardiovascular and all-cause mortality.**" (Leaf, 1999, the Lyon Diet Heart Study)

- in 2009, Harvard linked fish and omega-3 fats to type 2 diabetes. Following **195,204 adults** for 14 to 18 years, researchers reported in 2009 that they had found that **the more fish or long-chain**

omega-3 fatty acids participants consumed, the higher their risk of developing diabetes. (Kaushik et al, 2009)

- in 2010, 867 **elderly adults showed no benefit at all from** a fish-oil supplement **in tests for reaction time, spatial memory, and processing speed measurements.** (Dangour et al, 2010)

- A later 2010 JAMA report showed that **omega-3 supplements do not slow mental decline in Alzheimer's patients.** (Quinn et al, 2010)

- also in 2010, it was shown that babies get no benefit either from omega-3s. A JAMA report showed that **consumption of fish oil during pregnancy does not benefit babies' cognitive development.** (Makrides et al, 2010)

REFERENCES: Cocoa, chocolate and flavonoids and Omega-3 and fish oil section

(Dangour et al, 2010) (Dangour AD, Allen E, Elbourne D, et al. Effect of 2-y n23 long-chain polyunsaturated fatty acid supplementation on cognitive function in older people: a randomized, double-blind, controlled trial. Am J Clin Nutr. 2010;91:1725-1732)

(de Lorgeril et al, 1994) (de Lorgeril M, Renaud S, Mamelle N, et al. Mediterranean alpha-linolenic acid-rich diet in secondary prevention of coronary heart disease. Lancet 1994;343:1454-1459[Erratum, Lancet 1995;345:738.])

(Dijkstra et al, 2009, the Rotterdam Study) (Dijkstra SC, Brouwer IA, van Rooij FJA, Hofman A, Witteman JCM, Geleijnse JM. Intake of very long chain n-3 fatty acids from fish and the incidence of heart failure: the Rotterdam Study. Eur J Heart Fail. 2009;11:922-928)

(Hooper et al, 2006) (Hooper L, Thompson RL, Harrison RA, et al. Risks and benefits of omega-3 fats for mortality, cardiovascular disease, and cancer: systematic review. BMJ. 2006;332:752-760)

(Kaushik et al, 2009) (Kaushik M, Mozaffarian D, Spiegelman D, Manson JE, Willett WC, Hu FB. Long-chain omega-3 fatty acids, fish intake, and the risk of type 2 diabetes mellitus. Am J Clin Nutr. 2009;90:613-620)

(Kramer et al, 1982) (Kramer, JKG and others. Reduction of Myocardial Necrosis in Male Albino Rats by Manipulation of Dietary Fatty Acid Levels. Lipids 17, 372-382, 1982)

(Krombout et al, 2010) (Kromhout D, Giltay EJ, Geleijnse JM. n-3 fatty acids and cardiovascular events after myocardial infarction. N Engl J Med. 2010;363:2015-2026)

(Lau, 2005) (Lau FC, Shukitt-Hale B, Joseph JA. The beneficial effects of fruit polyphenols on brain aging. Neurobiol Aging 2005;26S:S128-32)

(Leaf, 1999, the Lyon Diet Heart Study) (Leaf A. Dietary prevention of coronary heart disease: the Lyon Diet Heart Study. Circulation 1999;99:733-735)

(MacLean et al, 2005) (MacLean CH, Newberry SJ, Mojica WA, et al. Effects of omega-3 fatty acids on cancer risk: a systematic review. JAMA. 2005;295:405-415)

(Makrides et al, 2010) (Makrides M, Gibson RA, McPhee AJ, et al. Effect of DHA Supplementation During Pregnancy on Maternal Depression and Neurodevelopment of Young Children. JAMA. 2010;304:1675-1683)

(Quinn et al, 2010) (Quinn JF, Rama R, Thomas RG, et al. Docosahexaenoic acid supplementation and cognitive decline in Alzheimer disease. JAMA. 2010;304:1903-1911)

(Ramsden et al, 2010) (Ramsden CE, Hibbeln JR, Majchrzak SF, Davis JM. n-6 Fatty acid-specific and mixed polyunsaturate dietary interventions have different effects on CHD risk: a meta-analysis of randomised controlled trials. Br J Nutr 2010;104:1586-1600)

(Ratnayake et al, 2000) (WMN Ratnayake and others. Influence of Sources of Dietary Oils on the Life Span of Stroke-Prone Spontaneously Hypertensive Rats. Lipids, 2000;35(4):409-420)

(Riatt et al, 2005) (Riatt MH, Connor WE, Morris C, et al. Fish oil supplementation and risk of ventricular tachycardia and ventricular fibrillation in patients with implantable defibrillators: a randomized controlled trial. JAMA. 2005;293:2884-2891)

(Sauer et al, 1997) (Sauer, FD and others. Additional vitamin E required in milk replacer diets that contain canola oil. Nutrition Research 17(2), 259-269, 1997)

(Trenholm et al, 1979) (HL Trenholm and others. An Evaluation of the Relationship of Deitary Fatty Acids to Incidence of Myocardial Lesions in Male Rats. Canadian Institute of Food Science Technology Journal, October 1979;12(4):189-193)

(Viles et al, 1978) (RO Vles and others. Nutritional Evaluation of Low-Erucic-Acid Rapeseed Oils. Toxicological Aspects of Food Safety, Archives of Toxicology, Supplement 1, 1978:23-32)

(Wallsundera et al, 2000) (MN Wallsundera and others. Vegetable Oils High in Phytosterols Make Erythrocytes Less Deformable and Shorten the Life Span of Stroke-Prone Spontaneously Hypertensive Rats. Journal of the American Society for Nutritional Sciences, May, 2000;130(5):1166-78)

CHAPTER TEN

AN EPIC CHRONOLOGY OF ANTIOXIDANT STUDY REPORTS

This is a selective listing of 316 interventional initial or follow up study reports, showing either marginal effect, negligible effects, no effects at all. Ninety of these reports show harmful effects of the antioxidant vitamins **A** (beta carotene), **C** (ascorbic acid) or **E** (alpha tocopherol), selenium, other antioxidants or combinations thereof. Total number of participants for all of the above studies is in excess of 15,000,000 subjects, some of which may have been repeats in follow up or parallel studies. Studies with multivitamins, which contain the antioxidant vitamins, were also included.

ANTIOXIDANT VITAMIN FAILURES AND DANGERS

I have 316 study reports, on over 15 million subjects, showing "disappointing" results with antioxidants in preventing or reversing disease. Thus, prudent questions are, "Why take them in the absence of a deficiency state?" "Why waste money for marginally effective or harmful pills?"

Dietary antioxidant supplements are taken as a supplementary component of the diet that improves the capability of something else or is added as a supplement to what seems insufficient. They are added in quantities calculated to make up for alleged deficiencies or to inhibit allegedly harmful normal chemical reactions.

Antioxidants are substances that inhibit oxidation or inhibit reactions promoted by oxygen, peroxides or EMODs. Basically, the term "antioxidant" could be substituted with the term "inhibitor" of oxidation.

Many of these study reports are intervention trials.

**316 Antioxidant Intervention Trial &
Analysis Study Reports:**
R.M. Howes M.D., Ph.D.

The following list will be arranged as: Paper (study) number; Title of Study or Paper; Author reference; and Chronological Year; and Number of participants or trials in the respective study.

1. **Failure of High-dose Vitamin C (ascorbic acid) Therapy to Benefit Patients with Advanced Cancer. A Controlled Trial.** (Creagan et al, 1979) (#159 patients with advanced cancer)
2. **The Multiple Risk Factor Intervention Trial (MRFIT).** (The Multiple Risk Factor Intervention Trial Research Group, 1982) (#360,000 middle aged men).
3. **High-dose Vitamin C versus Placebo in the Treatment of Patients with Advanced Cancer Who Have had no Prior Chemotherapy.** (Moertel et al, 1985) (#100 patients with advanced colorectal cancer)
4. **Skin Cancer Prevention Study** (Greenberg et al, 1990) (#1,805 men and women with recent nonmelanoma skin cancer)
5. **Diet in the Epidemiology of Postmenopausal Breast Cancer in the New York State Cohort** (Graham et al, 1992) (#18,586 postmenopausal women)
6. **Women's Health Study (WHS)** (Buring and Hennekens, 1992); (#39,876 healthy women)

7. **Isotretinoin-Basal Cell Carcinoma Study Group** (Tangrea et al, 1992, 1993) (#981 patients with two or more previously treated basal cell carcinomas)
8. **Prospective Study of the Intake of Vitamins C, E, and A and the Risk of Breast Cancer** (Hunter et al, 1993) (#89,494 women)
9. **Serum micronutrients and the subsequent risk of cervical cancer in a population-based nested case-control study.** (Batieha, 1993) (#15,161 women)
10. **A randomized trial of vitamin A and vitamin E supplementation for retinitis pigmentosa.** (Berson, 1993) (#601 patients aged 18 through 49 years with retinitis pigmentosa)
11. **Lack of an association between serum vitamin E and myocardial infarction in a population with high vitamin E levels.** (Hense et al. 1993) (#4,002 men and women)
12. **Nutrition intervention trials in Linxian, China: supplementation with specific vitamin/mineral combinations, cancer incidence, and disease-specific mortality in the general population.** (Blot et al, 1993) (#29,584
13. **α-Tocopherol, β-Carotene Cancer Prevention Study (ATBC study)** (Heinonen et al, 1994) (#29,133 men)
14. **Polyp Prevention Study** (Greenberg et al, 1994) (#864)
15. **Blood antioxidants and indices of lipid peroxidation in subjects with angina pectoris.** (Duthie et al. 1994) (#25 subjects with stable angina pectoris with 200 matched controls)
16. **Effect of vitamin C supplementation on lipoprotein cholesterol, apolipoprotein, and triglyceride concentrations** (Jaques et al, 1995) (#139)
17. **The effect of high-dose ascorbate supplementation on plasma lipoprotein(a) levels in patients with premature coronary heart disease** (Bostom et al, 1995) (#44 patients with premature CHD)
18. **Cholesterol Lowering Atherosclerosis Study (CLAS) (1995)** (Hodis et al, 1995) (#156 men)
19. **Effects of Vitamin E on susceptibility of low-density lipoprotein and low-density lipoprotein subfractions to**

oxidation and on protein glycation in **NIDDM.** (Reaven, 1995) (#21 men with NIDDM)

20. **Excretion of alpha-tocopherol into human seminal plasma after oral administration** (Moilanen and Hovatta, 1995) (#15 unselected male volunteers)

21. **Dietary factors and risk of prostate cancer: a case-control study in Ontario, Canada.** (Rohan et al. 1995) (#414)

22. **The β-Carotene and Retinol Efficacy Trial (CARET)** (Omenn et al, 1996) (#14,254 heavy smokers and 4,060 asbestos workers) (total #18,314)

23. **Cambridge Heart Antioxidant Study (CHAOS)** (Stephens et al., 1996) (#2,002 patients with coronary atherosclerosis)

24. **Physicians' Health Study (PHSI)** (Hennekens et al, 1996) (#22,071 US Physicians and Malignant Neoplasms or CVD)

25. **Dietary Antioxidant Vitamins and Death from Coronary Heart Disease in Postmenopausal Women** (Kushi et al, 1996) (#34,486 postmenopausal women with no cardiovascular disease)

26. **Mortality associated with low plasma concentration of beta carotene and the effect of oral supplementation** (Greenberg et al, 1996) (#1,720 men and women)

27. **The effect of antioxidant treatment on human spermatozoa and fertilization rate in an in vitro fertilization program** (Geva et al, 1996) (#Fifteen fertile normospermic male)

28. **Energy, nutrient intake and prostate cancer risk: a population-based case-control study in Sweden.** (Andersson et al. 1996) (#1,062)

29. **Effects of selenium supplementation for cancer prevention in patients with carcinoma of the skin. A randomized controlled trial. Nutritional Prevention of Cancer Study Group.** (Clark et al. 1996) (#1,312 patients)

30. **Alpha-Tocopherol and beta-carotene supplements and lung cancer incidence in the alpha-tocopherol, beta-carotene cancer prevention study: effects of base-line**

characteristics and study compliance. (Albanes et al, 1996) (#29,133 men, smokers

31. **Antioxidant Vitamin Effect on Traditional CVD Risk Factors** (Miller et al, 1997) (#297 retired teachers)
32. **ATBC Sub-Study Shows Increased CVD Deaths** (Rapola et al, 1997) (#1,862 men, with prior myocardial infarction)
33. **Effect of preoperative supplementation with alpha-tocopherol and ascorbic acid on myocardial injury in patients undergoing cardiac operations** (Westhuyzen et al, 1997) (#77 undergoing elective coronary artery bypass grafting)
34. **The influence of antioxidant nutrients on platelet function in healthy volunteers** (Calzada et al, 1997) (#40 healthy volunteers)
35. **The Multivitamins and Probucol Study** (Tardif et al, 1997) (#317 participants)
36. **Vitamin C intake and cardiovascular disease risk factors in persons with non-insulin-dependent diabetes mellitus** (Mayer-Davis et al, 1997) (#Insulin Resistance Atherosclerosis Study (IRAS, n = 520) **and from the San Luis Valley Diabetes Study (SLVDS, n = 422)** (total #942)
37. **Preformed Vitamin A Study Showed No Trend to Reduce Breast Cancer Risk.** (Longnecker, 1997) (#3,543 cases and 9,406 controls)
38. **No effect of supplementation with vitamin E, ascorbic acid, or coenzyme Q10 on oxidative DNA damage estimated by 8-oxo-7,8-dihydro-2'-deoxyguanosine excretion in smokers.** (Priemé et al. 1997) (#142 smoking men)
39. **Vitamin E Worsens Metabolic Parameters in Type 2 Diabetics.** (Skrha et al. 1997) (#12)
40. **Effect of Vitamin E and Beta Carotene on the Incidence of Primary Nonfatal Myocardial Infarction and Fatal Coronary Heart Disease** (Virtamo et al, 1998) (#27,271 Finnish male smokers)
41. **SU.VI.MAX** (Vasquez et al., 1998) (#13,017 French adults)
42. **A Sub-Study of SU.VI.MAX** (#1,162 subjects aged older than 50 years)

43. **The Nurses' Health Study and Folic Acid and Colon Cancer** (Giovannucci, 1998) (#88,756 women taking vitamin C and B-carotene, for 8 years)
44. **Effect of B-group vitamins and antioxidant vitamins on hyperhomocysteinemia** (Woodside, et al. 1998) (#101 men)
45. **Relationships of serum carotenoids, retinol, alpha-tocopherol, and selenium with breast cancer risk: results from a prospective study in Columbia, Missouri (United States).** (Dorgan, 1998) (#105 cases of histologically confirmed breast cancer)
46. **Incidence of cataract operations in Finnish male smokers unaffected by alpha tocopherol or beta carotene supplements.** (Teikari, 1998) (#28,934 male smokers)
47. **Effects of α tocopherol and β carotene supplements on symptoms, progression, and prognosis of angina pectoris.** (Rapola et al, 1998) (#1,795 male smokers aged 50–69 years who had angina pectoris)
48. **The effects of antioxidant supplementation during Percoll preparation on human sperm DNA integrity** (Hughes et al, 1998) (#150 patients)
49. **Dietary intake of antioxidant (pro)-vitamins, respiratory symptoms and pulmonary function: the MORGEN study** (Grievink et al. 1998) (#6,555 adults)
50. **Antioxidant nutrient intake and diabetic retinopathy: the San Luis Valley Diabetes Study.** (Mayer-Davis et al. 1998) (#387 participants with type 2 diabetes)
51. **Prospective association between lipid soluble antioxidants and coronary heart disease in men. The Multiple Risk Factor Intervention Trial.** (Evans et al. 1998) (#743)
52. **GISSI-Prevention Trial** (GISSI-Prevenzione Investigators; 1999) (#11,324 patients with recent MI)
53. **Women's Health Study** (Lee et al., 1999) (#39,876 healthy women); 50 mg β-carotene (alternate days)
54. **The Health Professionals Follow-Up Study** (Ascherio et al. 1999) (#43,738 men)

55. **Familial hypercholesterolemia, intima-to-media thickness (FH IMT study)** (Raal et al, 1999) (#15 with homozygous familial hypercholesterolemia)

56. **Beta carotene supplementation in prevention of basal-cell and squamous-cell carcinomas of the skin** (Green et al, 1999) (#1,383 participants)

57. **Vitamins A, C and E and the risk of breast cancer: results from a case-control study in Greece,** (Bohlke, 1999) (#830 patients with breast cancer plus 1,548 controls)

58. **Dietary antioxidants and risk of myocardial infarction in the elderly: the Rotterdam Study.** (Klipstein-Grobusch et al, 1999) (#4,802 participants of the Rotterdam Study aged 55–95 y who were free of MI)

59. **A prospective study of vitamin supplement intake and cataract extraction among US women.** (Chasan-Taber et al, 1999) (#47,152 female nurses)

60. **Antioxidant treatment of patients with asthenozoospermia or moderate oligoasthenozoospermia with high-dose vitamin C and vitamin E: a randomized, placebo-controlled, double-blind study** (Rolf et al, 1999) (#31 without genital infection but with asthenozoospermia)

61. **Antioxidant supplementation in vitro does not improve human sperm motility** (Donnelly et al, Fertil Steril. 1999) (#60 patients)

62. **The effect of ascorbate and alpha-tocopherol supplementation in vitro on DNA integrity and hydrogen peroxide-induced DNA damage in human spermatozoa** (Donnelly et al, Mutagenesis. 1999) (#Semen samples with normozoospermic and asthenozoospermic profiles (n = 15 for each control and antioxidant group)

63. **Heart Outcome Prevention Evaluation Study (HOPE)** (Yusuf et al, 2000) (#9,541 patients at high risk for cardiovascular events or diabetes)

64. **Meta-Analysis of Vitamin E in CVD, Ischemic Heart Disease (IHD) and Mortality** (Dagenais et al. 2000) (#51,000 participants)

65. **Vitamin C and the risk of acute myocardial infarction.** (Riemersma et al, 2000) (#180 males with a first AMI and 177 healthy volunteers)
66. **The effects of combined conventional treatment, oral antioxidants and essential fatty acids on sperm biology in subfertile men** (Comhaire et al, 2000) (#27 infertile men)
67. **Oral vitamin C and endothelial function in smokers: short-term improvement, but no sustained beneficial effect.** (Raitakari et al. 2000) (#20 healthy young adult smokers)
68. **Effects of vitamin E on chronic and acute endothelial dysfunction in smokers.** (Neunteufl et al, 2000) (#22 healthy male smokers)
69. **Smoking characteristics, antioxidant vitamins, and carotid artery wall thickness among life-long smokers.** (de Waart et al. 2000) (#158 male life-long cardiovascular disease (CVD)-free smokers)
70. **Controlled trial of alpha-tocopherol and beta-carotene supplements on stroke incidence and mortality in male smokers.** (Leppala et al. 2000) (#28,519 male cigarette smokers)
71. **A randomized, 12-year primary-prevention trial of beta carotene supplementation for nonmelanoma skin cancer in the Physician's Health Study.** (Frieling et al, 2000) (#22,071)
72. **Serum carotenoids and atherosclerosis. The Rotterdam Study.** (Klipstein-Grobusch et al, 2000) (#108 subjects with aortic atherosclerosis)
73. **Multivitamin use and mortality in a large prospective study.** (Watkins et al, 2000) (#1,063,023 adults)
74. **Dietary antioxidant vitamins, retinol, and breast cancer incidence in a cohort of Swedish women.** (Michels, 2001) (#59,036 women free of cancer)
75. **The Perth Carotid Ultrasound Disease Assessment Study (CUDAS)** (McQuillan et al 2001) (#1,111 subjects)
76. **Randomized Trial of Supplemental ß-Carotene to Prevent Second Head and Neck Cancer** (Mayne et al,

2001) (#264 patients who had been curatively treated for a recent early-stage squamous cell carcinoma of the oral cavity, pharynx, or larynx.)

77. **Age-Related Eye Disease Study Research Group (AREDS)** (AREDS, 2001) (#4,757 participants)
78. **The secondary prevention HDL Atherosclerosis Treatment study (HATS)** (Brown et al, 2001) (#160 participants)
79. **Vitamin C and Vitamin E Supplement Use and Colorectal Cancer Mortality in a Large American Cancer Society cohort. (Cancer Prevention Study II cohort - CPS-II)** (Jacobs, 2001) (#711,891 men and women in U.S.A.)
80. **Risk of Ovarian Carcinoma and Consumption of Vitamins A, C and E and Specific Carotenoids: a prospective analysis.** (Fairfield, 2001) (#80,326 women)
81. **Carotenoids, Alpha-tocopherols, and Retinol in Plasma and Breast Cancer Risk in Northern Sweden.** (Hulten, 2001) (#201 cases and 290 referents)
82. **Risk of ovarian carcinoma and consumption of vitamins A, C, and E and specific carotenoids: a prospective analysis** (Fairfield et al. 2001) (#80,326 participants in Nurses' Health Study)
83. **A controlled clinical trial of vitamin E supplementation in patients with congestive heart failure** (Keith et al, 2001) (#56 with advanced heart failure)
84. **Antioxidant vitamins and prevention of cardiovascular disease: epidemiological and clinical trial data.** (Marchioli et al. 2001) (#not available)
85. **The Vitamin E Atherosclerosis Prevention Study (VEAPS)** (Hodis et al, 2002) (#353 were randomized (176 placebo, 177 vitamin E)
86. **MRC/BHF** (MRC/BHF, 2002) (#20,536); (600 mg vitamin E, 250 mg vitamin C and 20 mg beta-carotene daily)
87. **Antioxidant Vitamins and US Physician CVD Mortality** (Muntwyler et al. 2002) (#83,639 male U.S.A. physicians)

88. **Women's Angiographic Vitamin and Estrogen (WAVE) Trial** (Waters et al, 2002) (#423 postmenopausal women, with at least one 15% to 75% coronary stenosis)
89. **Mega-dose vitamins and minerals in the treatment of non-metastatic breast cancer: an historical cohort study** (Lesperance et al, 2002) (#90 patients with non-metastatic breast cancer who received conventional treatment)
90. **The Roche European American Cataract Trial (REACT)** (Chylack et al. 2002) (#445 patients)
91. **Vitamin E supplementation and macular degeneration** (Taylor et al, 2002) (#1,193 subjects)
92. **Vitamin E on Cardiovascular and Microvascular Outcomes in High-Risk Patients With Diabetes. Results of the HOPE Study and MICRO-HOPE Substudy** (Lonn et al. 2002) (#3,654 with diabetes)
93. **Vitamin C and Vitamin E Supplement Use and Bladder Cancer Mortality in a Large Cohort of US Men and Women** (Cancer Prevention Study II (CPS-II) (Jacobs et al., 2002) (#991,522 US adults in the Cancer Prevention Study II (CPS-II) cohort.)
94. **Supplemental Vitamin C & E and Multivitamin use and Stomach cancer Mortality in U.S.A.** (Jacobs et al. Jan. 2002) (#1,045,923)
95. **Vitamin E and C Supplements and Risk of Dementia** (Laurin et al, 2002) (#3,734 Japanese men)
96. **Retinol intake and bone mineral density in the elderly: the Rancho Bernardo Study.** (Promislow et al, 2002) (#570 women and 388 men)
97. **Vitamin A intake and hip fractures among postmeno-pausal women.** (Feskanich et al, 2002) (#72,337 postmeno-pausal women)
98. **A prospective study on supplemental vitamin e intake and risk of colon cancer in women and men.** (Wu et al, 2002) (#87,998 females from the Nurses' Health Study and 47, 344 males from the Health Professionals Follow-up Study) (#135,332 total participants)

99. **The Collaborative Primary Prevention Project (PPP)** (Chiabrando et al., 2002) (#144 participants with CHD risk factors)
100. **Antioxidants to slow aging, facts and perspectives. (STUDY)** (Bonnefoy et al. 2002) (#not applicable)
101. **Selenium and vitamin E supplements for prostate cancer: evidence or embellishment?** (Moyad et al. 2002) (#not applicable)
102. **Vitamin E in cardiovascular disease: has the die been cast?** (Yusoff. K. 2002) (#not applicable)
103. **Prospective study of carotenoids, tocopherols, and retinoid concentrations and the risk of breast cancer.** (Sato et al, 2002) (#590)
104. **Lack of effect of oral vitamin C on blood pressure, oxidative stress and endothelial function in Type II diabetes** (Darko et al, 2002) (#35)
105. **Antioxidant vitamins and risk of cardiovascular disease. Review of large-scale randomised trials** (Clarke, Armitage, 2002) **(#70,000** people from 3 large-scale trials in healthy populations and on vitamin E supplementation involving 29,000 patients at high-risk of cardiovascular disease from 5 large-scale trials)
106. **Vitamins E & A fail to reduce incidence or mortality of lung cancer: Cochrane Database Syst Rev. 2003.** (Caraballoso et al., 2003) (#109,394 participants)
107. **Use of antioxidant vitamins for the prevention of cardiovascular disease: meta-analysis of randomized trials**. (Vivekananthan et al., 2003) (The vitamin E trials involved a total of #81,788 patients, and the beta-carotene trials involved #138,113)
108. **Antioxidant Vitamins Effect on Alzheimer's Disease: Washington Heights-Inwood Columbia Aging Project** (Luchsinger et al, 2003) (#980 elderly subjects)
109. **Neoplastic and Antineoplastic Effects of Beta Carotene on Colorectal Adenoma Recurrence: Results of a Randomized Trial** (Baron et al, 2003) (#864 subjects who had had an adenoma removed and were polyp-free)

110. **Routine Vitamin Supplementation To Prevent Cardiovascular Disease: A Summary of the Evidence for the U.S. Preventive Services Task Force** (Morris and Carson, 2003)

111. **Midlife Dietary Intake of Antioxidants and Risk of Late-Life Incident Dementia: The Honolulu-Asia Aging Study** (Laurin et al, 2003) (#2,459 men)

112. **Serum retinol levels and the risk of fracture.** (Michealsson, 2003) (#2,322 men)

113. **Impact of simvastatin, niacin, and/or antioxidants on cholesterol metabolism in CAD patients with low HDL.** (Matthan et al, 2003) (#123 HATS participants)

114. **A randomized trial of beta carotene and age-related cataract in US physicians.** (Christen et al, 2003) (#22,071 Male US physicians aged 40 to 84 years)

115. **Plasma carotenoids and tocopherols and risk of myocardial infarction in a low-risk population of US male physicians.** (Hak et al. 2003) (#531 physicians diagnosed with MI)

116. **Selenium supplementation and secondary prevention of nonmelanoma skin cancer in a randomized trial.** (Duffield-Lillico, 2003) (#1,312)

117. **Acetylcysteine, coronary procedure and prevention of contrast-induced worsening of renal function: which benefit for which patient?** (Kefer et al. 2003) (#108)

118. **Oral acetylcysteine does not protect renal function from moderate to high doses of intravenous radiographic contrast** (Boccalandro et al. 2003) (#106 consecutive patients)

119. **Supplemental vitamin C increase cardiovascular disease risk in women with diabetes** (Lee et al, 2004) (#1,923 postmenopausal women who reported being diabetic)

120. **Cochrane Database Syst Rev. 2004: Vitamins E & A fail to reduce incidence or mortality of gastrointestinal cancer.** (Cochrane Database Syst Rev. G. Bjelakovic et al, 2004) (#170,525 participants)

121. **ATBC 6-year followup study (2004)** (Thornwall et al., 2004) (#29,133 male smokers)

122. **HOPE study of vitamin E on renal insufficiency (2004)** (Mann et al, 2004) (#993 people with a serum creatinine > or =1.4 to 2.3 mg/dL. And renal insufficiency)

123. **Randomized trials of vitamin E in the treatment and prevention of cardiovascular disease (2004)** (Eidelman et al., 2004) (7 large-scale randomized trials)

124. **Effect of supplemental vitamin E for the prevention and treatment of cardiovascular disease** (Shekelle et al, 2004) (#Eighty-four eligible trials)

125. **SU.VI.MAX Study (2004)** (Hercberg et al, 2004) (#A total of 13,017 French adults)

126. **Meta-analysis: high-dosage vitamin E supplementation may increase all-cause mortality** (Miller et al., 2004) (#135,967 subjects)

127. **The role of vitamin E in the prevention of coronary events and stroke. Meta-analysis of randomized controlled trials** (Alkhenizan and Al-Omran, 2004) (#80,645 subjects)

128. **Oats, Antioxidants and Endothelial Function in Overweight, Dyslipidemic Adults** (Katz et al, 2004) (#30) (16 males ≥age 35; 14 postmenopausal females)

129. **Vitamin C Worsens Coronary Atherosclerosis in Those with Two Copies of the Haptoglobin 2 Gene.** (Levy, 2004) (#299 postmenopausal women)

130. **Vitamin C for preventing and treating the common cold. Cochrane Database Syst Rev. 2004;(4):CD000980.** (Douglas, 2004) (#11,350 study participants)

131. **Vitamin E supplementation and cataract: randomized controlled trial.** (McNeil, 2004) (#1,193 eligible subjects with early or no cataract)

132. **Antioxidant vitamins and coronary heart disease risk: a pooled analysis of 9 cohorts.** (Knekt et al, 2004) (#293,172 subjects free of CHD at baseline)

133. **Dietary carotenoids and risk of lung cancer in a pooled analysis of seven cohort studies.** (Mannisto et al, 2004) (#399,765 participants)

134. **Impact of antioxidants, zinc, and copper on cognition in the elderly: a randomized, controlled trial.** (Yaffe et al, 2004) (#2,166 elderly persons)

135. **No long-term effect of combined vitamins E and C on coronary and peripheral endothelial function** (Kinlay et al. 2004) (#not applicable)

136. **Age-related cataract in a randomized trial of beta-carotene in women.** (Christen et al. 2004) (#39,876)

137. **A review of the epidemiological evidence for the 'antioxidant hypothesis'. (The British Nutrition Foundation was recently commissioned by the Food Standards Agency) (an independent review of the scientific literature on the role of antioxidants in chronic disease prevention)** (Stanner et al. Pub Health Nutr. 2004) (#not applicable) (British Nutrition Foundation independent review)

138. **Fruit, vegetable, and antioxidant intake and all-cause, cancer, and cardiovascular disease mortality in a community-dwelling population in Washington County, Maryland (CLUE)** (Genkinger et al. 2004) (#6,151)

139. **Vitamin E and beta-carotene supplementation and hospital-treated pneumonia incidence in male smokers.** (Hemila et al, 2004) (#29,133 men aged 50 to 69 years, who smoked at least five cigarettes per day)

140. **Early Infant Multivitamin Supplementation Is Associated With Increased Risk for Food Allergy and Asthma** (Milner et al. 2004) (#over 8,000)

141. **Antioxidants and cardiovascular disease: Still a topic of interest.** (Nojiri et al, 2004)

142. **Effects of Long-Term Daily Low-Dose Supplementation With Antioxidant Vitamins and Minerals on Structure and Function of Large Arteries.** (Zureik et al, 2004) (#1,162)

143. **Use of multivitamins and prostate cancer mortality in a large cohort of US men.** (Stevens et al, 2005) (#475,726 men who were cancer-free)

144. **Vitamin A Supplementation for Reducing the Risk of Mother-to-child Transmission of HIV Infection.** (Wiysonge et al, 2005) (#3,033 females)

145. **The Alzheimer's Disease Cooperative Study (ADCS) Group** (Petersen et al, 2005) (#769 subjects)

146. **Vitamin E Supplementation in Alzheimer's Disease, Parkinson's Disease, Tardive Dyskinesia, and Cataract: Part 2** (Pham et al, 2005)

147. **Dementia and Alzheimer's Disease in Community-Dwelling Elders Taking Vitamin C and/or Vitamin E:** (Fillenbaum et al, 2005) (#626 elderly)

148. **HOPE-TOO Extension** (Lonn et al, 2005) (#3,994 original study enrollees)

149. **Women's Health Study (WHS)** (Lee et al, 2005) (#39,876 apparently healthy US women)

150. **A randomized trial of antioxidant vitamins to prevent second primary cancers in head and neck cancer patients** (Bairati et al, 2005 Apr 6) (#540 patients with stage I or II head and neck cancer treated by radiation therapy)

151. **Randomized trial of antioxidant vitamins to prevent acute adverse effects of radiation therapy in head and neck cancer patients** (Bairati et al, 2005 Aug 20) (#540 patients with stage I or II head and neck cancer treated by radiation therapy)

152. **Effect of intensive lipid lowering, with or without antioxidant vitamins, compared with moderate lipid lowering on myocardial ischemia** (Stone et al, 2005) (#300 patients with stable coronary disease)

153. **Vitamin C and vitamin E for Alzheimer's disease.** (Boothby and Doering, 2005)

154. **Effects of vitamins C and E on oxidative stress markers and endothelial function in patients with systemic lupus erythematosus: a double blind, placebo controlled pilot study.** (Tam et al. 2005) (#39 patients with SLE)

155. **Antioxidants for preventing pre-eclampsia. (Rumbold et al, Apr 18, 2005. CD004072)** (#35,812 women and 37,353 pregnancies)

156. **Vitamin E in the primary prevention of cardiovascular disease and cancer: the Women's Health Study: a randomized controlled trial.** (Lee et al. 2005) (#39,876 healthy women)

157. **Vitamin E supplementation in pregnancy.** (Rumbold et al, Apr 18, 2005. CD004072) (#566 women)

158. **Antioxidant vitamin supplementation in the prevention of cardiovascular disease (they are not recommended).** (Yuen et al. 2005)

159. **Effect of multivitamin and multimineral supplements on morbidity from infections in older people (MAVIS trial)** (Avenell et al. 2005) (910 men and women 65 or over)

160. **Vitamin K3 triggers human leukemia cell death through hydrogen peroxide generation and histone hyperacetylation** (Lin, Kang, Zheng, 2005) (#)

161. **Fruits and vegetables and ovarian cancer risk in a pooled analysis of 12 cohort studies.** (Koushik et al, 2005) (#560,441 women)

162. **Perioperative N-acetylcysteine to prevent renal dysfunction in high-risk patients undergoing cabg surgery: a randomized controlled trial** (Burns KE, et al. 2005) (#295 patients required elective or urgent CABG)

163. **Vitamin-mineral supplementation and the progression of atherosclerosis** (Bleys et al, 2006) (searched the MEDLINE, EMBASE, and CENTRAL databases)

164. **Multivitamin/mineral supplements and prevention of chronic disease.** (Huang et al, 2006 May)

165. **The Efficacy and Safety of Multivitamin and Mineral Supplement Use To Prevent Cancer and Chronic Disease in Adults: A Systematic Review for a National Institutes of Health State-of-the-Science Conference.** (Huang et al, 2006 Sept)

166. **Antioxidants Vitamin C and Vitamin E for the Prevention and Treatment of Cancer** (Coulter et al, 2006) (Thirty-eight studies; participant # not available)
167. **Vitamin C levels in Type 2 diabetes and low vitamin C levels does not improve endothelial dysfunction or insulin resistance** (Chen et al, 2006) (#32 type 2 diabetics)
168. **Meta-analysis: antioxidant supplements for primary and secondary prevention of colorectal adenoma (2006)** (Bjelakovic et al., 2006) (#17,620 participants)
169. **Australian Collaborative Trial of Supplements (ACTS)** (Rumbold et al, 2006) (#1,877 subjects)
170. **The Antioxidants in Prevention of Cataracts Study (APC Study): effects of antioxidant supplements on cataract progression in South India.** (Gritz, 2006) (#798)
171. **Vitamins in Pre-eclampsia (VIP) Trial Consortium** (Poston et al., 2006) (#2,410 women at increased risk for pre-eclampsia, analayzed 2,395)
172. **SU.VI.MAX (2006) Antioxidants do not affect fasting blood glucose** (Czernichow et al, 2006) (#3,146 subjects)
173. **Vitamin E and Risk of Type 2 Diabetes in the Women's Health Study** (Liu et al., 2006) (#38,716 apparently healthy U.S. women)
174. **Vitamin E supplementation and cognitive function in women: The Women's Health Study (2006)** (Kang et al., 2006) (#39,876 healthy US women)
175. **Supplemental and dietary vitamin E, beta-carotene, and vitamin C intakes and prostate cancer risk (PLCO Trial)** (Kirsh et al, 2006) (#29,361 men during up to 8 years of follow-up)
176. **Intakes of Vitamins A, C and E and Folate and Multivitamins and Lung Cancer: a pooled analysis of 8 prospective studies.** (Cho et al, 2006) (#430,281 persons over a maximum of 6-16 years in the studies)
177. **The Melbourne Atherosclerosis Vitamin E Trial (MAVET): a study of high dose vitamin E in smokers.** (Magliano et al, 2006) (#409 male and female smokers)

178. **Dietary supplementation with different vitamin C doses: no effect on oxidative DNA damage in healthy people.** (Herbert et al. 2006) (#160 volunteers)

179. **Carotenoids and cardiovascular health American Journal of Clinical Nutrition (not recommended).** (Voutilainen et al. 2006) (#not applicable)

180. **Vitamins C and E and the risks of pre-eclampsia and perinatal complications** (Rumbold et al, 2006) (#1,877 randomly assigned)

181. **Smoking, alcohol drinking, green tea consumption and the risk of esophageal cancer in Japanese men.** (Ishikawa et al, 2006) (#9,008 men in Cohort 1 and 17,715 men in Cohort 2)

182. **Intake of major carotenoids and the risk of epithelial ovarian cancer in a pooled analysis of 10 cohort studies.** (Koushik et al, 2006) (#521,911 women)

183. **Mortality in Randomized Trials of Antioxidant Supplements for Primary and Secondary Prevention; Systematic Review and Meta-analysis** (Bjelakovic et al, 2007) (# 232,606 participants)

184. **Multivitamin Use and Risk of Prostate Cancer in the National Institutes of Health–AARP Diet and Health Study** (Lawson et al, 2007) (#295,344 men)

185. **A Randomized Factorial Trial of Vitamins C and E and Beta Carotene in the Secondary Prevention of Cardiovascular Events in Women: Results From the Women's Antioxidant Cardiovascular Study** (Cook et al. 2007) (#8,171 female health professionals at increased risk)

186. **Use of Supplements of Multivitamins, Vitamin C, and Vitamin E in Relation to Mortality** (Pocobelli et al, 2007) (#77,719 subjects aged 50–76 years)

187. **Health Professionals Follow-up Study (2007): Effect of vitamins C, E, A and carotenoids and the occurrence of oral pre-malignant lesions** (Maserejian et al, 2007) (#42,340 men enrolled in the Health Professionals Follow-up Study) (#207 found with oral premalignant lesions)

188. **Antioxidant meta-analysis for the treatment of macular degeneration (2007)** (Chong et al, 2007) (#149,203 subjects)

189. **Effect of RRR-α-tocopherol supplementation on carotid atherosclerosis in patients with stable coronary artery disease (CAD)** (Devaraj et al, 2007) (#90 patients with CAD)

190. **Overview of the Women's Antioxidant Cardiovascular Study (WACS) (2007)** (Zaharris et al, 2007) (#8,171 women)

191. **Serum alpha-tocopherol, concurrent and past vitamin E intake, and mild cognitive impairment** (Dunn et al, 2007) (#526 subjects)

192. **The role of vitamin E in the prevention of cancer: a meta-analysis of randomized controlled trials.** (Alkhenizan and Hafez, 2007) (#167,025 subjects)

193. **Chemoprevention of Primary Liver Cancer: A Randomized, Double-Blind Trial in Linxian, China.** (Qu et al, 2007) (29,450 subjects)

194. **Risk of Mortality with Vitamin E Supplements: The Cache County Study.** (Hayden et al, 2007)

195. **Multivitamin-multimineral supplements and eye disease: age-related macular degeneration and cataract.** (Seddon, 2007) The Dietary Ancillary Study of the Eye Disease Case-Control Study (EDCCS)

196. **Antioxidant Supplementation Increases the Risk of Skin Cancers in Women but Not in Men.** (Hercberg et al, 2007) (#French adults, 7,876 women and 5,141 men. Total # = 13,017)

197. **Antioxidant Vitamin Supplement Use and Risk of Dementia or Alzheimer's Disease in Older Adults** (Gray et al, 2007) (#2,969)

198. **Beta carotene supplementation and age-related maculopathy in a randomized trial of US physicians** (Christen et al. 2007) (#22,071 apparently healthy US male physicians)

199. **Atherosclerosis and oxidant stress: the end of the road for antioxidant vitamin treatment?** (Thomson et al. 2007) (#not applicable)

200. **Effects of antioxidant supplementation on the aging process.** (Fusco et al. 2007) (#not applicable)

201. **Effect of high-dose alpha-tocopherol supplementation on biomarkers of oxidative stress and inflammation and carotid atherosclerosis in patients with coronary artery disease.** (Devaraj et al. 2007) (#90 patients with CAD)

202. **Multivitamins do not improve radiation therapy-related fatigue: results of a double-blind randomized crossover trial.** (de Souza et al. 2007) (randomized 40 patients to either placebo or Centrum Silver)

203. **Combined vitamin C and E supplementation during pregnancy for preeclampsia prevention: a systematic review** (Polyzos et al, 2007) (randomized 4,680 pregnant women to either the combination of vitamin C and vitamin E or placebo)

204. **Antioxidant therapy to prevent pre-eclampsia: a randomized controlled trial** (Spinnato et al. 2007) (707 of 739 randomly assigned patients)

205. **The effect of vitamin E on blood pressure in individuals with type 2 diabetes: a randomized, double-blind, placebo-controlled trial** (Ward et al, 2007) (#58 with type 2 diabetes randomized)

206. **National Institutes of Health State-of-the-Science Conference Statement:Multivitamin/Mineral Supplements and Chronic Disease Prevention** (NIH State-of-the Science Panel. 2007)

207. **Tumor-targeted induction of oxystress for cancer therapy** (Fang, Nakamura, Iyer, 2007)

208. **Diet and risk of ovarian cancer in the California Teachers Study cohort.** 000 (Chang et al, 2007) (#97,275)

209. **Dietary carotenoids and risk of colorectal cancer in a pooled analysis of 11 cohort studies.** (Mannisto et al, 2007) (#702,647 participants)

210. **Risk of Mortality with Vitamin E Supplements: The Cache County Study.** (Hayden et al, 2007)

211. **Phase II, randomized, controlled trial of high-dose N-acetylcysteine in high-risk cardiac surgery patients** (Haase M, et al. 2007) (#60 cardiac surgery patients at higher risk of postoperative renal failure)

212. **Antioxidant supplements for prevention of mortality in healthy participants and patients with various diseases.** (Bjelakovic, Nikolova, Gludd, Simonetti and Gludd, 2008 Apr) (#232,550 Cochrane Database Syst Rev.)

213. **Systematic review: primary and secondary prevention of gastrointestinal cancers with antioxidant supplements.** (Bjelakovic, Nikolova, Simonette and Gludd, 2008 Sept) (#211,818 participants)

214. **Vitamins E and C in the prevention of cardiovascular disease in men: the Physicians' Health Study II randomized controlled trial** (Sesso et al, 2008) (#14,641 US male physicians)

215. **Both {alpha}- and -Carotene, but Not Tocopherols and Vitamin C, Are Inversely Related to 15-Year Cardiovascular Mortality in Dutch Elderly Men** (Buijsse et al, 2008) (#559 men (mean age about 72 y) free of chronic diseases)

216. **Vitamin E and selenium supplementation and risk of prostate cancer in the Vitamins and lifestyle (VITAL) study cohort** (Peters et al, 2008) (#35,242 men)

217. **VITAL (VITamins And Lifestyle) study 2008** (Slatore et al, 2008) (#77,721 men and women)

218. **Vitamin E for Alzheimers and mild cognitive impairment. Cochrane Database Syst Rev. (2008)** (Isaac et al, 2008) (#769 participants)

219. **Efficacy of Antioxidant Supplementation in Reducing Primary Cancer Incidence and Mortality: Systematic Review and Meta-analysis** (Bardia et al, 2008) (#104,196 participants)

220. **Carotenoids and the risk of developing lung cancer: a systematic review.** (Gallicchio et al, 2008) (Six randomized clinical trials)

221. **Antioxidant enriched enteral nutrition and oxidative stress after major gastrointestinal tract surgery.** (van Stijn et al, 2008) (#21 undergoing major upper gastrointestinal tract surgery)

222. **Vitamin E supplementation may transiently increase tuberculosis risk in males who smoke heavily and have high dietary vitamin C intake.** (Hemila and Kaprio, 2008 Oct) (#29,023 males aged 50-69 years, smoking at baseline, with no tuberculosis)
223. **Vitamin E supplementation and pneumonia risk in males who initiated smoking at an early age: effect modification by body weight and dietary vitamin C.** (Hemila and Kaprio, 2008 Nov) (#21,657 ATBC Study participants who initiated smoking by the age of 20 years)
224. **Oral administration of vitamin C decreases muscle mitochondrial biogenesis and hampers training-induced adaptations in endurance performance.** (Gomez-Cabrera et al, 2008) (#14)
225. **Antioxidant vitamin and mineral supplements for preventing age-related macular degeneration.** (Evans and Henshaw, 2008) (#23,099 participants)
226. **Dietary antioxidants and the long-term incidence of age-related macular degeneration: the Blue Mountains Eye Study.** (Tan et al, 2008) (#2,454 Australian population-based cohort study)
227. **Multivitamin-multimineral supplement use and mammographic breast density.** (Berube et al, 2008) (#Premenopausal (777) and postmenopausal (783) women, total 1,560)
228. **Antioxidant supplements to prevent or slow down the progression of AMD: a systematic review and meta-analysis.** (Evans, 2008) (#23,099 people were randomized in three trials)
229. **High maternal plasma antioxidant concentrations associated with preterm delivery.** (Joshi et al. 2008) (#140 normotensive pregnant women)
230. **Vitamin E and age-related cataract in a randomized trial of women.** (Christen et al. 2008) (#39,876)
231. **Antioxidants in cardiovascular health and disease: key lessons from epidemiologic studies.** (Icox et al. 2008) (#not applicable)

232. **Both alpha and beta-Carotene, but Not Tocopherols and Vitamin C, Are Inversely Related to 15-Year Cardiovascular Mortality in Dutch Elderly Men - the Zutphen Elderly Study** (Buijsse et al. 2008) (#not applicable)

233. **Vitamin E and selenium supplementation and risk of prostate cancer in the Vitamins and lifestyle (VITAL) study cohort.** (Peters et al. 2008) (#35,242 men)

234. **Observational studies on the effect of dietary antioxidants on asthma: a meta-analysis.** (Gao et al. 2008) (#13,653)

235. **Dietary antioxidants and the long-term incidence of age-related macular degeneration: the Blue Mountains Eye Study** (Tan et al. 2008) (#Of 3,654 baseline (1992-1994) participants initially 49 years or older, 2,454 were reexamined after 5 years, 10 years, or both)

236. **Carotenoids and the risk of developing lung cancer: a systematic review.** (Gallicchio et al. 2008) (#Six randomized clinical trials examining the efficacy of beta-carotene supplements and 25 prospective observational studies assessing the associations between carotenoids and lung cancer were analyzed by using random-effects meta-analysis) (#6 RCTs and 25 prospective observational studies)

237. **Vitamins E and C in the prevention of cardiovascular disease in men: the Physicians' Health Study II randomized controlled trial** (Sesso et al, 2008) (#14,641 US male physicians)

238. **Mistletoe therapy in oncology** (Horneber et al, 2008) (#21 included studies, overall comprising 3,484 randomized cancer patients)

239. **Oxystress inducing antitumor therapeutics via tumor-targeted delivery of PEG-conjugated D-amino acid oxidase =** (Fang et al, 2008) (#)

240. **Clinical outcomes of contrast-induced nephropathy in patients undergoing percutaneous coronary intervention: a prospective, multicenter, randomized study to analyze the effect of hydration and acetylcysteine** (Chen SL, et al. 2008) (#936)

241. **NAC had no effect on disease progression in non-diabetic kidney failure patients** (Renke M. et al. 2008) (#20 non-diabetic patients with proteinuria)

242. **N-acetylcysteine to reduce renal failure after cardiac surgery: a systematic review and meta-analysis** (Naughton et al. 2008) (#Seven randomized controlled trials (RCTs, n = 1000)

243. **Is there a role for supplemented antioxidants in the prevention of atherosclerosis?** (Katsiki and Manes, 2009)

244. **Plasma Carotenoids, Retinol, and Tocopherols and Postmenopausal Breast Cancer Risk in theMultiethnic Cohort Study: a nested case-control study.** (Epplein, 2009) (#286 incident postmenopausal breast cancer cases were matched to 535 controls)

245. **Vitamins E and C in the prevention of prostate and total cancer in men: the Physicians' Health Study II randomized controlled trial (2009)** (Gaziano et al, 2009) (#14,641 male physicians)

246. **Vitamin E, vitamin C, beta carotene, and cognitive function among women with or at risk of cardiovascular disease: The Women's Antioxidant and Cardiovascular Study (2009)** (Kang et al, 2009) (#2,824 participants)

247. **Effects of vitamins C and E and beta-carotene on the risk of type 2 diabetes in women at high risk of cardiovascular disease: Women's Antioxidant Cardiovascular Study (2009)** (Song et al, 2009) (#8,171 female health professionals)

248. **Vitamins C and E and Beta Carotene Supplementation and Cancer Risk: Women's Antioxidant Cardiovascular Study (2009)** (Lin et al, 2009) (#7,627 female health professionals)

249. **Effect of selenium and vitamin E on risk of prostate cancer and other cancers: the Selenium and Vitamin E Cancer Prevention Trial (SELECT) (2009)** (Lippman, 2009) (#35,533 men)

250. **Multivitamin Use and Risk of Cancer and Cardiovascular Disease in the Women's Health Initiative Cohorts** (Neuhouser et al, 2009) (#161,808 participants)

251. **Effects of antioxidant supplements on cancer prevention: meta-analysis of randomized controlled trials** (Myung et al, 2009) (#161,045 total subjects)

252. **Decision Analysis Supports the Paradigm That Indiscriminate Supplementation of Vitamin E Does More Harm than Good** (Dotan et al, 2009) (over 300,000 participants)

253. **Effects of long-term antioxidant supplementation and association of serum antioxidant concentrations with risk of metabolic syndrome in adults (SU.VI.MAX)** (Czernichow et al, 2009) (#5,220 adults)

254. **Total and Cancer Mortality After Supplementation With Vitamins and Minerals: 10 year Follow-up of the Linxian General Population Nutrition Intervention Trial.** (Qiao et al, 2009) (#29,584 adults)

255. **Vitamin A and retinol intakes and the risk of fractures among participants of the Women's Health Initiative Observational Study.** (Caire-Juvera et al. 2009) (#75,747 women from the Women's Health Initiative Observational Study)

256. **Modification of the effect of vitamin E supplementation on the mortality of male smokers by age and dietary vitamin C.** (Hemila and Kaprio, 2009 Apr) (#29,023 males aged 50-69 years, smoking at baseline, with no tuberculosis)

257. **Vitamin E supplement use and the incidence of cardiovascular disease and all-cause mortality in the Framingham Heart Study: Does the underlying health status play a role?** (Dietrich et al, 2009) (#4,270 Framingham study participants)

258. **Antioxidants prevent health-promoting effects of physical exercise in humans.** (Ristow et al, 2009) (#39 healthy adults)

259. **Long-term use of beta-carotene, retinol, lycopene, and lutein supplements and lung cancer risk: results from**

the **VITamins And Lifestyle (VITAL) study.** (Satia et al, 2009) (#77,126 (VITAL) cohort Study in Washington State)

260. **Concentrations of antioxidant vitamins in maternal and cord serum and their effect on birth outcomes.** (Wang et al. 2009) (#143 mother-neonate pairs)

261. **WHO Vitamin C and Vitamin E trial group. World Health Organisation multicentre randomised trial of supplementation with vitamins C and E among pregnant women at high risk for pre-eclampsia in populations of low nutritional status from developing countries.** (Villar et al, 2009) (#687 women)

262. **Vitamin and mineral use and risk of prostate cancer: the case-control surveillance study.** (Zhang et al. 2009) (#1,706 prostate cancer cases and 2,404 matched controls, total 4,111)

263. **"Is the oxidative stress theory of aging dead?" (STUDY)** (Pérez et al. 2009) (#not applicable)

264. **The oxidative stress menace to coronary vasculature: any place for antioxidants?** (Briasoulis et al. 2009) (#not applicable)

265. **Oral antioxidant supplementation does not prevent acute mountain sickenss: Double blind, randomized placebo-controlled trial.** (Baillie et al. 2009) (#83)

266. **Total dietary antioxidant index and survival in patients with glioblastoma multiforme.** (Il'yasova et al. 2009) (#814 glioblastoma multiforme cases)

267. **Associations between alpha-tocopherol, beta-carotene, and retinol and prostate cancer survival** (Watters et al. 2009) (#29,133)

268. **Antioxidant Supplementation and Risk of Incident Melanomas. Results of a Large Prospective Cohort Study.** (Asgari et al, 2009) (#69,671 men and women)

269. **Folic acid and risk of prostate cancer: results from a randomized clinical trial.** (Figueiredo et al, 2009) (#643 randomly assigned men)

270. **Green tea, black tea consumption and risk of lung cancer: a meta-analysis (Tang et al, 2009)** (#meta-analysis included 22 studies)

271. **Green tea consumption and risk of stomach cancer: a meta-analysis of epidemiologic studies** (Myung, Int J Cancer. et al, 2009) **(#13 epidemiologic studies)**

272. **Green tea consumption and gastric cancer in Japanese: a pooled analysis of six cohort studies.** (Inoue et al, 2009) (# 219,080 subjects, 3,577 cases of gastric cancer)

273. **Green tea (Camellia sinensis) for the prevention of cancer. Chochrane Database Syst Rev. 2009 Jul 8;(3):CD005004).** (Boehm et al, 2009) (#Fifty-one studies with more than 1.6 million participants were included)

274. **Lack of genoprotective effect of phytosterols and conjugated linoleic acids on Caco-2 cells (and did not exhibit potential anti-carcinogenic activity)** (Daly et al, 2009) (#)

275. **Pre-radiotherapy plasma carotenoids and markers of oxidative stress are associated with survival in head and neck squamous cell carcinoma patients: a prospective study** (Sakhi et al, 2009) (#178 total; 78 HNSCC patients and 100 healthy controls)

276. **The protective effect of silibinin against mitomycin C-induced intrinsic apoptosis in human melanoma A375-S2 cells.** (Jiang et al, 2009)

277. **Are the health attributes of lycopene related to its antioxidant function?** (Erdman, Ford and Lindshield, 2009)

278. **Does antioxidant vitamin supplementation protect against muscle damage?** (McGinley et al. 2009)

279. **The oxidative stress menace to coronary vasculature: any place for antioxidants?** (Briasoulis et al. 2009)

280. **Use of Supplements of Multivitamins, Vitamin C, and Vitamin E in Relation to Mortality** (Pocobelli et al, 2009) (#77,719 subjects aged 50–76 years)

281. **N-acetylcysteine does not prevent contrast-induced nephropathy after cardiac catheterization in patients**

with diabetes mellitus and chronic kidney disease: a randomized clinical trial (Amini et al. 2009) (#90 patients)

282. **N-acetylcysteine in cardiovascular-surgery-associated renal failure: a meta-analysis** (Nigwekar SU, Kandula P. 2009) (#Twelve studies comprising 1,324 patients)

283. **Meta-analysis of N-acetylcysteine to prevent acute renal failure after major surgery** (Ho M, Morgan DJ. 2009) (#10 studies involving a total of 1,193 adult patients)

284. **Efficacy of N-acetylcysteine in preventing renal injury after heart surgery: a systematic review of randomized trials** (Adabag et al. 2009) (#1163)

285. **Vitamin C supplements and the risk of age-related cataract: a population-based prospective cohort study in women.** (Rautiainen et al, 2010) (#24,593 women)

286. **Micronutrient concentrations and subclinical atherosclerosis in adults with HIV.** (Falcone et al, 2010) (#298 Nutrition for Healthy Living participants)

287. **Multivitamin use and breast cancer incidence in a prospective cohort of Swedish women.** (Larsson et al, 2010) (#35,329 cancer-free women)

288. **Vitamins C and E to prevent complications of pregnancy-associated hypertension** (Roberts et al, 2010) (#10,154)

289. **Vitamin E and age-related macular degeneration in a randomized trial of women.** (Christen et al. 2010. women) (#39,876)

290. **Age-related cataract in a randomized trial of vitamins E and C in men.** (Christen et al. 2010. men) (#11,545)

291. **Vitamins C and E for prevention of pre-eclampsia in women with type 1 diabetes (DAPIT)** (McCance et al. 2010) (#762 women)

292. **Daily intake of antioxidants in relation to survival among adult patients diagnosed with malignant glioma.** (DeLorenze et al. 2010) (#not applicable)

293. **Effects of vitamin on stroke subtypes: meta-analysis of randomized controlled trials.** (Schurks et al. 2010) (#118,765)

Antioxidant Failures & Dangers

294. **Multivitamin/Mineral supplementation does not affect standardized assessment of academic performance in elementary school children** (Perlman et al. 2010) (#students in grades three through six, approximate age range=8 to 12 years old)

295. **Differential effects of concomitant use of vitamins C and E on trophoblast apoptosis and autophagy between normoxia and hypoxia-reoxygenation** (Hung et al, 2010)

296. **Green Tea Drinking and Subsequent Risk of Breast Cancer in a Population to Based Cohort of Japanese Women/No effect.** (Iwasaki et al, 2010) (#53,793)

297. **Green tea consumption and breast cancer risk or recurrence: a meta-analysis.** (Ogunleye et al, 2010) (#5,617 breast cancer cases)

298. **Long-Term Use of supplemental vitamin C, vitamin D and vitamin E does not reduce the risk of urothelial cell carcinoma of the bladder in the VITamins and Lifestyle Study.** (Hotaling 2010) (#77,050 eligible VITAL participants)

299. **The vitamin C: vitamin K3 system - enhancers and inhibitors of the anticancer effect.** (Lamson DW et al, 2010)

300. **Effects of vitamin E on plasma lipid status and oxidative stress in chinese women with metabolic syndrome.** (Wang et al, 2010) (#not available)

301. **Serum selenium and prognosis in cardiovascular disease: results from the AtheroGene study.** (Lubos et al, 2010) (#1,731)

302. **Low Total and Nonheme Iron Intakes Are Associated with a Greater Risk of Hypertension.** (Galan et al, 2010) (#2,895)

303. **Multivitamin use and the risk of myocardial infarction: a population-based cohort of Swedish women.** (Rautiainen et al, 2010) (#31,671 women with no history of cardiovascular disease (CVD) and 2,262 women with a history of CVD; total number - 33,933)

304. **Vitamin E and All-cause Mortality: A Meta-Analysis** (Abner EL, et al, 2011) (#246,371 subjects and 29,295 all-cause deaths)

305. **Impact of high-dose N-acetylcysteine versus placebo on contrast-induced nephropathy and myocardial reperfusion injury in unselected patients with ST-segment elevation myocardial infarction undergoing primary percutaneous coronary intervention. The LIPSIA-N-ACC (Prospective, Single-Blind, Placebo-Controlled, Randomized Leipzig Immediate PercutaneouS Coronary Intervention Acute Myocardial Infarction N-ACC) Trial** (Thiele H, et al. 2010) (#251)

306. **NAC had no effect on blood pressure and surrogate markers of cardiovascular injury in non-diabetic patients with CKD** (Renke et al. 2010) (#20 non-diabetic patients with albuminuria)

307. **Quercetin and Ferulic Acid Aggravate Renal Carcinoma in Long-Term Diabetic Victims.** (Chiu-Lan Hsieh et al, 2010)

308. **No beneficial effects of pine bark extract on cardiovascular disease risk factors** (Drieling et al. 2010) (#130 individuals with increased cardiovascular disease risk)

309. **Effect of vitamins C and E on antioxidant status of breast-cancer patients undergoing chemotherapy** (Suhail et al, 2011) (#forty untreated breast-cancer patients (stage II) and compared with those of healthy controls)

310. **The effect of supplemental vitamins and minerals on the development of prostate cancer: a systematic review and meta-analysis.** (Stratton, Godwin, 2011) (#Fourteen articles were included)

311. **The outcome of 5-ALA-mediated photodynamic treatment in melanoma cells is influenced by vitamin C and heme oxygenase-1.** (Grimm et al, 2011) (#)

312. **The protective effects of nutritional antioxidant therapy on Ehrlich solid tumor-bearing mice depend on the type of antioxidant therapy chosen: histology, genotoxicity and hematology evaluations.** (Miranda-Vilela AL, et al, 2011)

313. **Prenatal exposure to flavonoids: Implication for cancer risk.** (Vanhees et al, 2011)
314. **Peroxiredoxin 6 overexpression attenuates cisplatin-induced apoptosis in human ovarian cancer cells.** (Pak et al, 2011)
315. **Dietary phytocompounds and risk of lymphoid malignancies in the California Teachers Study cohort.** 000 (Chang et al, 2010) (#110,215)
316. **Black and green tea consumption and the risk of coronary artery disease: a meta-analysis** (Wang et al, 2011) (#18 studies)

This totals 316 study reports with over 15,000,000 participants.

**Radicophobes can ignore the truth or
they can reject the truth but
they can not change the magnificent truths
regarding the crucial role
of EMODs in the life process of all aerobes
or the inherent splendor of oxygen.**
R. M. Howes M.D., Ph.D.
1/26/09

REFERENCES

(Abner et al, 2011) (Abner EL, et al. Vitamin E and All-cause Mortality: A Meta-Analysis. Curr Aging Sci. 2011 Jan 14).

(Adabag et al. 2009) (Adabag AS, Ishane A, Bloomfield HE, Ngo AK, Wilt TJ. Efficacy of N-acetylcysteine in preventing renal injury after heart surgery: a systematic review of randomized trials. Eur Heart J. 2009 Aug;30(15):1910-7).

(Alkhenizan and Hafez, 2007) (Alkhenizan A, Hafez K. The role of vitamin E in the prevention of cancer: a meta-analysis of randomized controlled trials. Ann Saudi Med. 2007 Nov-Dec;27(6):409-14)

(Amini et al. 2009) (Amini M, Salarifar M, Amirgaigloo A, Masoudkabir F, Esfahane F. N-acetylcysteine does not prevent contrast-induced nephropathy after cardiac catheterization in patients with diabetes mellitus and chronic kidney disease: a randomized clinical trial. Trials. 2009 Jun 29;10:45).

(Anderssen et al. 1996) (Andersson SO, Wolk A, Bergström R, Giovannucci E, Lindgren C, Baron J, Adami HO. Energy, nutrient intake and prostate cancer risk: a population-based case-control study in Sweden. Int J Cancer. 1996 Dec 11;68(6):716-22).

(AREDS, 2001) (A randomized, placebo-controlled, clinical trial of high-dose supplementation with vitamins C and E and beta carotene for age-related cataract and vision loss: AREDS report no. 9. Age-Related Eye Disease Study Research Group. Arch Ophthalmol. 2001 Oct;119(10):1439-5)

(Arrowsmith et al, 1989) (JB Arrowsmith et al. Morbidity and mortality among low birth weight infants exposed to an intravenous vitamin E product, E-Ferol. Pediatrics. 1989 Feb;83(2):244-9)

(Ascherio et al. 1999) (Ascherio A, Rimm EB, Hernan MA, Giovannucci E, Kawachi I, Stampfer MJ, Willett WC. Relation of consumption of vitamin E, vitamin C, and carotenoids to risk for stroke among men in the United States. Ann Intern Med. 1999 Jun 15;130(12):963-70)

(Asgari et al. 2009) (Maryam M. Asgari, Sonia S. Maruti, Lawrence H. Kushi, Emily White. Antioxidant Supplementation and Risk of Incident Melanomas. Arch Dermatol. 2009;145(8):879-882)

(Asplund, 2002) (Asplund K. Antioxidant vitamins in the prevention of cardiovascular disease: a systematic review. J Intern Med. 2002 May;251(5):372-92)

(Avenell et al. 2005) (Avenell A. et al. Effect of multivitamin and mul-timineral supplements on morbidity from infections in older people (MAVIS trial): Pragmatic, randomized, double blind, placebo controlled trial. BMJ 2005;331:324)

(Baillie et al. 2009) . (Baillie JK, Thompson AA, Irving JB, Bates MG, Sutherland AI, Macnee W, Maxwell SR, Webb DJ. Oral antioxidant sup-plementation does not prevent acute mountain sickness: double blind, randomized placebo-controlled trial. QJM. 2009 May;102(5):341-8).

(Bairati et al, 2005 Apr 6) (Bairati I, Meyer F, Gélinas M, Fortin A, Nabid A, Brochet F, Mercier JP, Têtu B, Harel F, Mâsse B, Vigneault E, Vass S, del Vecchio P, Roy J. A randomized trial of antioxidant vitamins to pre-vent second primary cancers in head and neck cancer patients. J Natl Cancer Inst. 2005 Apr 6;97(7):481-8)

(Bairati et al, 2005 Aug 20) (Bairati I, Meyer F, Gélinas M, Fortin A, Nabid A, Brochet F, Mercier JP, Têtu B, Harel F, Abdous B, Vigneault E, Vass S, Del Vecchio P, Roy J. Randomized trial of antioxidant vitamins to prevent acute adverse effects of radiation therapy in head and neck cancer patients. J Clin Oncol. 2005 Aug 20;23(24):5805-13)

(Bardia et al, 2008) (Aditya Bardia, Imad M. Tleyjeh, James R. Cerhan, Amit K. Sood, Paul J. Limburg, Patricia J. Erwin and Victor M. Montori. Efficacy of Antioxidant Supplementation in Reducing Primary Cancer Incidence and Mortality: Systematic Review and Meta-analysis. January 2008 vol. 83 no. 1 23-34)

(Baron et al, 2003) (John A. Baron, Bernard F. Cole, Leila Mott, Robert Haile, Maria Grau, Timothy R. Church, Gerald J. Beck, E. Robert Greenberg. Neoplastic and Antineoplastic Effects of Beta Carotene on Colorectal Adenoma Recurrence: Results of a Randomized Trial. JNCI Journal of the National Cancer Institute 2003 95(10):717-722)

(Batieha et al, 1993) (Batieha AM, Armenian HK, Norkus EP, Morris JS, Spate VE, Comstock GW. Serum micronutrients and the subsequent risk of cervical cancer in a population-based nested case-control study. Cancer Epidemiol Biomarkers Prev. 1993 Jul-Aug;2(4):335-9)

(Berson et al, 1993) (Berson EL, Rosner B, Sandberg MA, et al. A randomized trial of vitamin A and vitamin E supplementation for retinitis pigmentosa. Arch Ophthalmol. 1993;111(6):761-772)

(Berube et al, 2008) (Sylvie Bérubé, Caroline Diorio and Jacques Brisson. Multivitamin-multimineral supplement use and mammographic breast density. American Journal of Clinical Nutrition, Vol. 87, No. 5, 1400-1404, May 2008)

(Bjelakovic et al, Cochrane Database Syst Rev. 2004) (Bjelakovic G, Nikolova D, Simonetti RG, Gluud C. Antioxidant supplements for preventing gastrointestinal cancers. Cochrane Database Syst Rev (2004) (4):CD004183)

(Bjelakovic et al, Lancet. 2004) (Bjelakovic G, Nikolova D, Simonetti RG, Gluud C. Antioxidant supplements for prevention of gastrointestinal cancers: a systematic review and meta-analysis. Lancet (2004) 364:1219–28)

(Bjelakovic et al., 2006) (Bjelakovic G, Nagorni A, Nikolova D, et al. Meta-analysis: antioxidant supplements for primary and secondary prevention of colorectal adenoma. Aliment Pharmacol Ther. 2006;24:281-91)

(Bjelakovic et al, 2007) (Goran Bjelakovic, Dimitrinka Nikolova, Lise Lotte Gluud, Rosa G. Simonetti, and Christian Gluud. "Mortality in Randomized Trials of Antioxidant Supplements for Primary and Secondary Prevention; Systematic Review and Meta-analysis." JAMA 2007;297:842-857.Vol. 297 No. 8, February 28, 2007)

(Bjelakovic pp. 842-57 et al, 2007) (Bjelakovic G, Nikolova D, Gluud LL, Simonetti RG, Gluud C. "Mortality in randomized trials of antioxidant supplements for primary and secondary prevention: systematic review and meta-analysis." JAMA 2007 28;297(8):84-57)

(Bjelakovic, Nikolova, Gludd, Simonetti and Gludd, 2008 Apr) (Bjelakovic G, Nikolova D, Gluud LL, Simonetti RG, Gluud C.. Antioxidant supplements for prevention of mortality in healthy participants and patients with various diseases. Cochrane Database Syst Rev. 2008 Apr 16;(2):CD007176)

(Bjelakovic, Nikolova, Simonette and Gludd, 2008 Sept) (Bjelakovic G, Nikolova D, Simonetti RG, Gluud C. Systematic review: primary and secondary prevention of gastrointestinal cancers with antioxidant supplements. Aliment Pharmacol Ther. 2008 Sep 15;28(6):689-703)

(Bleys et al, 2006) (Vitamin-mineral supplementation and the progression of atherosclerosis: a meta-analysis of randomized controlled trials. Joachim Bleys, Edgar R Miller, III, Roberto Pastor-Barriuso, Lawrence J Appel and Eliseo Guallar. American Journal of Clinical Nutrition, Vol. 84, No. 4, 880-887, October 2006)

(Boccalandro et al. 2003) (Boccalandro F, Amhad M, Smalling RW, Sdringola S. Oral acetylcysteine does not protect renal function from moderate to high doses of intravenous radiographic contrast. Catheter Cardiovasc Interv. 2003 Mar;58(3):336-41).

(Boehm et al, 2009) (Boehm et al. Green tea (Camellia sinensis) for the prevention of cancer. Chochrane Database Syst Rev. 2009 Jul 8;(3):CD005004).

(Bohlke et al, 1999) (K Bohlke, D Spiegelman, A Trichopoulou, K Katsouyanni and D Trichopoulos. Vitamins A, C and E and the risk of breast cancer: results from a case-control study in Greece. British Journal of Cancer (1999) 79, 23–29)

(Bonnefoy et al. 2002) (Bonnefoy M, Drai J, Kostka T. Antioxidants to slow aging, facts and perspectives. Presse Med. 2002 Jul 27;31(25):1174-84).

(Boothby and Doering, 2005) (Boothby LA, Doering PL. Vitamin C and vitamin E for Alzheimer's disease. Ann Pharmacother. 2005 Dec;39(12):2073-80)

(Bostom et al, 1995) (Bostom AG, Hume AL, Eaton CB, Laurino JP, Yanek LR, Regan MS, McQuade WH, Craig WY, Perrone G, Jacques PF. The effect of high-dose ascorbate supplementation on plasma lipoprotein(a) levels in patients with premature coronary heart disease. Pharmacotherapy 1995 Jul-Aug;15(4):458-64)

(Bove et al, 1985) (Vasculopathic hepatotoxicity associated with E-Ferol syndrome in low-birth-weight infants. K. E. Bove, N. Kosmetatos, K. E. Wedig, D. J. Frank, S. Whitlatch, V. Saldivar, J. Haas, C. Bodenstein and W. F. Balistreri. JAMA. Vol. 254. No. 17. November 1, 1985)

(Bradley et al, 1965) (Bradley, B.E., Jr., Vedros, N.A., Defalco, A.J., Lawson, D.W., Vineyard, G.C. and Urschel, H.C. The effect of intra-arterial hydrogen peroxide in rabbits infected with clostridum perfringens. J Trauma 1965; 6: 799)

Briasoulis et al. 2009) (Briasoulis A, Tousoulis D, Antoniades C, Stefanadis C. The oxidative stress menace to coronary vasculature: any place for antioxidants? Curr Pharm Des. 2009;15(26):3078-90).

(Braunersreuther, Jaquet, 2011) (Braunersreuther V, Jaquet V. Reactive Oxygen Species in Myocardial Reperfusion Injury: From Physiopathology to Therapeutic Approaches. Curr Pharm Biotechnol. 2011 Apr 6. [Epub ahead of print])

(Brown et al, 2001) (Brown BG, Zhao XQ, Chait A, Fisher LD, Cheung MC, Morse JS, Dowdy AA, Marino EK, Bolson EL, Alaupovic P, Frohlich J, Albers JJ. Simvastatin and niacin, antioxidant vitamins, or the combination for the prevention of coronary disease. N Engl J Med. 2001 Nov 29;345(22):1583-92)

(Brown et al, 2001) (Brown BG, Zhao XQ, Chait A, et al. Simvastatin and niacin, antioxidant vitamins, or the combination for the prevention of coronary disease. N Engl J Med 2001;345:1583-1592)

(Bruckdorfer, 2008) (Bruckdorfer KR. Antioxidants and CVD. Proc Nutr Soc. 2008 May;67(2):214-22)

(Buettner and Jurkiewicz, 1996) (Buettner GR. & Jurkiewicz BA, (1996) Catalytic metals, ascorbate and free radicals: combinations to avoid. Radiat. Res. 145, 532-541)

(Buijsse et al, 2008) (B. Buijsse, E. J. M. Feskens, L. Kwape, F. J. Kok, and D. Kromhout. Both {alpha}- and -Carotene, but Not Tocopherols and Vitamin C, Are Inversely Related to 15-Year Cardiovascular Mortality in Dutch Elderly Men. J. Nutr., February 1, 2008; 138(2): 344 – 350)

(Buring and Hennekens, 1992) (Buring JE, Hennekens CH. The women's health study: summary of the design. J Myocardial Ischemia 1992; 4:27-9)

(Burns KE, et al. 2005) (Burns KE, et al. Perioperative N-acetylcysteine to prevent renal dysfunction in high-risk patients undergoing cabg surgery: a randomized controlled trial. JAMA. 2005 Jul 20;294(3):342-50).

(Caire-Juvera et al, 2009) (Caire-Juvera G,, Ritenbaugh C, Wactawski-Wende J, Snetselaar LG, Chen Z. Vitamin A and retinol intakes and the risk of fractures among participants of the Women's Health Initiative Observational Study. Am J Clin Nutr. 2009 Jan;89(1):323-30)

(Calzada et al, 1997) (Calzada C, Bruckdorfer KR, Rice-Evans CA. The influence of antioxidant nutrients on platelet function in healthy volunteers. Atherosclerosis 1997 Jan 3;128(1):97-105)

(Caraballoso et al., 2003) (Drugs for preventing lung cancer in healthy people. M. Caraballoso et al. Cochrane Database Syst Rev. 2003;(2):CD002141)

(Celik et al, 2010) (Celik T, Yuksel C, Iyisoy A. Alpha tocopherol use in the management of diabetic cardiomyopathy: lessons learned from randomized clinical trials. J Diabetes Complications. 2010 Jul-Aug;24(4):286-8)

(Chang et al, 2007) (Chang ET, et al. Diet and risk of ovarian cancer in the California Teachers Study cohort. Am J Epidemiol. 2007 Apr 1;165(7):802-13).

(Chang et al, 2010) (Chang ET, et al. Dietary phytocompounds and risk of lymphoid malignancies in the California Teachers Study cohort. Cancer Causes Control. 2011 Feb;22(2):237-49).

(Chasan-Taber et al, 1999) (Chasan-Taber L, Willett W C, Seddon J M. et al A prospective study of vitamin supplement intake and cataract extraction among US women. Epidemiology 1999. 10679–684)

(Chen et al, 2006) (Chen et al. High-dose oral vitamin C partially replenishes vitamin C levels in patients with Type 2 diabetes and low vitamin C levels but does not improve endothelial dysfunction or insulin resistance. Am. J. Physiol. Heart Circ. Physiol. 2006;290:H137-H145)

(Chen SL, et al. 2008) (Chen SL, et al. Clinical outcomes of contrast-induced nephropathy in patients undergoing percutaneous coronary intervention: a prospective, multicenter, randomized study to analyze the effect of hydration and acetylcysteine. Int J Cardiol. 2008 Jun 6;126(3):407-13).

(Chiabrando et al., 2002) (Long-term vitamin E supplementation fails to reduce lipid peroxidation in people at cardiovascular risk: analysis of underlying factors. Chiabrando C, Avanzini F, Rivalta C, Colombo F, Fanelli R, Palumbo G, Roncaglioni MC; PPP Collaborative Group on the antioxidant effect of vitamin E. Curr Control Trials Cardiovasc Med. 2002 Mar 19;3(1):5)

(Chiu-Lan Hsieh et al, 2010) (Chiu-Lan Hsieh, Chiung-chi Peng, Yu-Ming Cheng, et al. Quercetin and Ferulic Acid Aggravate Renal Carcinoma in Long-Term Diabetic Victims. J. Agric. Food Chem., 2010, 58 (16), pp 9273–9280)

(Cho et al, 2006) (Cho E. et al. Intakes of vitamins A, C and E and folate and multivitamins and lung cancer: a pooled analysis of 8 prospective studies. Int J Cancer. 2006 Feb 15;118(4):970-8)

(Chong et al, 2007) ("Dietary antioxidants and primary prevention of age related macular degeneration: systematic review and meta-analysis" Elaine W-T Chong, Tien Y Wong, Andreas J Kreis, Julie A Simpson, Robyn H Guymer. British Medical Journal (BMJ)., doi:10.1136/bmj.39350.500428.47 (published 8 October 2007)

(Christen et al, 2003) (Christen WG; Manson JE; Glynn RJ; Gaziano JM; Sperduto RD; Buring JE; Hennekens CH. A randomized trial of beta carotene and age-related cataract in US physicians. Arch Ophthalmol. 2003; 121(3):372-8)

(Christen et al. 2007) (Christen WG, Manson JE, Glynn RJ, Gaziano JM, Chew EY, Buring JE, Hennekens CH. Beta carotene supplementation and age-related maculopathy in a randomized trial of US physicians. Arch Ophthalmol. 2007 Mar;125(3):333-9).

(Christen et al. 2008) (Christen WG, Glynn RJ, Chew EY, Buring JE. Vitamin E and age-related cataract in a randomized trial of women. Ophthalmology. 2008 May;115(5):822-829.e1).

(Christen et al. 2010) (Christen WG, Glynn RJ, Chew EY, Buring JE. Vitamin E and age-related macular degeneration in a randomized trial of women. Ophthalmology. 2010 Jun;117(6):1163-8).

(Christen et al. 2010) (Christen WG, et al. Age-related cataract in a randomized trial of vitamins E and C in men. Arch Ophthalmol. 2010 Nov;128(11):1397-405).

(Chylack et al. 2002) (Chylack LT Jr, Brown NP, Bron A, Hurst M, Kopcke W, Thien U, Schalch W. The Roche European American Cataract Trial (REACT): a randomized clinical trial to investigate the efficacy of an oral antioxidant micronutrient mixture to slow progression of age-related cataract. Ophthalmic Epidemiol. 2002 Feb;9(1):49-80)

(Clarke, Armitage, 2002) (Clarke R, Armitage J. Antioxidant vitamins and risk of cardiovascular disease. Review of large-scale randomised trials. Cardiovasc Drugs Ther. 2002 Sep;16(5):411-5)

(Comhaire et al, 2000) (Comhaire FH, Christophe AB, Zalata AA, Dhooge WS, Mahmoud AM, Depuydt CE. The effects of combined conventional treatment, oral antioxidants and essential fatty acids on sperm biology in subfertile men. Prostaglandins Leukot Essent Fatty Acids. 2000 Sep;63(3):159-65)

(Cook et al, 2007) (A Randomized Factorial Trial of Vitamins C and E and Beta Carotene in the Secondary Prevention of Cardiovascular Events in Women: Results From the Women's Antioxidant

Cardiovascular Study. Nancy R. Cook, ScD; Christine M. Albert, MD; J. Michael Gaziano, MD; Elaine Zaharris, BA; Jean MacFadyen, BA; Eleanor Danielson, MIA; Julie E. Buring, ScD; JoAnn E. Manson, MD, DrPH. Arch Intern Med. 2007;167(15):1610-1618)

(Coulter et al, 2006) (Antioxidants Vitamin C and Vitamin E for the Prevention and Treatment of Cancer. Coulter, Ian D.; Hardy, Mary L.; Morton, Sally C.; Hilton, Lara G.; Tu, Wenli; Valentine, Di; Shekelle, Paul G. Journal of General Internal Medicine, Volume 21, Number 7, July 2006, pp. 735-744(10))

(Creagan et al, 1979) (Creagan ET, Moertel CG, O'Fallon JR, Schutt AJ, O'Connell MJ, Rubin J, Frytak S. Failure of high-dose vitamin C (ascorbic acid) therapy to benefit patients with advanced cancer. A controlled trial. N Engl J Med. 1979 Sep 27;301(13):687-90)

(Czernichow et al, 2006) (Antioxidant supplementation does not affect fasting plasma glucose in the Supplementation with Antioxidant Vitamins and Minerals (SU.VI.MAX) study in France: association with dietary intake and plasma concentrations. S. Czernichow, A. Couthouis, S. Bertrais, A.-C. Vergnaud, L. Dauchet, P. Galan, and S. Hercberg. Am. J. Clinical Nutrition, August 1, 2006; 84(2): 395 - 399)

(Czernichow et al, 2009) (Czernichow, S. et al. Effects of long-term antioxidant supplementation and association of serum antioxidant concentrations with risk of metabolic syndrome in adults. Am. J. Clinical Nutrition, Vol. 90, No. 2, 329-335, August 2009)

(Dagenais etg al, 2000) (Dagenais GR, Marchioli R, Yusuf S, Tognoni G. Beta-carotene, vitamin C, and vitamin E and cardiovascular diseases. Curr Cardiol Rep. 2000 Jul;2(4):293-9)

(Daly et al, 2009) (Daly TJ, et al. Lack of genoprotective effect of phytosterols and conjugated linoleic acids on Caco-2 cells. Food and Chemical toxicology Vol. 47. Issue 8. August 2009. pp. 1791-1796).

(Darko et al, 2002) (Darko D, Dornhorst A, Kelly FJ, et al, Lack of effect of oral vitamin C on blood pressure, oxidative stress and endothelial function in Type II diabetes. Clin Sci (Lond). 2002 Oct;103(4): 339-44)

(Delcourt et al, 2003) (Delcourt C, Carriere I, Delange M, et al. Associations of cataract with antioxidant inzymes and other risk factors: the Frinch Age-Related Eye Diseases (POLA) Study. Ophthalmology 2003;110:2318-26)

(DeLorenze et al. 2010) (DeLorenze GN, McCoy L, Tsai AL, Quesenberry CP Jr, Rice T, Il'yasova D, Wrensch M. Daily intake of antioxidants in relation to survival among adult patients diagnosed with malignant glioma. BMC Cancer. 2010 May 19;10:215).

(Demling et al. 2002) (Demling R, Ikegami K, Picard L, Lalonde C. Administration of large doses of vitamin C does not decrease oxidant-induced lung lipid peroxidation caused by bacterial-independent acute peritonitis Inflammation. 2002. Volume 18, Number 5, 499-510, DOI: 10.1007/BF01560697). ANIMAL STUDY

(de Souza et al. 2007) (de Souza et al. Multivitamins do not improve radiation therapy-related fatigue: results of a double-blind randomized crossover trial. 2007. Am J Clin Oncol. 2007 Aug;30(4):432-6)

(Devaraj et al, 2007) (S. Devaraj, R. Tang, B. Adams-Huet, A. Harris, T. Seenivasan, J. A de Lemos, and I. Jialal. Effect of high-dose {alpha}-tocopherol supplementation on biomarkers of oxidative stress and inflammation and carotid atherosclerosis in patients with coronary artery disease. Am. J. Clinical Nutrition, November 1, 2007; 86(5): 1392 – 1398)

(Devereux, Seaton, 2005) (Devereux G, Seaton A. Diet as a risk factor for atopy and asthma. J Allergy Clin Immunol. 2005 Jun;115(6):1109-17; quiz 1118)

(de Waart et al. 2000) (de Waart FG, Smilde TJ, Wollersheim H, Stalenhoef AF, Kok FJ. Smoking characteristics, antioxidant vitamins, and carotid artery wall thickness among life-long smokers. J Clin Epidemiol. 2000 Jul;53(7):707-14)

(Dickinson et al, 2009) (Dickinson A, Boyon N, Shao A. Physicians and nurses use and recommend dietary supplements: report of a survey. Nutr J. 2009 Jul 1;8:29)

(Dietrich et al, 2009) (M. Dietrich, P. Jacques, M. Pencina, K. Lanier, M. Keyes, G. Kaur, P. Wolf, R. D'Agostino, R. Vasan. Vitamin E supplement use and the incidence of cardiovascular disease and all-cause mortality in the Framingham Heart Study: Does the underlying health status play a role? Atherosclerosis, 2009. Volume 205, Issue 2, Pages 549-553)

(Donnelly et al, Fertil Steril. 1999) (Donnelly ET, McClure N, Lewis SE. Antioxidant supplementation in vitro does not improve human sperm motility. Fertil Steril. 1999 Sep;72(3):484-95)

(Donnelly et al, Mutagenesis. 1999) (Donnelly ET, McClure N, Lewis SE. The effect of ascorbate and alpha-tocopherol supplementation in vitro on DNA integrity and hydrogen peroxide-induced DNA damage in human spermatozoa. Mutagenesis. 1999 Sep;14(5):505-12)

(Dorgan et al, 1998) (Dorgan JF, Sowell A, Swanson CA, et al. Relationships of serum carotenoids, retinol, alpha-tocopherol, and selenium with breast cancer risk: results from a prospective study in Columbia, Missouri (United States). Cancer Causes Control. 1998;9(1):89-97)

(Dotan, Lichtenberg and Pinchuk, 2009) (Dotan Y, Lichtenberg D, Pinchuk I. No evidence supports vitamin E indiscriminate supplementation. Biofactors. 2009 Nov-Dec;35(6):469-73)

(Douglas et al, 2004) (Douglas RM, Hemila H, D'Souza R, Chalker EB, Treacy B. Vitamin C for preventing and treating the common cold. Cochrane Database Syst Rev. 2004(4):CD000980)

(Drieling et al. 2010) (Drieling RL, Gardner CD, Ma J, Ahn DK, Stafford RS. No beneficial effects of pine bark extract on cardiovascular disease risk factors. Arch Intern Med. 2010 Sep 27;170(17):1541-7)

(Duffield-Lillico, 2003) (Duffueld-Lillico AJ, et al, Selenium supplementation and secondary prevention of nonmelanoma skin cancer in a randomized trial. J Natl Cancer Inst. 2003 Oct 1;95(19):1477-81).

(Dunn et al, 2007) (Julie E. Dunn, Sandra Weintraub, Anne M. Stoddard and Sarah Banks. Serum α-tocopherol, concurrent and past vitamin E intake, and mild cognitive impairment. Neurology. 200768: 670-676)

(Duthie et al. 1994) (Duthie GG, Beattie JA, Arthur JR, Franklin M, Morrice PC, James WP. Blood antioxidants and indices of lipid peroxidation in subjects with angina pectoris. Nutrition. 1994 Jul-Aug;10(4):313-6).

(Eidelman et al., 2004) (Eidelman RS, Hollar D, Hebert PR, Lamas GA, Hennekens CH. Randomized trials of vitamin E in the treatment and prevention of cardiovascular disease. Arch Intern Med. 2004;164: 1552-1556)

(Epplein, 2009) (Meira Epplein et al. Plasma carotenoids, retinol, and tocopherols and postmenopausal breast cancer risk in the Multiethnic Cohort Study: a nested case-control study. Breast Cancer Research. 2009, 11:R49)

(Erdman, Ford and Lindshield, 2009) (Erdman JW Jr, Ford NA, Lindshield BL. Are the health attributes of lycopene related to its antioxidant function? Arch Biochem Biophys. 2009 Mar 15;483(2):229-35)

(Evans et al. 1998) (Evans RW, Shaten BJ, Day BW, Kuller LH. Prospective association between lipid soluble antioxidants and coronary heart disease in men. The Multiple Risk Factor Intervention Trial. Am J Epidemiol. 1998; 147: 180–186).

(Evans, 2008) (Evans J. Antioxidant supplements to prevent or slow down the progression of AMD: a systematic review and meta-analysis. Eye (Lond). 2008 Jun;22(6):751-60)

(Evans and Henshaw, 2008) (Evans JR, Henshaw K. Antioxidant vitamin and mineral supplements for preventing age-related macular degeneration. Cochrane Database Syst Rev. 2008 Jan 23;(1):CD000253)

(Fairfield et al, 2001) (Fairfield KM, Hankinson SE, Rosner BA, Hunter DJ, Colditz GA, Willett WC. Risk of ovarian carcinoma and consumption of vitamins A, C, and E and specific carotenoids: a prospective analysis. Cancer. 2001 Nov 1;92(9):2318-26)

(Falcone et al, 2010) (E Liana Falcone, Alexandra Mangili, Alice M Tang, Clara Y Jones, Margo N Coods, Joseph F Polak and Christine A Wanke . Micronutrient concentrations and subclinical atherosclerosis in adults with HIV. Am J Clin Nutr 91: 1213-1219, 2010. Vol. 91, No. 5, 1213-1219, May 2010)

(Fang et al, 2008) (Fang J, Deng D, Nakamura H, et al. Oxystress inducing antitumor therapeutics via tumor-targeted delivery of PEG-conjugated D-amino acid oxidase. Int J Cancer. 2008 Mar 1;122(5): 1135-44).

(Feskanich et al, 2002) (Feskanich D, Singh F, Willett WC, Colditz GA. Vitamin A intake and hip fractures among postmenopausal women. J Am Med Assoc 2002;287:47-54)

(Figueiredo et al, 2009) (Figueiredo JC et al. Folic acid and risk of prostate cancer: results from a randomized clinical trial. J Natl Cancer Inst. 2009 Mar 18;101(6):432-5).

(Fillenbaum et al, 2005) (Fillenbaum et al. Dementia and Alzheimer's Disease in Community-Dwelling Elders Taking Vitamin C and/ or Vitamin E. The Annals of Pharmacotherapy: Vol. 39, No. 12, pp. 2009-2014)

(Foronjy et al, 2008) (Foronjy R, Wallace A, D'Armiento J. The phar-mokinetic limitations of antioxidant treatment for COPD. Pulm Pharmacol Ther. 2008;21(2):370-9)

(Frieling et al, 2000) (Frieling UM, et al. A randomized, 12-year pri-mary-prevention trial of beta carotene supplementation for nonmela-noma skin cancer in the physician's health study. Arch Dermatol. 2000 Feb;136(2):179-84).

(Fusco et al. 2007) (Fusco D, Colloca G, Lo Monaco MR, Cesari M. Effects of antioxidant supplementation on the aging process. Clin Interv Aging. 2007;2(3):377-87).

(Gallicchio et al, 2008) (Gallicchio L. et al. Carotenoids and the risk of developing lung cancer: a systematic review. Am J Clin Nutr. 2008 Aug;88(2):372-83)

(Gao et al. 2008) (Gao J, Gao X, Li W, Zhu Y, Thompson PJ. Observational studies on the effect of dietary antioxidants on asthma: a meta-analy-sis. Respirology. 2008 Jun;13(4):528-36).

(Gaziano et al, 2009) (Vitamins E and C in the prevention of prostate and total cancer in men: the Physicians' Health Study II randomized controlled trial. Gaziano, JM., JAMA. 2009 Jan 7;301(1):52-62. Epub 2008 Dec 9)

(Genkinger et al. 2004) (Genkinger JM, Platz EA, Hoffman SC, Comstock GW, Helzlsouer KJ. Fruit, vegetable, and antioxidant intake and all-cause, cancer, and cardiovascular disease mortality in a com-munity-dwelling population in Washington County, Maryland. Am J Epidemiol. 2004 Dec 15;160(12):1223-33).

(Geva et al, 1996) (Geva E, Bartoov B, Zabludovsky N, Lessing JB, Lerner-Geva L, Amit A. The effect of antioxidant treatment on human spermatozoa and fertilization rate in an in vitro fertilization program. Fertil Steril. 1996 Sep;66(3):430-4)

(Gibbons et al, 2003) (Gibbons RJ, Abrams J, Chatterjee K, et al; American College of Cardiology/American Heart Association Task Force on Practice Guidelines. Committee on the Management of Patients With Chronic Stable Angina. ACC/AHA 2002 guideline update for the management of patients with chronic stable angina—summary article: a report of the American College of Cardiology/American Heart Association Task Force on Practice Guidelines (Committee on the Management of Patients With Chronic Stable Angina). *Circulation.* 2003; 107: 149–158)

(Giovannucci, et al., 1998) (Giovannucci, E., M.J. Stampfer, G.A. Colditz, D.J. Hunter, C. Fuchs, B.A. Rosner, F.E. Speizer & W.C. Willett. 1998. Multivitamin use, folate, and colon cancer in women in the Nurses' Health Study. Ann. Intern. Med. 129(7):517-524)

(GISSI-Prevenzione Investigators;1999) (Dietary supplement with n-3 polyunsaturated acids and vitamin E after myocardial infarction: results of the GISSI-Prevention trial. Gruppo, Italiano per lo Studio Sopravvivenza nell'Infarto miocardico. Lancet, 1999; 354: 447-455)

(Gomez-Cabrera et al, 2008) (Gomez-Cabrera MC, Domenech E, Romagnoli M, Arduini A, Borras C, Pallardo FV, Sastre J, Viña J. Oral administration of vitamin C decreases muscle mitochondrial biogenesis and hampers training-induced adaptations in endurance performance. Am J Clin Nutr. 2008 Jan;87(1):142-9)

(Graham et al, 1992) (Graham S, Sielezny M, Marshall J, Priore R, Freudenheim J, Brasure J, Haughey B, Nasca P, Zdeb M. Diet in the epidemiology of Postmenopausal Breast Cancer in the New York State Cohort. Am J Epidemiol 1992;136:3127-37)

(Gray et al, 2007) (Gray SL et al. Antioxidant Vitamin Supplement Use and Risk of Dementia or Alzheimer's Disease in Older Adults. Journal of the American Geriatric Society. 2007. Volume 56 Issue 2, Pages 291 - 295)

(Greenberg et al, 1994) (Greenberg ER, Baron JA, Tosteson TD, Freeman DH Jr, Beck GJ, Bond JH, Colacchio TA, Coller JA, Frankl HD, Haile RW, et al. A clinical trial of antioxidant vitamins to prevent colorectal adenoma. Polyp Prevention Study Group. N Engl J Med. 1994 Jul 21;331(3):141-7)

(Greenberg et al, 1996) (Greenberg ER, Baron JA, Karagas MR, Stukel TA, Nierenberg DW, Stevens MM, Mandel JS, Haile RW. Mortality associated with low plasma concentration of beta carotene and the effect of oral supplementation. JAMA. 1996; 275: 699-703)

(Green et al, 1999) (Green A, Williams G, Neale R, et al.: Daily sunscreen application and beta carotene supplementation in prevention of basal-cell and squamous-cell carcinomas of the skin: a randomised controlled trial. Lancet 354 (9180): 723-9, 1999)

(Grievink et al. 1998) (Grievink L, Smit HA, Ocké MC, van 't Veer P, Kromhout D. Dietary intake of antioxidant (pro)-vitamins, respiratory symptoms and pulmonary function: the MORGEN study. Thorax. 1998 Mar;53(3):166-71).

(Grimm et al, 2011) (Grimm S, Mvondo d, Grune T, Bruesing N. The outcome of 5-ALA-mediated photodynamic treatment inmelamoma-cells is influenced by vitamin C and heme oxygenase-1. Biofactors. 2011 Jan;37(1):17-24).

(Gritz, 2006) (Gritz DC, Srinivasan M, Smith SD, et al. The Antioxidants in Prevention of Cataracts Study: effects of antioxidant supplements on cataract progression in South India. Br J Ophthalmol. 2006;90(7):847-851)

(Haase M, et al. 2007) (Haase M, et al. Phase II, randomized, controlled trial of high-dose N-acetylcysteine in high-risk cardiac surgery patients. Crit Care Med. 2007 May;35(5):1324-31).

(Hadi, Suwaidi, 2007) (Hadi HA, Suwaidi JA. Endothelial dysfunction in diabetes mellitus. Vasc Health Risk Manag. 2007;3(6):853-76)

(Hak et al. 2003) (Hak AE, Stampfer MJ, Campos H, Sesso HD, Gaziano JM, Willett W, Ma J: Plasma carotenoids and tocopherols and risk of myocardial infarction in a low-risk population of US male physicians. Circulation 108:802 :E9012 –E9013,2003).

(Hayden et al,2007) (K. Hayden, K. Welsh-Bohmer, H. Wengreen, P. Zandi, C. Lyketsos, J. Breitner. Risk of Mortality with Vitamin E Supplements: The Cache County Study. Am J Med. 2007 Feb;120(2):180-4)

(Heinonen et al, 1994) (Heinonen, O.P., J.K. Huttunen, D. Albanes & ATBC cancer prevention study group. 1994. The effect of vitamin E and beta carotene on the incidence of lung cancer and other cancers in male smokers. N. Engl. J. Med. 330:1029-1035)

(Hemila et al. 2004) (Hemilä H, Virtamo J, Albanes D, Kaprio J. Vitamin E and beta-carotene supplementation and hospital-treated pneumonia incidence in male smokers. Chest. 2004 Feb;125(2):557-65).

(Hemila and Kaprio, 2008 Oct) (Hemilä H, Kaprio J. Vitamin E supplementation may transiently increase tuberculosis risk in males who smoke heavily and have high dietary vitamin C intake. Br J Nutr. 2008 Oct;100(4):896-902)

(Hemila and Kaprio, 2008 Nov) (Hemilä H, Kaprio J. Vitamin E supplementation and pneumonia risk in males who initiated smoking at an early age: effect modification by body weight and dietary vitamin C. Nutr J. 2008 Nov 19;7:33)

(Hemila and Kaprio, 2009 Apr) (Hemilä H, Kaprio J. Modification of the effect of vitamin E supplementation on the mortality of male smokers by age and dietary vitamin C. Am J Epidemiol. 2009 Apr 15;169(8):946-53)

(Hennekens et al, 1996) (Hennekens CH, Buring JE, Manson JE, et al. Lack of effect of long-term supplementation with beta carotene on the incidence of malignant neoplasms and cardiovascular disease. N Engl J Med. 1996;334:1145-1149)

(Hense et al. 1993) (Hense HW, Stender M, Bors W, Keil U. Lack of an association between serum vitamin E and myocardial infarction in a population with high vitamin E levels.Atherosclerosis. 1993 Oct;103(1): 21-8).

(Herbert et al. 2006) (Herbert KE, Fletcher S, Chauhan D, Ladapo A, Nirwan J, Munson S, Mistry P. Dietary supplementation with different vitamin C doses: no effect on oxidative DNA damage in healthy people. Eur J Nutr. 2006 Mar;45(2):97-104).

(Hercberg et al, 2004) (Hercberg S, Galan P, Preziosi P, Bertrais S, Mennen L, Malvy D, Roussel A-M, Favier A, Briançon S. SU.VI.MAX Study. Arch Intern Med 2004;164:2335–42)

(Hercberg et al, 2007) (Serge Hercberg et al. Antioxidant Supplementation Increases the Risk of Skin Cancers in Women but Not in Men. American Society for Nutrition J. Nutr. 137:2098-2105, September 2007)

(Ho M, Morgan DJ. 2009) (Ho M, Morgan DJ. Meta-analysis of N-acetylcysteine to prevent acute renal failure after major surgery. Am J Kidney Dis. 2009 Jan;53(1):33-40).

(Hodis et al, 1995) (Hodis HN, Mack WJ, LaBree L, et al. Serial coronary angiographic evidence that antioxidant vitamin intake reduces progression of coronary artery atherosclerosis. JAMA 1995;273(23): 1849-54)

(Hodis et al, 2002) (Hodis HN, Mack WJ, LaBree L, Mahrer PR, Sevanian A, Liu CR, Liu CH, Hwang J, Selzer RH, Azen SP; VEAPS Research Group. Alpha-tocopherol supplementation in healthy individuals reduces low-density lipoprotein oxidation but not atherosclerosis: the Vitamin E Atherosclerosis Prevention Study (VEAPS). Circulation. 2002; 106: 1453–1459)

(Honarbakhsh, Schachter, 2009) (Honarbakhsh S, Schachter M. Vitamins and cardiovascular disease. Br J Nutr. 2009 Apr;101(8):1113-31)

(Horneber et al, 2008) (Horneber et al. Mistletoe therapy in oncology. Cochrane Database Syst Rev. 2008 Apr 16;(2):CD003297).

(Hotaling et al, 2010) ((James Hotaling, Jonathan Wright, Gaia Pocobelli, Michael Porter and Emily White. Long-Term Use of supplemental vitamin C, vitamin D and vitamin E does not reduce the risk of urothelial cell carcinoma of the bladder in the VITamins and Lifestyle Study. The Journal of Urology. Vol. 183, Issue 4, Supplement 1, April 2010, Page e450)

(Howes R.M. 2009, Am J Cosm Surg) (Howes, RM. Antioxidant Vitamins: A Review of Policy Statements and Recommendations. *The American Journal of Cosmetic Surgery.* 2009;26(2):63-78)

(Huang et al, 2006 May) (Huang HY, Caballero B, Chang S, Alberg A, Semba R, Schneyer C, Wilson RF, Cheng TY, Prokopowicz G, Barnes GJ 2nd, Vassy J, Bass EB. Multivitamin/mineral supplements and prevention of chronic disease. Evid Rep Technol Assess (Full Rep). 2006 May;(139):1-117)

(Huang et al, 2006 Sept) (H. Huang et al. The Efficacy and Safety of Multivitamin and Mineral Supplement Use To Prevent Cancer and Chronic Disease in Adults: A Systematic Review for a National

Institutes of Health State-of-the-Science Conference. September 5, 2006, Vol. 145. Issue 5. Pages 372-385)

(Hughes et al, 1998) (Hughes CM, Lewis SE, McKelvey-Martin VJ, Thompson W. The effects of antioxidant supplementation during Percoll preparation on human sperm DNA integrity. Hum Reprod. 1998 May;13(5):1240-7)

(Hulten, 2001) (Hulten K, Van Kappel AL, Winkvist A, et al. Carotenoids, alpha-tocopherols, and retinol in plasma and breast cancer risk in northern Sweden. Cancer Causes Control. 2001;12(6):529-537)

(Hung et al, 2010) (Hung TH, Chen SF, Li MJ, Yeh YL, Hsieh TT. Differential effects of concomitant use of vitamins C and E on tro-phoblast apoptosis and autophagy between normoxia and hypoxia-reoxygenation. PLoS One. 2010 Aug 16;5(8):e12202).

(Hunter et al, 1993) (A Prospective Study of the Intake of Vitamins C, E, and A and the Risk of Breast Cancer. David J. Hunter, JoAnn E. Manson, Graham A. Colditz, Meir J. Stampfer, Bernard Rosner, Charles H. Hennekens, Frank E. Speizer, and Walter C. Willett. The New England Journal of Medicine. Vol. 329:234-240. No. 4. July 22, 1993)

(Icox et al. 2008) (Icox BJ, Curb JD, Rodriguez BL. Antioxidants in car-diovascular health and disease: key lessons from epidemiologic stud-ies. Am J Cardiol. 2008 May 22;101(10A):75D-86D).

(Il'yasova et al. 2009) (Il'yasova D, Marcello JE, McCoy L, Rice T, Wrensch M. Total dietary antioxidant index and survival in patients with glioblastoma multiforme. Cancer Causes Control. 2009 Oct;20(8): 1255-60).

(Inoue et al, 2009) (Inoue et al. Green tea consumption and gastric cancer in Japanese: a pooled analysis of six cohort studies. Gut. 2009 Oct;58(10):1323-32).

(Isaac et al, 2008) (Vitamin E for Alzheimers and mild cognitive impairment. Isaac MG, Quinn R, Tabet N. .Cochrane Database Syst Rev. 2008 Jul 16;(3):CD002854)

(Ishikawa et al, 2006) (Ishikawa et al. Smoking, alcohol drinking, green tea consumption and the risk of esophageal cancer in Japanese men. J Epidemiol. 2006 Sep;16(5):185-92).

(Iwasaki et al, 2010) (Motoki Iwasaki; Manami Inoue; Shizuka Sasazuki; Norie Sawada; Taiki Yamaji; Taichi Shimazu; Walter C Willett; Shoichiro Tsugane. Green Tea Drinking and Subsequent Risk of Breast Cancer in a Population to Based Cohort of Japanese Women. Breast Cancer Research. 2010;12(5:R88).

(Jacobs, 2001) (Jacobs EJ, Connell CJ, Patel AV, Chao A, Rodriguez C, Seymour J, McCullough ML, Calle EE, Thun MJ. Vitamin C and vitamin E supplement use and colorectal cancer mortality in a large American Cancer Society cohort. Cancer Epidemiol Biomarkers Prev. 2001 Jan;10(1):17-23)

(Jacob and Ames, 2003) (Robert A. Jacob et al. and Bruce Ames. Moderate Antioxidant Supplementation Has No Effect on Biomarkers of Oxidant Damage in Healthy Men with Low Fruit and Vegetable Intakes. J. Nutr. 133:740-743, March 2003)

(Jacobs et al., 2002) (Jacobs EJ, Henion AK, Briggs PJ, Connell CJ, McCullough ML, Jonas CR, Rodriguez C, Calle EE, Thun MJ. Vitamin C and vitamin E supplement use and bladder cancer mortality in a large cohort of US men and women. American Journal of Epidemiology 2002;156: 1002-10)

(Jacobs et al. Jan. 2002) (Jacobs EJ, Connell CJ, McCullough ML, Chao A, Jonas CR, Rodriguez C, Calle EE, Thun MJ. Vitamin C, vitamin E, and multivitamin supplement use and stomach cancer mortality in the Cancer Prevention Study II cohort. Cancer Epidemiol Biomarkers Prev. 2002 Jan;11(1):35-41)

(Jiang et al, 2009) (Jiang YY, et al. The protective effect of silibinin against mitomycin C-induced intrinsic apoptosis in human melanoma A375-S2 cells. J Pharmacol Sci. 2009 Oct;111(2):137-46).

(Jaques et al, 1995) (Jacques PF, Sulsky SI, Perrone GE, Jenner J, Schaefer EJ Epidemiology Program, US Department of Agriculture Human Nutrition Research Center on Aging, Tufts University, Boston, MA 02111, USA. (Jacques PF, Sulsky SI, Perrone GE, Jenner J, Schaefer EJ. Effect of vitamin C supplementation on lipoprotein cholesterol, apolipoprotein, and triglyceride concentrations. Ann Epidemiol 1995 Jan;5(1):52-9)

(Joshi et al. 2008) (Joshi SR, Mehendale SS, Dangat KD, Kilari AS, Yadav HR, Taralekar VS. High maternal plasma antioxidant concentrations associated with preterm delivery. Ann Nutr Metab. 2008;53(3-4):276-82).

(Kang et al., 2006) (A randomized trial of vitamin E supplementation and cognitive function in women. Kang JH, Cook N, Manson J, Buring JE, Grodstein F. Arch Intern Med. 2006 Dec 11-25;166(22):2462-8)

(Kang et al, 2009) (Vitamin E, vitamin C, beta carotene, and cognitive function among women with or at risk of cardiovascular disease: The Women's Antioxidant and Cardiovascular Study. Kang JH, Cook NR, Manson JE, Buring JE, Albert CM, Grodstein F. Circulation. 2009 Jun 2;119(21):2772-80. Epub 2009 May 18)

(Katsiki and Manes, 2009) (Katsiki N, Manes C. Is there a role for supplemented antioxidants in the prevention of atherosclerosis? Clin Nutr. 2009 Feb;28(1):3-9)

(Katz et al, 2004) (D. L. Katz, M. A. Evans, W. Chan, H. Nawaz, B. P. Comerford, M. L. Hoxley, V.Y. Njike, and P. M. Sarrel. Oats, Antioxidants and Endothelial Function in Overweight, Dyslipidemic Adults. J. Am. Coll. Nutr., October 1, 2004; 23(5): 397 – 403)

(Kefer et al. 2003) (Kefer JM, Hanet CE, Boitte S, Wilmotte L, DeKock M. Acetylcysteine, coronary procedure and prevention of contrast-induced worsening of renal function: which benefit for which patient? Acta Cardiol. 2003 Dec;58(6):555-60).

(Keith et al, 2001) (Keith ME, Jeejeebhoy KN, Langer A, Kurian R, Barr A, O'Kelly B, Sole MJ. A controlled clinical trial of vitamin E supplementation in patients with congestive heart failure. Am J Clin Nutr. 2001 Feb;73(2):219-24)

(Kinlay et al. 2004) (Kinlay S, Behrendt D, Fang JC, Delagrange D, Morrow J, Witztum JL, Rifai N, Selwyn AP, Creager MA, Ganz P. Long-term effect of combined vitamins E and C on coronary and peripheral endothelial function. J Am Coll Cardiol. 2004 Feb 18;43(4):629-34).

(Kirsh et al, 2006) (Kirsh VA, Hayes RB, Mayne ST, Chatterjee N, Subar AF, Dixon LB, et al. Supplemental and Dietary Vitamin E, Beta-Carotene, and Vitamin C Intakes and Prostate Cancer Risk. PCLO. J Natl Cancer Inst 2006;98:245-254)

(Kizhakekuttu, Widlansky, 2010) (Kizhakekuttu TJ, Widlansky ME. Natural antioxidants and hypertension: promise and challenges. Cardiovasc Ther. 2010 Aug;28(4):e20-32. Epub 2010 Mar 29)

(Klipstein-Grobusch et al, 1999) (K. Klipstein-Grobusch, J. M Geleijnse, J. H den Breeijen, H. Boeing, A. Hofman, D. E Grobbee, and J. C. Witteman. Dietary antioxidants and risk of myocardial infarction in the elderly: the Rotterdam Study. Am. J. Clinical Nutrition, February 1, 1999; 69(2): 261 - 266)

(Knekt et al, 2004) (P. Knekt, J. Ritz, M. A Pereira, E. J O'Reilly, K. Augustsson, G. E Fraser, U. Goldbourt, B. L Heitmann, G. Hallmans, S. Liu, et al. Antioxidant vitamins and coronary heart disease risk: a pooled analysis of 9 cohorts. Am. J. Clinical Nutrition, December 1, 2004; 80(6): 1508 - 1520)

(Koushik et al, 2005) (Koushik A, et al. Fruits and vegetables and ovarian cancer risk in a pooled analysis of 12 cohort studies. Cancer Epidemiol Biomarkers Prev. 2005 Sep;14(9):2160-7).

(Kushi et al, 1996) (L. H. Kushi et al. Dietary Antioxidant Vitamins and Death from Coronary Heart Disease in Postmenopausal Women. New England Journal of Medicine. Vol. 334. No. 18. May 2, 1996. pp. 1156-1162)

(Lamson, Brignall, 1999) (Lamson DW, Brignall MS. Antioxidants in cancer therapy; their actions and interactions with oncologic therapies. Altern Med Rev 1999;4(5):304-29).

(Larsson et al, 2010) (Susanna C Larsson, Agneta Åkesson, Leif Bergkvist, and Alicja Wolk. Multivitamin use and breast cancer incidence in a prospective cohort of Swedish women. Am J Clin Nutr Published online 24 March 2010. Am J Clin Nutr Vol. 91, No. 5, 1268-1272, May 2010)

(Laurin et al, 2002) (Laurin D, Foley DJ, Masaki KH, et al. Vitamin E and C supplements and risk of dementia. JAMA 2002;288:2266–8)

(Laurin et al, 2003)(Midlife Dietary Intake of Antioxidants and Risk of Late-Life Incident Dementia: The Honolulu-Asia Aging Study)

(Lawson et al, 2007) (Lawson KA, Wright ME, Subar A, Mouw T, Schatzkin A, Leitzmann MF. Multivitamin use and risk of prostate cancer in the National Institutes of Health–AARP Diet and Health Study. J Natl Cancer Inst (2007) 99:754–64)

(Lee et al., 1999) (Lee IM, Cook NR, Manson JE, Buring JE, Hennekens CH. Beta-carotene supplementation and incidence of cancer and cardiovascular disease: the Women's Health Study. J Natl Cancer Inst. 1999 Dec 15;91(24):2102-6)

(Lee et al, 2004) (Duk-Hee Lee, Aaron R Folsom, Lisa Harnack, Barry Halliwell and David R Jacobs, Jr. Does supplemental vitamin C increase cardiovascular disease risk in women with diabetes? American Journal of Clinical Nutrition, Vol. 80, No. 5, 1194-1200, November 2004)

(Lee et al, 2005) (Vitamin E in the primary prevention of cardiovascular disease and cancer: the Women's Health Study: a randomized controlled trial. Lee IM, Cook NR, Gaziano JM, Gordon D, Ridker PM, Manson JE, et al. JAMA. 2005;294:56–65)

(Leppala et al. 2000) (Leppala JM, Virtamo J, Fogelholm R, Huttunen JK, Albanes D, Taylor PR, et al. Controlled trial of alpha-tocopherol and beta-carotene supplements on stroke incidence and mortality in male smokers. Arterioscler Thromb Vasc Biol2000;20:230-5).

(Lesperance et al, 2002) (Lesperance ML, Olivotto IA, Forde N, et al. Mega-dose vitamins and minerals in the treatment of non-metastatic breast cancer: an historical cohort study. Breast Cancer Res Treat (2002) 76(2):137–143)

(Levy et al, 2004) (Levy AP, Friedenberg P, Lotan R, et al. The effect of vitamin therapy on the progression of coronary artery atherosclerosis varies by haptoglobin type in postmenopausal women. Diabetes Care. 2004;27(4):925-930)

(Lichtenstein, 2009) (Lichtenstein AH. Nutrient supplements and cardiovascular disease: a heartbreaking story. J Lipid Res. 2009 Apr;50 Suppl:S429-33)

(Lin, Kang, Zheng, 2005) (Lin C, Kang J, Zheng R. Vitamin K3 triggers human leukemia cell death through hydrogen peroxide generation and histone hyperacetylation. Pharmazie. 2005 Oct;60(10):765-71).

(Lin et al, 2009) (Jennifer Lin, Nancy R. Cook, Christine Albert, Elaine Zaharris, J. Michael Gaziano, Martin Van Denburgh, Julie E. Buring, JoAnn E. Manson. Vitamins C and E and Beta Carotene Supplementation and

Cancer Risk: A Randomized Controlled Trial. JNCI Journal of the National Cancer Institute 2009 101(1):14-23)

(Lippman et al, 2009) (Effect of selenium and vitamin E on risk of prostate cancer and other cancers: the Selenium and Vitamin E Cancer Prevention Trial (SELECT). Lippman, SM. JAMA. 2009 Jan 7;301(1):39-51. Epub 2008 Dec 9)

(Liu et al., 2006) (S. Liu, I-M. Lee, Y. Song, M. Van Denburgh, N. R. Cook, J. E. Manson, and J. E. Buring. Vitamin E and Risk of Type 2 Diabetes in the Women's Health Study Randomized Controlled Trial. Diabetes, October 1, 2006; 55(10): 2856 – 2862)

(Longnecker, 1997) (Longnecker MP, Newcomb PA, Mittendorf R, Greenberg ER, Willett WC. Intake of carrots, spinach, and supplements containing vitamin A in relation to risk of breast cancer. Cancer Epidemiol Biomarkers Prev. 1997;6(11):887-892)

(Lonn et al. 2002) (Eva Lonn et al. Effects of Vitamin E on Cardiovascular and Microvascular Outcomes in High-Risk Patients With Diabetes. Results of the HOPE Study and MICRO-HOPE Substudy. Diabetes Care 25:1919-1927, 2002)

(Lonn et al, 2005) (Effects of long-term vitamin E supplementation on cardiovascular events and cancer: a randomized controlled trial. E. Lonn et al. JAMA. 2005 Mar 16;293(11):1338-47)

(Luchsinger et al, 2003) (Luchsinger et al. Antioxidant Vitamin Intake and Risk of Alzheimer Disease. Arch Neurol 2003;60:203-208)

(Magliano et al, 2006) (Magliano, Dianna; McNeil, John; Branley, Pauline; Shiel, Louise; Demos, Lisa; Wolfe, Rory; Kotsopoulos, Dimitra; McGrath, Barry. The Melbourne Atherosclerosis Vitamin E Trial (MAVET): a study of high dose vitamin E in smokers. European Journal of Cardiovascular Prevention & Rehabilitation. June 2006 - Volume 13 - Issue 3 - pp 341-347)

(Mann et al, 2004) (Effects of vitamin E on cardiovascular outcomes in people with mild-to-moderate renal insufficiency: results of the HOPE study. J.F. Mann et al. Kidney Int. 2004 Apr;65(4):1375-80)

(Mannisto et al, 2004) (Mannisto S, et al. Dietary carotenoids and risk of lung cancer in a pooled analysis of seven cohort studies. Cancer Epidemiol Biomarkers Prev. 2004 Jan;13(1):40-8).

(Mannisto et al, 2007) (Mannisto S, et al. Dietary carotenoids and risk of colorectal cancer in a pooled analysis of 11 cohort studies. Am J Epidemiol. 2007 Feb 1;165(3):246-55).

(Marchioli et al. 2001) (Marchioli R, Schweiger C, Levantesi G, Tavazzi L, Valagussa F. Antioxidant vitamins and prevention of cardiovascular disease: epidemiological and clinical trial data. Lipids. 2001;36 Suppl:S53-63).

(Matthan et al, 2003) (Matthan N.R. et al. Impact of simvastatin, niacin, and/or antioxidants on cholesterol metabolism in CAD patients with low HDL. Journal of Lipid Research, Vol. 44, 800-806, April 2003)

(Maserejian et al, 2007) (Maserejian NW, Giovanncci E, Rosner B, Joshipura K. Prospective Study of Vitamins C, E, A, and Carotenoids and Risk of Oral Premalignant Lesions in Men. International J of Cancer. 120(5):970-7; 2006)

(Mayer-Davis et al, 1997) (Mayer-Davis EJ, Monaco JH, Marshall JA, Rushing J, Juhaeri. Vitamin c intake and cardiovascular disease risk factors in persons with non-insulin-dependent diabetes mellitus. From the Insulin Resistance Atherosclerosis Study and the San Luis Valley Diabetes Study. Prev Med 1997 May-Jun;26(3):277-83)

(Mayer-Davis et al. 1998) (Mayer-Davis EJ, Bell RA, Reboussin BA, Rushing J, Marshall JA, Hamman RF. Antioxidant nutrient intake and diabetic retinopathy: the San Luis Valley Diabetes Study. Ophthalmology. 1998 Dec;105(12):2264-70).

(Mayne et al, 2001) (Susan T. Mayne et al. Randomized Trial of Supplemental ß-Carotene to Prevent Second Head and Neck Cancer. Cancer Research 61, 1457-1463, February 15, 2001)

(McCance et al. 2010) (McCance DR, Holmes VA, Maresh MJ, Patterson CC, Walker JD, Pearson DW, Young IS; Diabetes and Pre-eclampsia Intervention Trial (DAPIT) Study Group. Vitamins C and E for prevention of pre-eclampsia in women with type I diabetes (DAPIT): a randomised placebo-controlled trial. Lancet. 2010 Jul 24;376(9737):259-66).

(McGinley et al. 2009) (McGinley C, Shafat A, Donnelly AE. Does anti-oxidant vitamin supplementation protect against muscle damage? Sports Med. 2009;39(12):1011-32).

(McNeil et al, 2004) (McNeil JJ, Robman L, Tikellis G, Sinclair MI, McCarty CA, Taylor HR. Vitamin E supplementation and cataract: randomized controlled trial. Ophthalmology. 2004;111(1):75-84)

(McQuillan et al. 2001) (McQuillan BM, Hung J, Beilby JP, Nidorf M, Thompson PL. Antioxidant vitamins and the risk of carotid atherosclerosis. The Perth Carotid Ultrasound Disease Assessment study (CUDAS). J Am Coll Cardiol. 2001 Dec;38(7):1788-94)

(Miller et al., 1997) (Miller, E.R. 3rd, L.J. Appel, O.A. Levander & D.M. Levine. 1997. The effect of antioxidant vitamin supplementation on traditional cardiovascular risk factors. J. Cardiovasc. Risk. 4(1):19-24)

(Miller et al., 2004) (Miller ER 3d, Pastor-Barriuso R, Dalal D, Riemersma RA, Appel LJ, Guallar E. Meta-analysis: high-dosage vitamin E supplementation may increase all-cause mortality. Ann Intern Med 2005;142:37-46)

(Milner et al. 2004) (Joshua D. Milner, Daniel M. Stein, Robert McCarter, Rachel Y. Moon. Early Infant Multivitamin Supplementation

Is Associated With Increased Risk for Food Allergy and Asthma. PEDIATRICS Vol. 114 No. 1 July 2004, pp. 27-32).

(Miranda-Vilela et al, 2011) (Miranda-Vilela AL, et al. The protective effects of nutritional antioxidant therapy on Ehrlich solid tumor-bearing mice depend on the type of antioxidant therapy chosen: histology, genotoxicity and hematology evaluations. J Nutr Biochem. 2011 Jan 25. [Epub ahead of print]).

(Moertel et al, 1985) (Moertel CG, Fleming TR, Creagan ET, Rubin J, O'Connell MJ, Ames MM. High-dose vitamin C versus placebo in the treatment of patients with advanced cancer who have had no prior chemotherapy. A randomized double-blind comparison. N Engl J Med. 1985 Jan 17;312(3):137-41)

(Moilanen and Hovatta, 1995) (Moilanen J, Hovatta O. Excretion of alpha-tocopherol into human seminal plasma after oral administration. Andrologia. 1995 May-Jun;27(3):133-6)

(Morris and Carson, 2003) (Routine Vitamin Supplementation To Prevent Cardiovascular Disease: A Summary of the Evidence for the U.S. Preventive Services Task Force. Morris and Carson. ANN INTERN MED 2003;139:56-70. Review)

(Moyad et al. 2002) (Moyad MA. Selenium and vitamin E supplements for prostate cancer: evidence or embellishment? Urology. 2002 Apr;59(4 Suppl 1):9-19).

(MRC/BHF, 2002) (MRC/BHF Heart Protection Study of antioxidant vitamin supplementation in 20,536 high-risk individuals: a randomized placebo-controlled trial. Lancet. 2002 Jul 6;360(9326):23-33)

(Munteanu, Zingg, 2007) (Munteanu A, Zingg JM. Cellular, molecular and clinical aspects of vitamin E on atherosclerosis prevention. Mol Aspects Med. 2007 Oct-Dec;28(5-6):538-90)

(Muntwyler et al. 2002) (Muntwyler J, Hennekens CH, Manson JE, Buring JE, Gaziano JM. Vitamin supplement use in a low-risk population of US male physicians and subsequent cardiovascular mortality. Arch Intern Med. 2002 Jul 8;162(13):1472-6)

(Munzel et al, 2010) (Munzel T, et al. Is oxidative stress a therapeutic target in cardiovascular disease? Eur Heart J. 2010 Nov;31(22):2741-8. Epub 2010 Oct 25)

(Myung et al, 2009) (Myung, S.-K., Kim, Y., Ju, W., Choi, H. J., Bae, W. K. (2010). Effects of antioxidant supplements on cancer prevention: meta-analysis of randomized controlled trials. Ann Oncol 21: 166-179).

(Myung, Int J Cancer. et al, 2009) (Myung et al, Green tea consumption and risk of stomach cancer: a meta-analysis of epidemiologic studies. Int J Cancer. 2009 Feb1;124(3):670-7).

(Naughton et al. 2008) (Naughton F, Wijeysundera D, Karkouti K, Tait G, Beattie WS. N-acetylcysteine to reduce renal failure after cardiac surgery: a systematic review and meta-analysis. Can J Anaesth. 2008 Dec;55(12):827-35).

(Neuhouser et al, 2009) (Multivitamin Use and Risk of Cancer and Cardiovascular Disease in the Women's Health Initiative Cohorts. Marian L. Neuhouser et al. Arch Intern Med. 2009;169(3):294-304)

(Neunteufl et al. 2000) (Neunteufl T, Priglinger U, Heher S, Zehetgruber M, Söregi G, Lehr S, Huber K, Maurer G, Weidinger F, Kostner K. Effects of vitamin E on chronic and acute endothelial dysfunction in smokers. J Am Coll Cardiol. 2000 Feb;35(2):277-83).

(Nigwekar SU, Kandula P. 2009) (Nigwekar SU, Kandula P. N-acetylcysteine in cardiovascular-surgery-associated renal failure: a meta-analysis. Ann Thorac Surg. 2009 Jan;87(1):139-47).

(NIH State-of-the Science Panel. 2007). (National Institutes of Health State-of-the-Science Conference Statement: Multivitamin/Mineral

Supplements and Chronic Disease Prevention. NIH State-of-the-Science Panel. ANN INTERN MED 2006;145:364-371).

(Ogunleye et al, 2010) (Ogunleye AA, Xue F, Michels KB: Green tea consumption and breast cancer risk or recurrence: a meta-analysis. Breast Cancer Res Treat 2010, 119:477–484).

(Omenn et al., 1996) (Risk factors for lung cancer and for intervention effects in CARET, the Beta-Carotene and Retinol Efficacy Trial. G.S. Omenn et al. J Natl Cancer Inst. 1996 Nov 6;88(21):1550-9)

(Omenn et al, NEJM. 1996) (Omenn GS, Goodman GE, Thornquist MD, et al. Effects of a combination of beta carotene and vitamin A on lung cancer and cardiovascular disease. N Engl J Med. 1996;334: 1150-1155)

(Padayatty et al, 2010) (Padayatty SJ, Sun AY, Chen Q, Espey MG, Drisko J, Levine M. Vitamin C: intravenous use by complementary and alternative medicine practitioners and adverse effects. PLoS One. 2010 Jul 7;5(7):e11414)

(Pak et al, 2011) (Pak JH et al. Peroxiredoxin 6 overexpression attenuates cisplatin-induced apoptosis in human ovarian cancer cells. Cancer Invest. 2011 Jan;29(1):21-8).

(Pérez et al. 2009) (Pérez VI, Bokov A, Van Remmen H, Mele J, Ran Q, Ikeno Y, Richardson A. "Is the oxidative stress theory of aging dead?". Biochimica et Biophysica Acta (BBA) - General Subjects (2009). 1790 (10): 1005–1014).

(Perlman et al. 2010) (Perlman et al. Multivitamin/Mineral supplementation does not affect standardized assessment of academic performance in elementary school children. 2010. J Am Diet Assoc. 2010 Jul;110(7):1089-93).

(Peters et al, 2008) (U. Peters et al. Vitamin E and selenium supplementation and risk of prostate cancer in the Vitamins and lifestyle (VITAL) study cohort. Cancer Causes Control. 2008 Feb;19(1):75-87)

(Pham et al, 2005) (Pham DQ, Plakogiannis R. Vitamin E supplementation in Alzheimer's disease, Parkinson's disease, tardive dyskinesia, and cataract: Part 2. Ann Pharmacother. 2005 Dec;39(12):2065-72)

(Pocobelli et al, 2007) (Gaia Pocobelli, Ulrike Peters, Alan R. Kristal and Emily White. Use of Supplements of Multivitamins, Vitamin C, and Vitamin E in Relation to Mortality. American Journal of Epidemiology 2009 170(4):472-483)

(Polyzos et al, 2007) (Polyzos NP et al. Combined vitamin C and E supplementation during pregnancy for preeclampsia prevention: a systematic review. Obstet Gynecol Surv. 2007 Mar;62(3):202-6).

(Poston et al., 2006) (L. Poston et al. Vitamin C and vitamin E in pregnant women at risk for pre-eclampsia (VIP trial): randomised placebo-controlled trial. The Lancet, Volume 367, Issue 9517, Pages 1145 - 1154, 8 April 2006)

(Priemé et al. 1997) (Priemé H, Loft S, Nyyssönen K, Salonen JT, Poulsen HE. No effect of supplementation with vitamin E, ascorbic acid, or coenzyme Q10 on oxidative DNA damage estimated by 8-oxo-7,8-dihydro-2'-deoxyguanosine excretion in smokers. Am J Clin Nutr. 1997 Feb;65(2):503-7).

(Promislow et al, 2002) (Promislow JH, Goodman-Gruen D, Slymen DJ, Barrett-Connor E. Retinol intake and bone mineral density in the elderly: the Rancho Bernardo Study. J Bone Miner Res. 2002;17(8):1349-1358)

(Qiao et al, 2009) (Y.-L. Qiao, S. M. Dawsey, F. Kamangar, J.-H. Fan, C. C. Abnet, X.-D. Sun, L. L. Johnson, M. H. Gail, Z.-W. Dong, B. Yu, et al. Total and Cancer Mortality After Supplementation With Vitamins and Minerals: Follow-up of the Linxian General Population Nutrition Intervention Trial. J Natl Cancer Inst, April 1, 2009; 101(7): 507 - 518)

(Qu et al, 2007) (Chen-Xu Qu et al, Chemoprevention of Primary Liver Cancer: A Randomized, Double-Blind Trial in Linxian, China. Journal of the National Cancer Institute. Vol 99, Issue 16. August 15, 2007. pp. 1240-1247)

(Raal et al, 1999) (Efficacy of vitamin E compared with either simvastatin or atorvastatin in preventing the progression of atherosclerosis in homozygous familial hypercholesterolemia. Raal FJ, Pilcher GJ, Veller MG, Kotze MJ, Joffe BI. Am J Cardiol. 1999 Dec 1;84(11):1344-6, A7)

(Raitakari, et al. 2000) (Raitakari OT, Adams MR, McCredie RJ, Griffiths KA, Stocker R, Celermajer DS. Oral vitamin C and endothelial function in smokers: short-term improvement, but no sustained beneficial effect. J Am Coll Cardiol. 2000 May;35(6):1616-21).

(Rapola et al, 1997) (Rapola, J.M., J. Virtamo, S. Ripatti, J.K. Huttunen, D. Albanes, P.R. Taylor & O. P. Heinonen. 1997. Randomized trial of alpha-tocopherol and beta-carotene supplements on incidence of major coronary events in men with previous myocardial infarction. Lancet. 349(9067):1715-1720)

(Rapola et al, 1998) (J M Rapola, J Virtamo, S Ripatti, J K Haukka, J K Huttunen, D Albanes, P R Taylor, and O P Heinonen. Effects of alpha tocopherol and beta carotene supplements on symptoms, progression, and prognosis of angina pectoris Heart, May 1, 1998; 79(5): 454 - 458)

(Rautiainen et al, 2010) (Susanne Rautiainen, Birgitta Ejdervik Lindblad, Ralf Morgenstern and Alicja Wolk. Vitamin C supplements and the risk of age-related cataract: a population-based prospective cohort study in women. Am J Clin Nutr 91: 487-493, 2010)

(Reaven, 1995) (Reaven PD, Herold DA, Barnett J, Edelman S. Effects of Vitamin E on susceptibility of low-density lipoprotein and low-density lipoprotein subfractions to oxidation and on protein glycation in NIDDM. Diabetes Care. 1995;18(6):807-816)

(Renke M. et al. 2008) (Renke M. et al. The effect of N-acetylcysteine on proteinuria and markers of tubular injury in non-diabetic patients with chronic kidney disease. A placebo-controlled, randomized, open, cross-over study. Kidney Blood Press Res. 2008;31(6):404-10).

(Renke et al. 2010) (Renke M. et al. The effect of N-acetylcysteine on blood pressure and markers of cardiovascular risk in non-diabetic patients with chronic kidney disease: a placebo-controlled, randomized, cross-over study. Med Sci Monit. 2010;16(7):PI13-8).

(Riemersma et al, 2000) (R. A Riemersma, K. F Carruthers, R. A Elton, and K. A. Fox. Vitamin C and the risk of acute myocardial infarction. Am. J. Clinical Nutrition, May 1, 2000; 71(5): 1181 - 1186)

(Ristow et al, 2009) (Ristow M, Zarse K, Oberbach A, Klöting N, Birringer M, Kiehntopf M, Stumvoll M, Kahn CR, Blüher M. Antioxidants prevent health-promoting effects of physical exercise in humans. Proc Natl Acad Sci U S A. 2009 May 26;106(21):8665-70)

(Roberts et al, 2010) (Roberts JM et al, Vitamins C and E to prevent complications of pregnancy-associated hypertension. N Engl J Med (2010) 362: 1282-91)

(Rohan et al. 1995) (Rohan TE, Howe GR, Burch JD, Jain M. Cancer Causes Control. Dietary factors and risk of prostate cancer: a case-control study in Ontario, Canada. 1995 Mar;6(2):145-54).

(Rolf et al, 1999) (Rolf C, Cooper TG, Yeung CH, Nieschlag E. Antioxidant treatment of patients with asthenozoospermia or moderate oligoasthenozoospermia with high-dose vitamin C and vitamin

E: a randomized, placebo-controlled, double-blind study. Hum Reprod. 1999 Apr;14(4):1028-33)

(Rumbold et al, Apr 18, 2005. CD004072) (Rumbold A, Middleton P, Crowther CA. Vitamin supplementation for preventing miscarriage. Cochrane Database Syst Rev. 2005 Apr 18;(2):CD004069).

(Rumbold et al, Apr 18, 2005. CD004072) (Rumbold A, Crowther CA. Vitamin E supplementation in pregnancy. Cochrane Database Syst Rev. 2005 Apr 18;(2):CD004069).

(Rumbold et al, 2006) (Vitamins C and E and the Risks of Preeclampsia and Perinatal Complications Alice R. Rumbold, Ph.D., Caroline A. Crowther, Ross R. Haslam, Gustaaf A. Dekker, and Jeffrey S. Robinson, for the ACTS Study Group. N Engl J Med. 2006 Apr 27;354(17):1796-806)

(Sakhi et al, 2009) (Sakhi AK et al, Pre-radiotherapy plasma carotenoids and markers of oxidative stress are associated with survival in head and neck squamous cell carcinoma patients: a prospective study. BMC Cancer. 2009 Dec 21;9:458).

(Satia et al, 2009) (Satia JA, Littman A, Slatore CG, Galanko JA, White E. Long-term use of beta-carotene, retinol, lycopene, and lutein supplements and lung cancer risk: results from the VITamins And Lifestyle (VITAL) study. Am J Epidemiol. 2009 Apr 1;169(7):815-28)

(Sato et al, 2002) (Sato R, et al, Prospective study of carotenoids, tocopherols, and retinoid concentrations and the risk of breast cancer. Cancer Epidemiol Biomarkers Prev. 2002 May;11(5):451-7).

(Schurks et al. 2010) (Effects of vitamin on stroke subtypes: meta-analysis of randomised controlled trials. Markus Schurks, Robert J Glynn, Pamela M. Rist, Christophe Tzourio, Tobias Kurth. BMJ 2010; 341:c5702 (online 11-4-10).

(Seddon, 2007) (Johanna M Seddon. Multivitamin-multimineral supplements and eye disease: age-related macular degeneration and cataract. American Journal of Clinical Nutrition, Vol. 85, No. 1, 304S-307S, January 2007)

(Seifried et al, 2003) (Seifried HE, McDonald SS, Anderson DE, Greenwald P, Milner JA. The antioxidant conundrum in cancer. Cancer Res (2003) 63:4295–8)

(Selman et al. 2006) (Selman C, McLaren JS, Meyer C, Duncan JS, Redman P, Collins AR, Duthie GG, Speakman JR. Life-long vitamin C supplementation in combination with cold exposure does not affect oxidative damage or lifespan in mice, but decreases expression of antioxidant protection genes. Mech Ageing Dev. 2006 Dec;127(12):897-904). ANIMAL STUDY

(Sesso, 2006) (Sesso HD. Carotenoids and cardiovascular disease: what research gaps remain? Curr Opin Lipidol. 2006 Feb;17(1):11-6)

(Sesso et al, 2008) (Vitamins E and C in the prevention of cardiovascular disease in men: the Physicians' Health Study II randomized controlled trial. Sesso, HD. et al. JAMA. 2008 Nov 12;300(18):2123-33. Epub 2008 Nov 9)

(Shekelle et al, 2004) (Shekelle PG, Morton SC, Jungvig LK, et al. Effect of supplemental vitamin E for the prevention and treatment of cardiovascular disease. J Gen Intern Med 2004;19:380-389)

(Skrha et al. 1997) (J. Skrha, G. Sindelka and J. Hilgertova. The effect of fasting and vitamin E on insulin action in obese type 2 diabetes mellitus. Annals of the New York Academy of Sciences, Vol 827, Issue 1 556-560, 1997).

(Slatore et al, 2008) (Christopher G. Slatore, Alyson J. Littman, David H. Au, Jessie A. Satia, and Emily White Long-Term Use of Supplemental Multivitamins, Vitamin C, Vitamin E, and Folate Does Not Reduce the

Risk of Lung Cancer. Am. J. Respir. Crit. Care Med. 177: 524-530. First published online Nov. 7, 2007 as doi:10.1164/rccm.200709-1398OC. Published in print March 1, 2008)

(Song et al. 2009) (Song Y, Cook NR, Albert CM, Van Denburgh M, Manson JE. Effects of vitamins C and E and beta-carotene on the risk of type 2 diabetes in women at high risk of cardiovascular disease: a randomized controlled trial. Am J Clin Nutr 2009 Aug;90(2):429-37)

(Spinnato et al. 2007) (Spinnato JA et al. Antioxidant therapy to prevent preeclampsia: a randomized controlled trial. Obstet Gynecol. 2007 Dec;110(6):1311-8).

(Stanner et al, 2004) (Stanner SA, Hughes J, Kelly CN, Buttriss J (2004). "A review of the epidemiological evidence for the 'antioxidant hypothesis'". Public Health Nutr 7 (3): 407–22)

(Stephens et al., 1996) (Stephens, NG et al. Randomized controlled trial of vitamin E in patients with coronary artery disease: Cambridge Heart Antioxidant Study (CHAOS)," Lancet, March 23, 1996; 347:781-786.)

(Stevens et al, 2005) (Stevens VL, McCullough MI, Diver WR, Rodriguez C, Jacobs EJ, Thun MJ, Calle EE. Use of multivitamins and prostate cancer mortality in a large cohort of US men. Cancer Causes Control. 2005 Aug; 16(6):643-50)

(Stocker, 2007) (Stocker R. Vitamin E. Novartis Found Symp. 2007;282:77-87; discussion 87-92, 212-8)

(Stone et al, 2005) (P.H. Stone et al. Effect of intensive lipid lowering, with or without antioxidant vitamins, compared with moderate lipid lowering on myocardial ischemia in patients with stable coronary artery disease: the Vascular Basis for the Treatment of Myocardial Ischemia Study. Circulation. 2005 Apr 12;111(14):1747-55)

(Stratton, Godwin, 2011) (Stratton J, Godwin M. The effect of supplemental vitamins and minerals on the development of prostate cancer: a systematic review and meta-analysis. Fam Pract. 2011. Jan 27. [Epub ahead of print]).

(Suhail et al, 2011) (Suhail N et al. Effect of vitamins C and E on antioxidant status of breast-cancer patients undergoing chemotherapy. J Clin Pharm Ther. 2011 Jan 4).

(Tam et al. 2005) (Tam LS, Li EK, Leung VY, Griffith JF, Benzie IF, Lim PL, Whitney B, Lee VW, Lee KK, Thomas GN, Tomlinson B. Effects of vitamins C and E on oxidative stress markers and endothelial function in patients with systemic lupus erythematosus: a double blind, placebo controlled pilot study. J Rheumatol. 2005 Feb;32(2):275-82).

(Tan et al, 2008) (Tan JS, Wang JJ, Flood V, Rochtchina E, Smith W, Mitchell P. Dietary antioxidants and the long-term incidence of age-related macular degeneration: the Blue Mountains Eye Study. Ophthalmology. 2008 Feb;115(2):334-41)

(Tang et al, 2009) (Tang et al, Green tea, black tea consumption and risk of lung cancer: a meta-analysis. Lung Cancer. 2009 Sep;65(3):274-83).

(Tangrea et al, 1992) (Tangrea JA, Edwards BK, Taylor PR, et al.: Long-term therapy with low-dose isotretinoin for prevention of basal cell carcinoma: a multicenter clinical trial. Isotretinoin-Basal Cell Carcinoma Study Group. J Natl Cancer Inst 84 (5): 328-32, 1992)

(Tangrea et al, 1993) (Tangrea JA, Adrianza E, Helsel WE, et al.: Clinical and laboratory adverse effects associated with long-term, low-dose isotretinoin: incidence and risk factors. The Isotretinoin-Basal Cell Carcinomas Study Group. Cancer Epidemiol Biomarkers Prev 2 (4): 375-80, 1993 Jul-Aug)

(Taylor et al, 2002) (Taylor HR, Tikellis G, Robman LD, McCarty CA, McNeil JJ. Vitamin E supplementation and macular degeneration: randomised controlled trial. BMJ. 2002 Jul 6;325(7354):11)

(Tardif et al, 1997) (Tardif, J.C., Cote, G and Lesperance, J., et al. Probucol and multivitamins in the prevention of restenosis after coronary angioplasty. Multivitamins and Probucol Study Group. N Engl J Med 1997; 337(6): 365-372)

(Teikari et al, 1998) (Teikari JM, Rautalahti M, Haukka J, et al. Incidence of cataract operations in Finnish male smokers unaffected by alpha tocopherol or beta carotene supplements. J Epidemiol Community Health. 1998;52(7):468-472)

(Thiele H, et al. 2010) (Thiele H, et al. Impact of high-dose N-acetylcysteine versus placebo on contrast-induced nephropathy and myocardial reperfusion injury in unselected patients with ST-segment elevation myocardial infarction undergoing primary percutaneous coronary intervention. The LIPSIA-N-ACC (Prospective, Single-Blind, Placebo-Controlled, Randomized Leipzig Immediate PercutaneouS Coronary Intervention Acute Myocardial Infarction N-ACC) Trial. J Am Coll Cardiol. 2010 May 18;55(20):2201-9).

(Thomson et al. 2007) (Thomson MJ, Puntmann V, Kaski JC. Atherosclerosis and oxidant stress: the end of the road for antioxidant vitamin treatment? Cardiovasc Drugs Ther. 2007 Jun;21(3):195-210).

(Thörnwall et al., 2004) (Effect of α-tocopherol and ß-carotene supplementation on coronary heart disease during the 6-year post-trial follow-up in the ATBC study. Markareetta E. Törnwall et al. European Heart Journal 2004 25(13):1171-1178)

(van Stijn et al, 2008) (van Stijn MF, Ligthart-Melis GC, Boelens PG, Scheffer PG, Teerlink T, Twisk JW, Houdijk AP, van Leeuwen PA. Antioxidant enriched enteral nutrition and oxidative stress after major gastrointestinal tract surgery. World J Gastroenterol. 2008 Dec 7;14(45):6960-9)

(Vasquez et al., 1998) (The SUVIMAX (France) study: the role of antioxidants in the prevention of cancer and cardiovascular disease. Vasquez, Martínez C, Galán P, Preziosi P, Ribas L, Serra LL, Hercberg S. Rev Esp Salud Publica. 1998 May-Jun;72(3):173-83.)

(Villar et al, 2009) (Villar J, Purwar M, Merialdi M, Zavaleta N, Thi Nhu Ngoc N, Anthony J, De Greeff A, Poston L, Shennan A; WHO Vitamin C and Vitamin E trial group. World Health Organisation multicentre randomised trial of supplementation with vitamins C and E among pregnant women at high risk for pre-eclampsia in populations of low nutritional status from developing countries. BJOG. 2009 May;116(6):780-8).

(Virtamo et al, 1998) (Virtamo J, Rapola JM, Ripatti S, et al. Effect of vitamin E and beta carotene on the incidence of primary nonfatal myocardial infarction and fatal coronary heart disease. Arch Intern Med. 1998;158:668-675)

(Vivekananthan et al., 2003) (Vivekananthan DP, Penn MS, Sapp SK, Hsu A, Topol EJ. Use of antioxidant vitamins for the prevention of cardiovascular disease: meta-analysis of randomised trials 2003 Lancet 2003 June 14; 361: 2017–23)

(Voutilainen et al. 2006) (Sari Voutilainen, Tarja Nurmi, Jaakko Mursu and Tiina H Rissanen. Carotenoids and cardiovascular health American Journal of Clinical Nutrition, Vol. 83, No. 6, 1265-1271, June 2006).

(Wang et al. 2090) (Wang YZ, Ren WH, Liao WQ, Zhang GY. Concentrations of antioxidant vitamins in maternal and cord serum and their effect on birth outcomes. J Nutr Sci Vitaminol (Tokyo). 2009 Feb;55(1):1-8).

(Ward et al, 2007) (Ward NC, Wu JH, Clarke MW, Puddey IB, Burke V, Croft KD, Hodgson JM. The effect of vitamin E on blood pressure in individuals with type 2 diabetes: a randomized, double-blind, placebo-controlled trial. J Hypertens. 2007 Jan;25(1):227-34).

(Waters et al, 2002) (Waters DD, Alderman EL, Hsia J, Howard BV, Cobb FR, Rogers WJ, Ouyang P, Thompson P, Tardif JC, Higginson L, Bittner V, Steffes M, Gordon DJ, Proschan M, Younes N, Verter JI. Effects of hormone replacement therapy and antioxidant vitamin supplements on coronary atherosclerosis in postmenopausal women: a randomized controlled trial. JAMA. 2002; 288: 2432–40)

(Watters et al. 2009) (Watters JL, Gail MH, Weinstein SJ, Virtamo J, Albanes D. Associations between alpha-tocopherol, beta-carotene, and retinol and prostate cancer survival. Cancer Res. 2009 May 1;69(9):3833-41).

(Westhuyzen et al, 1997) (Westhuyzen J, Cochrane AD, Tesar PJ, Mau T, Cross DB, Frenneaux MP, Khafagi FA, Fleming SJ. Effect of preoperative supplementation with alpha-tocopherol and ascorbic acid on myocardial injury in patients undergoing cardiac operations. J Thorac Cardiovasc Surg 1997 May;113(5):942-8)

(Wiysonge et al, 2005) (Wiysonge CS, Shey MS, Sterne JA, Brocklehurst P. Vitamin A supplementation for reducing the risk of mother-to-child transmission of HIV infection. Cochrane Database Syst Rev. 2005;(4):CD003648)

(Woodside, et al. 1998) (Woodside JV, Yarnell JW, McMaster D, Young IS, Harmon DL, McCrum EE, Patterson CC, Gey KF, Whitehead AS, Evans A. Effect of B-group vitamins and antioxidant vitamins on hyperhomocysteinemia: a double-blind, randomized, factorial-design, controlled trial. Am J Clin Nutr. 1998 May;67(5):858-66).

(Wu et al, 2002) (Wu K, Willett WC, Chan JM, Fuchs CS, Colditz GA, Rimm EB, Giovannucci EL. A prospective study on supplemental vitamin E intake and risk of colon cancer in women and men. Cancer Epidemiol Biomarkers Prev 2002;11:1298-304)

(Yaffe et al, 2004) (Yaffe K, Clemons TE, McBee WL, Lindblad AS; Age-Related Eye Disease Study Research Group. Impact of antioxidants, zinc, and copper on cognition in the elderly: a randomized, controlled trial. Neurology. 2004 Nov 9;63(9):1705-7)

(Yuen et al. 2005) (Yuen B, Furrer L, Ballmer PE. Antioxidant vitamin supplementation in the prevention of cardiovascular disease. Ther Umsch. 2005 Sep;62(9):615-8).

(Yusoff, 2002) (Yusoff K. Vitamin E in cardiovascular disease: has the die been cast? Asia Pac J Clin Nutr. 2002;11 Suppl 7:S443-7).

(Yusuf et al. 2000) (Yusuf, S., Dagenais, G., Progue, J. et al. Vitamin E supplementation and cardiovascular evens, in high-risk patients the Heart Outcomes Prevention Evaluation Study Investigators. N Engl J Med. 2000; 342; 154-160)

(Zaharris et al, 2007) (A Randomized Factorial Trial of Vitamins C and E and Beta Carotene in the Secondary Prevention of Cardiovascular Events in Women: Results From the Women's Antioxidant Cardiovascular Study. Zaharris, J. MacFadyen, E. Danielson, J. E. Buring, and J. E. Manson. Arch Intern Med, August 13, 2007; 167(15): 1610 – 1618)

(Zhang et al. 2009) (Zhang Y, Coogan P, Palmer JR, Strom BL, Rosenberg L. Vitamin and mineral use and risk of prostate cancer: the case-control surveillance study. Cancer Causes Control. 2009 Jul;20(5):691-8).

(Zhou et al, 2008) (Zhou C, Huang Y, Przedborski S. Oxidative stress in Parkinson's disease: a mechanism of pathogenic and therapeutic significance. Ann N Y Acad Sci. 2008 Dec;1147:93-104)

CHAPTER ELEVEN

Antioxidant Vitamin Studies (Expanded Data): Negligible or failed studies and analysis reports
SUMMARY
R.M. Howes MD, PhD

RCTs (randomized, controlled trials) and cohort studies have complementary strengths and weaknesses and share problems and challenges. Considerable care is appropriate in observational studies to minimize potential biases such as residual confounding bias and measurement error bias. RCTs are costly and logistically require careful preliminary development that may consider multiple sources of information, including observational studies, trials with intermediate outcomes and basic science research. An RCT may be justified when the preliminary research is strong and the public health and safety implications are sufficiently great.

Arguably, RCTs are considered the "gold standard" for evaluation of these types of studies.

The human diet is a very complex chemical mixture of foods and nutrients, with nearly countless interactive elements and studies attributing results to singular components of foods (such as fresh fruits and vegetables) must be viewed with caution. The dose of nutrients from diets and supplements can be highly influential and affect overall results of supplement studies. Extrapolation of results from one dose to another, and presumably from one agent to a mixture of agents, should be taken into account, along with all of the caveats I have already pointed out.

Studies with multivitamins, which contain the antioxidant vitamins, were also included.

This is a selective review of studies primarily showing marginal effectiveness, negligible effects, total ineffectiveness or the harmful potential and consequences of multivitamins, antioxidants and the antioxidant vitamins A, C and E.

Harmful effects have been italicized.

Circa 1979

Failure of high-dose vitamin C (ascorbic acid) therapy to benefit patients with advanced cancer. A controlled trial. (Creagan et al , 1979) (#159 patients with advanced cancer) **One hundred and fifty patients with advanced cancer** participated in a controlled double-blind study to evaluate the effects of high-dose vitamin C on symptoms and survival. Patients were divided randomly into a group that **received vitamin C (10 g per day)** and one that received a comparably flavored lactose placebo. Sixty evaluable patients received vitamin C and 63 received a placebo. Both groups were similar in age, sex, site of primary tumor, performance score, tumor grade and previous chemotherapy. The two groups showed no appreciable difference in changes in symptoms, performance status, appetite or weight. The median survival for all patients was about seven weeks, and the survival curves essentially overlapped. In this selected group of patients, **we were unable to show a therapeutic benefit of high-dose vitamin C treatment.**

CIRCA 1982

The Multiple Risk Factor Intervention Trial (MRFIT). (The Multiple Risk Factor Intervention Trial Research Group, 1982) (#360,000 middle aged men). The Multiple Risk Factor Intervention Trial **(MRFIT)** recruited over 360,000 middle aged men to participate in a diet-heart study. After 7 years, there was only a 2% difference in the blood cholesterol between the studied group and the controls and the study showed that **diet is basically a worthless preventative measure for heart attacks**.

The Multiple Risk Factor Intervention Trial was a **randomized primary prevention trial** to test the effect of a multifactor intervention program on mortality from coronary heart disease (CHD) in **12,866 high-risk men aged 35 to 57 years**. Men were randomly assigned either to a special intervention (SI) program consisting of stepped-care treatment for hypertension, counseling for cigarette smoking, and dietary advice for lowering blood cholesterol levels, or to their usual sources of health care in the community (UC). Over an average follow-up period of seven years, risk factor levels declined in both groups, but to a greater degree for the SI men. Mortality from CHD was 17.9 deaths per 1,000 in the SI group and 19.3 per 1,000 in the UC group, **a statistically nonsignificant difference** of 7.1% (90% confidence interval, — 15% to 25%). Total mortality rates were 41.2 per 1,000 (SI) and 40.4 per 1,000 (UC). Three possible explanations for these findings are considered: (1) the overall intervention program, under these circumstances, does not affect CHD mortality; (2) the intervention used does affect CHD mortality, but the benefit was not observed in this trial of seven years' average duration, with lower-than-expected mortality and with considerable risk factor change in the UC group; and (3) measures to reduce cigarette smoking and to lower blood cholesterol levels may have reduced CHD mortality within subgroups of the SI cohort, with a possibly unfavorable response to antihypertensive drug therapy in certain but not all hypertensive subjects.

(The Multiple Risk Factor Intervention Trial Research Group, 1982) 000

Circa 1985

High-dose vitamin C versus placebo in the treatment of patients with advanced cancer who have had no prior chemotherapy. A randomized double-blind comparison. (Moertel et al, 1985) (#100 patients with advanced colorectal cancer)

It has been claimed that high-dose vitamin C is beneficial in the treatment of patients with advanced cancer, especially patients who have had no prior chemotherapy. In a double-blind study **100 patients with advanced colorectal cancer** were randomly assigned to treatment with either **high-dose vitamin C (10 g daily)** or placebo. Overall, these patients were in very good general condition, with minimal symptoms. None had received any previous treatment with cytotoxic drugs. Vitamin C therapy showed no advantage over placebo therapy with regard to either the interval between the beginning of treatment and disease progression or patient survival. Among patients with measurable disease, none had objective improvement. On the basis of this and our previous randomized study, it can be concluded that **high-dose vitamin C therapy is not effective against advanced malignant disease regardless of whether the patient has had any prior chemotherapy.**

Circa 1990

Skin Cancer Prevention Study (Greenberg et al, 1990) (#1,805 men and women with recent nonmelanoma skin cancer); "A clinical trial of **beta carotene to prevent basal-cell and squamous-cell cancers of the skin.** The Skin Cancer Prevention Study Group." a randomized, double-blind, placebo-controlled intervention; men and women with recent nonmelanoma skin cancer; β-carotene; **No effect** on occurrence of new nonmelanoma skin cancers.

Circa 1992

Diet in the Epidemiology of Postmenopausal Breast Cancer in the New York State Cohort (Graham et al, 1992) (#18,586 postmenopausal women); **did not associate a greater vitamin E intake with a reduced risk of developing breast cancer.**

Women's Health Study (WHS) (Buring and Hennekens, 1992) (#39,876 healthy women); randomized, double-blind; placebo-controlled intervention; **600 IU of natural-source vitamin E taken every other day provided no overall benefit for major cardiovascular events or cancer, did not affect total mortality, and decreased cardiovascular mortality in healthy women. These data do not support recommending vitamin E supplementation for cardiovascular disease or cancer prevention among healthy women.** The WHS data suggest that **vitamin E provides**

no protection against cancer in women. In addition, **vitamin E offered no overall protection against CVD**.

Isotretinoin-Basal Cell Carcinoma Study Group (Tangrea et al, 1992, 1993) (#981 patients with two or more previously treated basal cell carcinomas) randomly assigned; low-dose regimen of isotretinoin not only is **ineffective in reducing the occurrence of basal cell carcinoma at new sites** in patients with two or more previously treated basal cell carcinomas but *also is associated with significant adverse systemic effects.* The **toxicity associated with the long-term administration of isotretinoin, even at the low dose** (10 mg/day) **used in this trial, must be weighted in planning future prevention trials.** Isotretinoin is a modified vitamin A molecule used to treat severe acne vulgaris.
(adverse mucocutaneous effects and serum triglyceride elevations)

Circa 1993

Prospective Study of the Intake of Vitamins C, E, and A and the Risk of Breast Cancer (Hunter et al, 1993) (#89,494 women); prospective study; **Large intakes of vitamin C or E did not protect women in our study from breast cancer. A low intake of vitamin A may increase the risk of this disease.**

Serum micronutrients and the subsequent risk of cervical cancer in a population-based nested case-control study. (Batieha et al, 1993) (#15,161 women) A nested case-control study was conducted in Washington County, MD, to determine whether low serum micronutrients are related to the subsequent risk of cervical cancer. Among the **15,161 women** who donated blood for future cancer research during a serum collection campaign in 1974, **18 developed invasive cervical cancer and 32 developed carcinoma in situ** during the period January 1975 through May 1990. For each of these 50 cases, two matched controls were selected from the same cohort. The frozen sera of the cases and their matched controls were analyzed for a number of nutrients. The mean serum levels of total carotenoids, alpha-carotene, beta-carotene, cryptoxanthin, and lycopene were lower among cases than they were among controls. **When examined by tertiles, the risk of cervical**

cancer was significantly higher among women in the lower tertiles of total carotenoids, alpha-carotene, and beta-carotene as compared to women in the upper tertiles and the trends were statistically significant. Cryptoxanthin was significantly associated with a lower risk of cervical cancer when examined as a continuous variable. Retinol, lutein, alpha- and gamma-tocopherol, and selenium were not related to cervical cancer risk. Smoking was also strongly associated with cervical cancer. These findings are suggestive of a protective role for total carotenoids, alpha-carotene and beta-carotene in cervical carcinogenesis and possibly for cryptoxanthin and lycopene as well.

A randomized trial of vitamin A and vitamin E supplementation for retinitis pigmentosa. (Berson et al, 1993) (#601 patients aged 18 through 49 years with retinitis pigmentosa) Objective. To determine whether supplements of vitamin A or vitamin E alone or in combination affect the course of retinitis pigmentosa. Design. Randomized, controlled, double-masked trial with 2x2 factorial design and duration of 4 to 6 years. Electroretinograms, visual field area, and visual acuity were measured annually. Setting. Clinical research facility. Patients. 601 patients aged 18 through 49 years with retinitis pigmentosa meeting preset eligibility criteria. Ninety-five percent of the patients completed the study. There were no adverse reactions. Intervention. Patients were assigned to one of four treatment groups receiving 15 000 IU/d of vitamin A, 15 000 IU/d of vitamin A plus 400 IU/d of vitamin E, trace amounts of both vitamins, or 400 IU/d of vitamin E. Main Outcome Measure. Cone electroretinogram amplitude. Results. The two groups receiving 15 000 IU/d of vitamin A had on average a slower rate of decline of retinal function than the two groups not receiving this dosage ($P=.01$). Among 354 patients with higher initial amplitudes, the two groups receiving 15 000 IU/d of vitamin A were 32% less likely to have a decline in amplitude of 50% or more from baseline in a given year than those not receiving this dosage ($P=.01$), while the two groups receiving 400 IU/d of vitamin E were 42% more likely to have a decline in amplitude of 50% or more from baseline than those not receiving this dosage ($P=.03$). While not statistically significant, similar trends were observed for rates of decline of visual field area. Visual acuity declined about 1 letter per year in all groups. Conclusions. These results support a beneficial effect of 15 000 IU/d of vitamin A and suggest an adverse effect of

400 IU/d of vitamin E on the course of retinitis pigmentosa. Supplementation with 400 IU/day of vitamin E has been found to accelerate the progression of retinitis pigmentosa that is not associated with vitamin E deficiency.

Lack of an association between serum vitamin E and myocardial infarction in a population with high vitamin E levels. (Hense et al. 1993) (#4,002 men and women)

The antioxidant effects of vitamin E may protect low density lipoproteins from peroxidation and thus inhibit the development of arteriosclerosis. Inverse associations between vitamin E levels and coronary heart disease have been reported from cross-sectional and ecologic studies. In the population-based **MONICA Augsburg cohort (2023 men, 1999 women, total 4,022, age 25-64 years at baseline in 1984, 93% of whom were reexamined in 1987/88)** we investigated the relationship between serum vitamin E concentrations and the risk of subsequent myocardial infarction (MI). Between 1984 and 1991, 46 cases of fatal and non-fatal myocardial infarction from this cohort were recruited for a nested case-control study. Four controls were sampled from the cohort for each case of MI with matching for age, sex, and total cholesterol. **There were no marked differences between cases and their matched controls in the means of vitamin E concentrations (33.9 mumol/l vs. 32.8 mumol/l, P = 0.37) or in the mean vitamin E/total cholesterol ratios (4.89 mumol/mmol vs. 4.82 mumol/mmol, P = 0.75).** The covariate adjusted relative risk (RR) for fatal plus non-fatal MI in the lowest tertile of vitamin E relative to the upper two tertiles was 0.72 (90% confidence interval: 0.33-1.57). Likewise, for the lowest tertile of the ratio (vitamin E/total cholesterol) the RR was 0.81 (0.42-1.56). The association was not modified by history of previous coronary heart disease, fatality of MI, temporal distance of MI onset from vitamin E determinations, or season.

Nutrition intervention trials in Linxian, China: supplementation with specific vitamin/mineral combinations, cancer incidence, and disease-specific mortality in the general population. (Blot et al, 1993) (#29,584)

Linxian Cancer Prevention Study; Country: China; Study Type: Primary prevention; randomized, double-blind, placebo-controlled intervention; Study Population: **29,584** poorly nourished men and women, aged 40-69 years; Duration of Treatment Years: 5.25; Daily Dose: 15 mg β-carotene, 30 mg α-tocopherol, 50 mg selenium; Primary Disease Outcome: Cancer; Results: **9% reduction in total morality, 13% decrease in cancer mortality; 21% decrease in stomach cancer deaths;** 10% decrease in cerebrovascular mortality (nonsignificant).

The **Chinese Cancer Prevention Trial** (the only completed RCT of vitamin C) studied **29,584 patients** who were taking more than 250 mg of **vitamin C** per day. **This study demonstrated no effect on prevention of cardiovascular disease mortality** (Blot et al, 1993) (Blot WJ, Li JY, Taylor PR, Guo W, Dawsey S, Wang GQ, et al. Nutrition intervention trials in Linxian, China: supplementation with specific vitamin/mineral combinations, cancer incidence, and disease-specific mortality in the general population. J Natl Cancer Inst 1993;85:1483-92). 000

Circa 1994

α-Tocopherol, β-Carotene Cancer Prevention Study (ATBC study) (Heinonen et al, 1994) (#29,133 men); randomized, double-blind, placebo-controlled intervention; no effect of vitamin E on lung cancer. Men with known coronary artery disease given 50 mg of a synthetic vitamin E **had no reduction in fatal heart attacks. *50% increase in hemorrhagic stroke deaths* among vitamin E group; *11% increase in ischemic heart disease deaths* among β-carotene group; *18% increase in lung cancer among β-carotene group.***

This study was stopped 21 months earlier than planned.

The incidence of lung cancer was 18% higher among men who took the beta-carotene supplement and eight percent more men in this group died, as compared to those receiving other treatments or placebo. (Albanes et al, 1996)

The negative (and harmful) results of the 2 beta carotene intervention trials were completely **unexpected and counterintuitive**, according to predominant thinking of the time, which was based totally on the free radical theory.

However, results from the **Alpha-Tocopherol Beta Carotene Prevention Study** (Heinonen et al, 1994) and the **Carotene and Retinol Efficiency Trial (CARET)** (Omenn et al, 1996) **showed an increase in lung cancer** among smokers or asbestos-exposed workers after beta carotene supplementation. **The Physician's Health Study**, in which only a small percentage of subjects were smokers (11%), showed no significant effect of beta carotene supplementation on lung cancer (Hennekens et al, 1996).

Polyp Prevention Study (Greenberg et al, 1994) (#864); randomized, double-blind, placebo-controlled intervention; **There was no evidence that either beta carotene or vitamins C and E reduced the incidence of adenomas.** CONCLUSION. **The lack of efficacy of these vitamins argues against the use of supplemental beta carotene and vitamins C and E to prevent colorectal cancer.**

Blood antioxidants and indices of lipid peroxidation in subjects with angina pectoris. (Duthie et al. 1994) (#25 subjects with stable angina pectoris with 200 matched controls) They tested the antioxidant hypothesis of coronary heart disease (CHD) by comparing blood antioxidants, indices of lipid peroxidation and classic (CHD) risk factors of 25 subjects with stable angina pectoris with 200 matched controls. Angina subjects had significantly increased plasma concentrations of total cholesterol, low density lipoproteins and triglycerides although body mass index, plasma cotinine concentration and blood pressure were similar to those of the control group. **Plasma concentrations of vitamin A, vitamin C and cholesterol- adjusted vitamin E did not differ between the groups** although **subjects with angina had significantly decreased plasma uric acid concentrations and elevated indices of lipid peroxidation.** Although the results are compatible with the antioxidant hypothesis, **it is unclear whether the increased oxidative stress in angina sufferers is a cause or consequence of the disease.**

Circa 1995

Effect of vitamin C supplementation on lipoprotein choles-terol, apolipoprotein, and triglyceride concentrations (Jaques et al, 1995) (#139); the overall results of this trial were negative.

The effect of high-dose ascorbate supplementation on plasma lipoprotein(a) levels in patients with premature cor-onary heart disease (Bostom et al, 1995) (#44 patients with pre-mature CHD); findings do not support a clinically important lowering effect of high-dose ascorbate on plasma Lp(a) in patients with premature CHD.

Cholesterol Lowering Atherosclerosis Study (CLAS) (1995) (Hodis et al, 1995) (#156 men); a randomized, placebo-controlled, serial angiographic clinical trial; Supplementary and dietary vitamin E and C intake (non-randomized) in association with cholesterol-lowering diet and either colestipol-niacin or placebo (randomized). Result: Overall, subjects with supplementary vitamin E intake of 100 IU per day or greater **demonstrated less coronary artery lesion progression** than did subjects with supplementary vitamin E intake less than 100 IU per day for all lesions and for mild/moderate lesions. Within the drug group, benefit of supplementary vitamin E intake was found for all lesions and mild/moderate lesions. Within the placebo group, benefit of sup-plementary vitamin E intake was not found. **No benefit was found for use of supplementary vitamin C exclusively or in conjunction with supple-mentary vitamin E, use of multivitamins, or increased dietary intake of vitamin E or vitamin C.**

Yet, they concluded that, "These results indicate an association between supple-mentary vitamin E intake and angiographically demonstrated reduction in coro-nary artery lesion progression." But this was only true in the group also receiving colestipol-niacin. **Otherwise, neither vitamins E nor C had a positive effect.** (Hodis et al, 1995).

Effects of Vitamin E on susceptibility of low-density lipo-protein and low-density lipoprotein subfractions to oxida-tion and on protein glycation in NIDDM. (Reaven et al, 1995)

(#21 men with NIDDM) To evaluate the effect of vitamin E supplementation on the susceptibility of low-density lipoprotein (LDL) and LDL subfractions to oxidation and on protein glycation in non-insulin-dependent diabetes mellitus (NIDDM). RESEARCH DESIGN --**Twenty-one men with NIDDM** (HbAlc = 6-10%), ages 50-70, were randomly assigned to either 1,600 IU/day of vitamin E or placebo for 10 weeks after a 4-week placebo period. LDL and LDL subfractions were isolated after 4 weeks of placebo and after 6 and 10 weeks of therapy. Susceptibility of LDL to copper-mediated oxidation was measured by conjugated diene formation (lag time) and formation of thiobarbituric acid-reactive substances (TBARS). Fasting serum glucose, mean weekly blood glucose, HbAlc, and glycated plasma protein concentrations were also determined at these time points. RESULTS--Vitamin E content in plasma and LDL increased 4.0- and 3.7-fold, respectively, in the vitamin E-treated group. Vitamin E decreased the susceptibility of LDL to oxidation in comparison with placebo. Vitamin E content also increased significantly in both buoyant and dense LDL subfractions, and their oxidation was dramatically reduced. The lag time of LDL oxidation correlated well with the content of vitamin E in both LDL and its subfractions. Glycemic indexes did not change significantly in either group during the study. **Protein glycation, including glycated hemoglobin, glycated albumin, glycated total plasma proteins, and glycated LDL were unchanged in the vitamin E group.** CONCLUSIONS--Supplementation of vitamin E in NIDDM leads to enrichment of LDL and LDL subfractions and reduced susceptibility to oxidation. **Despite a greater percentage increase in vitamin E content in small dense LDL, it remained substantially more susceptible to oxidation than was buoyant LDL.** This suggests that dense, LDL may gain less protection against oxidation from antioxidant supplementation than does larger, more buoyant LDL. **In contrast to previous reports, vitamin E supplementation did not reduce glycation of intracellular or plasma proteins.**

Excretion of alpha-tocopherol into human seminal plasma after oral administration (Moilanen and Hovatta, 1995) (#15 unselected male volunteers) In this open controlled study, we investigated the blood and seminal plasma concentrations of alpha-tocopherol in **15 unselected male volunteers**, who received either 600, 800 or 1200 mg d-alpha-tocopherol per day for 3 weeks. During the intervention, both the blood and seminal plasma vitamin E concentrations increased significantly, although the increase did not correlate with the dose administered. The highest median blood and seminal plasma concentrations were achieved with 800 mg vitamin E per day, but the differences between the group medians were not significant, except in the blood plasma concentration after the first week of treatment between men receiving 600 and 800 mg per day (P < 0.05). **No significant improvement was noted in the movement characteristics of spermatozoa, hypo-osmolar swelling of spermatozoa, or the velocity of deterioration in the parameters mentioned above.** The seminal plasma vitamin E concentrations achieved during the treatment remained low (< 1 mumol l-1) compared to the concentrations found effective in protecting spermatozoa from peroxidative damage in vitro.

Dietary factors and risk of prostate cancer: a case-control study in Ontario, Canada. (Rohan et al. 1995) (#414)

The relationship between risk of prostate cancer and dietary intake of energy, fat, vitamin A, and other nutrients was investigated in a case-control study conducted in Ontario, Canada. Cases were men with a recent, histologically confirmed diagnosis of adenocarcinoma of the prostate notified to the Ontario Cancer Registry between April 1990 and April 1992. Controls were selected randomly from assessment lists maintained by the Ontario Ministry of Revenue, and were frequency-matched to the cases on age. The study included **207 cases (51.4 percent of those eligible) and 207 controls** (39.4 percent of those eligible), and information on dietary intake was collected from them by means of a quantitative diet history. There was a positive association between energy intake and risk of prostate cancer, such that men at the uppermost quartile level of energy intake had a 75 percent increase in risk. In contrast, there was no clear association between the non-energy effects of total fat and monounsaturated fat intake and prostate cancer risk. There was some evidence for an inverse association with saturated fat intake, although the dose-response pattern was irregular. There was a weak (statistically nonsignificant) positive association between polyunsaturated fat intake and risk of prostate cancer. Relatively high levels of retinol intake were associated with reduced risk, but **there was essentially no association between dietary beta-carotene intake and risk. There was no alteration in risk in association with dietary fiber, cholesterol, and vitamins C and E.** Although these patterns were evident both overall and within age-strata, and persisted after adjustment for a number of potential confounding factors, **they could reflect (in particular) the effect of nonrespondent bias.**

Circa 1996

The β-Carotene and Retinol Efficacy Trial (CARET) (Omenn et al, 1996) (#14,254 heavy smokers and 4,060 asbestos workers) (total #18,314 men and women); randomized, double-blind, placebo-controlled intervention; Duration of Treatment Years: 4; Daily Dose: 30 mg β-carotene, 25,000 IU retinol (as retinyl palmitate); *28% increase in lung cancer; 26% increase in CVD (nonsignificant); 17% increase in total mortality* among treatment group. This study was stopped 21 months earlier than planned.

RMH Note: A 6-year follow-up of a large, randomized trial in people with a history of smoking has found that the overall harm associated with beta-carotene supplementation on cardiovascular disease mortality disappeared quickly after participants stopped taking the supplements. However, the risk of lung cancer may persist, especially in females and former smokers, according to the study in the December 1, 2004 issue of the Journal of the National Cancer Institute. Gary E. Goodman, M.D., of the Fred Hutchinson Cancer Research Center in Seattle, and colleagues followed the more than 18,000 participants in CARET for 6 years after the trial was stopped, until the end of 2001. **The increased risk of cardiovascular disease mortality quickly disappeared after participants stopped taking the supplements.** However, *women had a higher risk of death from cardiovascular disease or from any cause than men. In addition, **the incidence of lung cancer and deaths from all causes decreased but did not disappear completely after the supplementation ceased.*** The excess risk of lung cancer was restricted primarily to females and former smokers.

"When chemoprevention agents are administered to large, healthy populations, it is necessary to document long-term safety, efficacy and, importantly, the duration of the beneficial (or adverse) effect," the authors write. "**This is especially true when the basic underlying molecular and genetic mechanism of the agent is unclear.** The results of CARET and ATBC emphasize that chemoprevention trials require careful monitoring of all disease endpoints ... even after the study intervention is discontinued." (Goodman et al, 2004).

197

Based on the free radical theory, beta carotene was known to be an effective anti-oxidant and a precursor of vitamin A, and was therefore believe to be a plausible mechanism to block lung cancer. However, a series of beta carotene intervention trials were conducted that **categorically dispelled the notion that supplemental beta carotene could effectively reduce lung cancer risk** (Hennekens et al, 1996) (Heinonen et al, 1994) (Omenn et al, NEJM. 1996).

The lessons from the β-carotene studies in relation to chronic disease include the possibility that antioxidants may have surprising short and long term health consequences, even if generally regarded as safe, which could be important in determining the balance of benefits and risks for the individual.

Cambridge Heart Antioxidant Study (CHAOS) (Stephens et al., 1996) (#2,002 patients with coronary atherosclerosis); randomized, double-blind, placebo-controlled intervention; patients with coronary atherosclerosis; in patients with angiographically proven symptomatic coronary atherosclerosis, alpha-tocopherol treatment substantially reduces the rate of non-fatal MI, with beneficial effects apparent after 1 year of treatment (**77% decrease in risk of subsequent nonfatal MI); no benefit on cardiovascular mortality, *even though paralleled by a not significant 22% increase in all deaths.***

This landmark study was begun in the fall of 1982 to test the benefits and risks of aspirin and beta carotene in the primary prevention of cardiovascular disease and cancer. The original randomized trial, the Physicians' Health Study-I, ended in 1995. Its finding that daily low-dose aspirin decreased the risk of a first myocardial infarction by 44% helped focus on the role of aspirin in primary prevention of coronary heart disease. In 1996, it also **showed no benefit or harm from beta carotene**.

Physicians' Health Study (PHSI) (Hennekens et al, 1996) (#22,071 US Physicians and Malignant Neoplasms or CVD); randomized, double-blind, placebo-controlled intervention; β-carotene in US Physicians and Malignant Neoplasms or CVD; among healthy men, 12 years of supplementation with **beta carotene produced neither benefit nor harm in terms of the incidence of malignant neoplasms, cardiovascular disease, or death from all causes.**

Dietary Antioxidant Vitamins and Death from Coronary Heart Disease in Postmenopausal Women (Kushi et al, 1996) (#34,486 postmenopausal women with no cardiovascular disease); completed a questionnaire that assessed, among other factors, their intake of vitamins A, E, and C from food sources and supplements; vitamin E consumption appeared to be inversely associated with the risk of death from coronary heart disease. This association was particularly striking in the subgroup of 21,809 women who did not consume vitamin supplements.

There was little evidence that the intake of vitamin E from supplements was associated with a decreased risk of death from coronary heart disease, but the effects of high-dose supplementation and the duration of supplement use could not be definitively addressed. **Intake of vitamins A and C did not appear to be associated with the risk of death from coronary heart disease.**

Mortality associated with low plasma concentration of beta carotene and the effect of oral supplementation (Greenberg et al, 1996) (#1,720 men and women) **Cohort study** of plasma concentrations; randomized, controlled clinical trial of supplementation. A **# 1,720 total** of **1188 men and 532 women** with mean age of 63.2 years. Beta carotene, 50 mg per day for a median of 4.3 years. **Patients randomly assigned to beta carotene supplementation showed no reduction in relative mortality rates from all causes or from cardiovascular disease.** CONCLUSION: **These analyses provide no support for a strong effect of supplemental beta carotene in reducing mortality from cardiovascular disease or other causes.**

The following study by Clark et al, 1996 is included because of the studies claiming the antioxidant properties of selenium. No vitamins A, C or E were in this study and it is not part of the numbered studies.

The Nutritional Prevention of Cancer Trial (Clark et al. 1996) (#1,312 men and women with a history of basal or squamous cell carcinoma) A multicenter, double-blind, randomized, placebo-controlled cancer prevention trial. **Selenium treatment did not protect against development of basal or squamous cell carcinomas of the skin.**

The effect of antioxidant treatment on human spermatozoa and fertilization rate in an in vitro fertilization program (Geva et al, 1996) (#Fifteen fertile normospermic male)

Investigators studied the possible influence of antioxidant treatment on human spermatozoa and the fertilization rate in an IVF program. DESIGN: Prospective study. SETTING: In Vitro Fertilization Unit, Serlin Maternity Hospital, and the Laboratory of Male Fertility, Bar-Ilan University, Ramat-Gan, Israel. PATIENTS: **Fifteen fertile normospermic male** volunteers who had low fertilization rates in their previous IVF cycles. INTERVENTIONS: Vitamin E (alpha-tocopherol) 200 mg daily by mouth for 3 months. MAIN OUTCOME MEASURES: Lipid peroxidation potential (amount of malondialdehyde [MDA]), quantitative ultramorphologic analysis of spermatozoa, and fertilization rate per cycle. RESULTS: The high MDA levels significantly decreased from 12.6 +/- 9.4 nmol/10(8) spermatozoa to normal levels of 7.8 +/- 4.2 nmol/10(8) spermatozoa after 1 month of treatment. The fertilization rate per cycle increased significantly from 19.3 +/- 23.3 to 29.1 +/- 22.2 after 1 month of treatment. No additional effects on MDA levels and fertilization rate were observed after completion of treatment. **With regard to the quantitative ultramorphologic analysis, none of the sperm cell subcellular organelles were affected significantly by vitamin E treatment.** CONCLUSION: Vitamin E may improve the fertilization rate of fertile normospermic males with low fertilization rates after 1 month of treatment, possibly by reducing the lipid peroxidation potential, and with no change of the quantitative ultramorphologic analysis of subcellular organelles.

Energy, nutrient intake and prostate cancer risk: a population-based case-control study in Sweden. (Andersson et al. 1996) (#1,062)

The role of diet in the etiology of prostate cancer remains unclear, because results from several case-control and cohort studies on fat intake and risk of prostate cancer have been inconsistent; few of the studies have adjusted the results for caloric intake. To examine the relationship between energy, intake of several nutrients and risk of prostate cancer (all stages combined and advanced stages separately), we conducted a population-based case-control study in Orebro County, Sweden, from 1989 through 1994. A total of **526 patients with newly diagnosed prostate cancer and 536 controls, (total 1,062) randomly** selected from the population register and frequency-matched by age, were included in the analyses. Information about dietary intake was obtained from a self-administered semi-quantitative food frequency questionnaire. Odds ratios with 95% confidence intervals were estimated by unconditional logistic regression. *In age-adjusted analyses, there were positive associations of prostate cancer (all stages combined) risk with total energy intake as well as intake of total fat (saturated and monounsaturated), protein, retinol and zinc.* The **positive association with energy intake was stronger for advanced cancer, with an excess risk of 70% for the highest quartile vs. the lowest.** *After adjustment for energy intake, there was no apparent association of prostate cancers (all stages combined) with any of the investigated nutrients. However, a weak positive association between intake of retinol and advanced cancer was observed.* We conclude that our results provide some evidence that **total energy intake is a risk factor for prostate cancer.**

Effects of selenium supplementation for cancer prevention in patients with carcinoma of the skin. A randomized controlled trial. Nutritional Prevention of Cancer Study Group. (Clark et al. 1996) (#1,312 patients) **OBJECTIVE:** To determine whether a nutritional supplement of selenium will decrease the incidence of cancer. **DESIGN:** A multicenter, double-blind, randomized, placebo-controlled cancer prevention trial. **SETTING:** Seven dermatology clinics in the eastern United States. **PATIENTS:** A total of 1312 patients (mean age, 63 years; range, 18-80 years) with a history of basal cell or squamous cell carcinomas of the skin were randomized from 1983 through 1991. Patients were treated for a mean (SD) of 4.5 (2.8) years and had a total follow-up of 6.4 (2.0) years. **INTERVENTIONS:** Oral administration of 200 microg of selenium per day or placebo.

MAIN OUTCOME MEASURES: The primary end points for the trial were the incidences of basal and squamous cell carcinomas of the skin. The secondary end points, established in 1990, were all-cause mortality and total cancer mortality, total cancer incidence, and the incidences of lung, prostate, and colorectal cancers.

201

RESULTS: After a total follow-up of 8271 person-years, selenium treatment did not significantly affect the incidence of basal cell or squamous cell skin cancer. There were 377 new cases of basal cell skin cancer among patients in the selenium group and 350 cases among the control group, and 218 new squamous cell skin cancers in the selenium group and 190 cases among the controls. Analysis of secondary end points revealed that, compared with controls, patients treated with selenium had a nonsignificant reduction in all-cause mortality (108 deaths in the selenium group and 129 deaths in the control group and significant reductions in total cancer mortality (29 deaths in the selenium treatment group and 57 deaths in controls, total cancer incidence (77 cancers in the selenium group and 119 in controls, and incidences of lung, colorectal, and prostate cancers. Primarily because of the apparent reductions in total cancer mortality and total cancer incidence in the selenium group, the blinded phase of the trial was stopped early. No cases of selenium toxicity occurred.

CONCLUSIONS: Selenium treatment did not protect against development of basal or squamous cell carcinomas of the skin. However, results from secondary end-point analyses support the hypothesis that **supplemental selenium may reduce the incidence of, and mortality from, carcinomas of several sites**. These effects of selenium **require confirmation** in an independent trial of appropriate design before new public health recommendations regarding selenium supplementation can be made. (Clark et al. 1996). 000

Alpha-Tocopherol and beta-carotene supplements and lung cancer incidence in the alpha-tocopherol, beta-carotene cancer prevention study: effects of base-line characteristics and study compliance. (Albanes et al, 1996) (#29,133 men, smokers) === Experimental and epidemiologic investigations suggest that alpha-tocopherol (the most prevalent chemical form of vitamin E found in vegetable oils, seeds, grains, nuts, and other foods) and beta-carotene (a plant pigment and major precursor of vitamin A found in many yellow, orange, and dark-green, leafy vegetables and some fruit) might reduce the risk of cancer, particularly lung cancer. The initial findings of the Alpha-Tocopherol, Beta-Carotene Cancer Prevention Study (ATBC Study) indicated, however, that lung cancer incidence was increased among participants who received beta-carotene as a supplement. Simi-

lar results were recently reported by the Beta-Carotene and Retinol Efficacy Trial (CARET), which tested a combination of beta-carotene and vitamin A.

PURPOSE: We examined the effects of alpha-tocopherol and beta-carotene supplementation on the incidence of lung cancer across subgroups of participants in the ATBC Study defined by base-line characteristics (e.g., age, number of cigarettes smoked, dietary or serum vitamin status, and alcohol consumption), by study compliance, and in relation to clinical factors, such as disease stage and histologic type. Our primary purpose was to determine whether the pattern of intervention effects across subgroups could facilitate further interpretation of the main ATBC Study results and shed light on potential mechanisms of action and relevance to other populations.

METHODS: A total of **29,133 men aged 50-69 years who smoked five or more cigarettes** daily were **randomly assigned** to receive **alpha-tocopherol (50 mg), beta-carotene (20 mg), alpha-tocopherol and beta-carotene, or a placebo daily for 5-8 years** (median, 6.1 years). Data regarding smoking and other risk factors for lung cancer and dietary factors were obtained at study entry, along with measurements of serum levels of alpha-tocopherol and beta-carotene. Incident cases of lung cancer (n = 894) were identified through the Finnish Cancer Registry and death certificates. Each lung cancer diagnosis was independently confirmed, and histology or cytology was available for 94% of the cases. Intervention effects were evaluated by use of survival analysis and proportional hazards models. All P values were derived from two-sided statistical tests.

RESULTS: **No overall effect was observed for lung cancer from alpha-tocopherol supplementation.** *beta-Carotene supplementation was associated with increased lung cancer risk*. The beta-carotene effect appeared stronger, but not substantially different, in participants who smoked at least 20 cigarettes daily compared with those who smoked five to 19 cigarettes daily and in those with a higher alcohol intake compared with those with a lower intake.

CONCLUSIONS: **Supplementation with alpha-tocopherol or beta-carotene does not prevent lung cancer in older men who smoke.** *beta-Carotene supplementation at pharmacologic levels may modestly increase* *lung cancer incidence in cigarette smokers*, and this effect may be associated with heavier smoking and higher alcohol intake.

IMPLICATIONS: While the most direct way to reduce lung cancer risk is not to smoke tobacco, **smokers should avoid high-dose beta-carotene supplementation**. (Albanes et al, 1996).

Circa 1997

Antioxidant Vitamin Effect on Traditional CVD Risk Factors

(Miller et al, 1997) (#297 retired teachers); **the combined antioxidant supplement of vitamin C, E and beta carotene had no significant effect on the systolic and diastolic blood pressures, fasting serum lipids (total cholesterol, high-density lipoprotein cholesterol, and LDL cholesterol) and fasting glucose, with unadjusted and adjusted analyses.**

ATBC Sub-Study Shows Increased CVD Deaths (Rapola et al, 1997) (#1,862 men, with prior myocardial infarction); there **were no significant differences** in major coronary events but *significantly more deaths from fatal coronary heart disease.* There were no significant differences in the number of major coronary events between any supplementation group and the placebo group. There were *significantly more deaths from fatal coronary heart disease in the beta-carotene and combined alpha-tocopherol and beta-carotene groups than in the placebo group. The risk of fatal coronary heart disease increased in the groups that received either beta-carotene or the combination of alpha-tocopherol and beta-carotene.* They do not recommend the use of alpha-tocopherol or beta-carotene supplements in this group of patients.

Effect of preoperative supplementation with alpha-tocopherol and ascorbic acid on myocardial injury in patients undergoing cardiac operations (Westhuyzen et al, 1997) (#77 undergoing elective coronary artery bypass grafting) **There were no significant differences between the groups with respect to release of creatine kinase MB isoenzyme over 72 hours, nor in the reduction of the myocardial perfusion defect determined by thallium 201 uptake.**

Electrocardiography provided no evidence of a benefit from antioxidant supplementation. Thus the supplementation regimen provided no measurable reduction in myocardial injury after the operation.

The influence of antioxidant nutrients on platelet function in healthy volunteers (Calzada et al, 1997) (#40 healthy volunteers) Supplementation of healthy volunteers with *vitamin E decreased platelet function* whereas supplementation with vitamin c or beta-carotene had no significant effects.

The Multivitamins and Probucol Study (Tardif et al, 1997) (#317 participants); (Beta carotene 30 000 IU, vitamin C 500 mg and E 700 IU); the combination of vitamins C and E and beta carotene had no effect in reducing the rate of restenosis in patients after angioplasty.

Probucol reduced restenosis rates from 38.9% to 20.7% and rates of repeated PTCA from 24.4% to 11.2%; multivitamins alone had no effect. *Probucol has been pulled off the market due harmful effects and the likelihood of cardiac arrhythmias.*

Vitamin C intake and cardiovascular disease risk factors in persons with non-insulin-dependent diabetes mellitus (Mayer-Davis et al, 1997) (#Insulin Resistance Atherosclerosis Study (IRAS, n = 520**) and from the San Luis Valley Diabetes Study (SLVDS, n = 422**) (total #942); across a wide range of intake, vitamin C does not appear to be associated with improved CVD risk factor status among community-dwelling persons with diabetes.

Preformed Vitamin A Study Showed No Trend to Reduce Breast Cancer Risk. (Longnecker, 1997) (#3,543 cases and 9,406 controls) Intake of fruits, vegetables, vitamin A, and related compounds are associated with a decreased **risk of breast cancer** in some studies, but additional data are needed. To estimate intake of beta-carotene and vitamin A, the authors included nine questions on food and supplement use in a population-based case-control study of breast cancer risk conducted in Maine, Massachusetts, New Hampshire, and Wisconsin in 1988-1991. Multivariate-adjusted models were fit to data for **3,543 cases and 9,406 controls**. Eating carrots or spinach more than twice weekly, compared with no intake, was associated with an odds ratio of 0.56 (95% confidence interval 0.34-0.91). **Estimated intake of preformed vitamin A from all evaluated foods and supplements showed no trend or monotonic decrease in risk across categories of intake.** These data do not allow us to distinguish among several potential explanations for the protective association observed between intake of carrots and spinach and risk of breast cancer. The findings are, however, consistent with a diet rich in these foods having a modest protective effect.

The majority of epidemiological studies have failed to find significant associations between retinol intake and breast cancer risk in women.

No effect of supplementation with vitamin E, ascorbic acid, or coenzyme Q10 on oxidative DNA damage estimated by 8-oxo-7,8-dihydro-2'-deoxyguanosine excretion in smokers. (Priemé et al. 1997) (#142 smoking men).

The protective effect of fruit and vegetables against cancer has been related to their high antioxidant content. However, **results from intervention trials have not been conclusive on the protective effect of antioxidant supplementation.** In a randomized placebo-controlled trial we investigated the effect of dietary supplementation with antioxidants on a biomarker of oxidative DNA damage with mechanistic relation to carcinogenesis. One hundred forty-two smoking men aged 35-65 y were randomly assigned to one of the following seven treatments for 2 mo: 100 mg D-alpha-tocopheryl acetate plus 250 mg slow-release ascorbic acid twice a day (n = 20), 100 mg D-alpha-tocopheryl acetate twice a day (n = 20), 250 mg ascorbic acid twice a day (n = 21), 250 mg slow-release ascorbic acid twice a day (n = 21), 30 mg coenzyme Q10 in oil three times a day (n = 20), 30 mg coenzyme Q10 as granulate three times a day (n = 20), or placebo twice a day (n = 20). The trial outcome was the urinary excretion rate of 8-oxo-7, 8-dihydro-2'-deoxyguanosine (8-oxodG)-a repair product of oxidative DNA damage. **Two months of supplementation did not result in significant changes in the urinary excretion rate of 8-oxodG in any group. The lack of effect of antioxidant supplementation on the excretion rate of 8-oxodG, despite substantial increases in plasma antioxidant concentrations, agrees with the results from recent large intervention stud-**

ies with cancer as an endpoint. The cancer-protective effect of fruit and vegetables seems to rely not on the effect of single antioxidants but rather on other anticarcinogenic compounds or on a concerted action of several micronutrients present in these foods.

Vitamin E Worsens Metabolic Parameters in Type 2 Diabetics. (Skrha et al. 1997) (#12)

The influence of either short-term fasting or vitamin E administration on insulin action was studied in two groups of obese Type 2 diabetic patients. Twelve patients underwent 7 days of fasting (group A), whereas 600 mg of vitamin E was administered daily during 3 months in 9 diabetic patients (group B). Insulin action was examined by using hyperinsulinemic isoglycemic clamps (insulin infusion rate, 1.0 mU/kg/min) and insulin receptors on erythrocytes before and after respective regimens. An increase of glucose disposal rate and an increase of metabolic clearance rate of glucose were observed in group A after fasting. On the contrary, *decreases of glucose disposal rate, metabolic clearance rate of glucose, and insulin receptor number were found after vitamin E administration as compared with pretreated values. A worsening of diabetes control as observed by an increase of HbAlC was present* in the latter group. In summary, they found an improvement of insulin action after short-term fasting in contrast with the worsening of metabolic parameters after vitamin E administration in obese Type 2 diabetic patients.

Circa 1998

In 1998, The Medical Letter concluded: (Medical letter, 1998)

- The benefits of taking high doses of vitamin E remain to be established.
- There is no convincing evidence that taking supplements of vitamin C prevents any disease.
- No one should take beta carotene supplements

A second randomized trial, the Physician's Health Study-II, was started in 1997 to test the balance of benefits and risks of three other widely used, but as yet unproven, supplements for the primary prevention of cardiovascular disease, cancer, age-related eye disease, and cognitive decline--vitamin E, vitamin C, and a multivitamin. **The vitamin C and vitamin E components, which ended as** planned in 2007, found that these vitamin supplements do not prevent major cardiovascular events or cancer.

Nutritional Prevention of Cancer Study (Clark et al, 1996) (#1,312 men and women with a history of basal or squamous cell carcinoma); Daily Dose: 200 pg **selenium**; Primary Disease Outcome: Skin cancer, prostate cancer; Results: **No effect on incidence of skin cancer**; 63% reduction in prostate cancer incidence; reduction in total cancer mortality and total cancer incidence. *Reference to this particular study was included, even though it did not test vitamins A, C or E, due to the antioxidant claims of selenium.*

RMH Note: **Intakes of dietary or supplemental antioxidants were not associated with a decreased risk of prostate cancer among men in the Prostate (CaP), Lung, Colorectal, and Ovarian (PLCO) Cancer Screening Trial**, according to a study in the February 15, 2006 issue of the *Journal of the National Cancer Institute*. Kirsh and Hayes, at the National Cancer Institute, and colleagues assessed the risk of prostate cancer for **29,361 men ages** 55 to 74 enrolled in the PLCO Cancer Screening Trial, based on their daily intake of beta-carotene, vitamin E, and vitamin C. The researchers looked at intake of antioxidants from both dietary sources and from supplements. The authors found that, **overall, dietary or supplemental intake of vitamin E, vitamin C, or beta-carotene was not associated with prostate cancer incidence** in this group of PLCO trial participants. **In short,** in the 29,361 men in the trial, 1,338 cases of CaP were identified over the 8 years of follow-up. In general, **there was no clear CaP risk reduction resulting from dietary or supplemental intake of vitamins E and C or b-carotene** (Kirsh et al, 2006).

Effect of Vitamin E and Beta Carotene on the Incidence of Primary Nonfatal Myocardial Infarction and Fatal Coronary Heart Disease (Virtamo et al, 1998) (#27,271 Finnish male smokers) effect of vitamin E (alpha tocopherol) and beta carotene supplementation on major coronary events in the Alpha-Tocopherol, Beta-Carotene Cancer Prevention Study. The incidence of primary major coronary events decreased 4% among recipients of vitamin E and increased 1% among recipients of beta carotene compared with the respective nonrecipients. Neither agent affected the incidence of nonfatal myocardial infarction. Supplementation with vitamin E decreased the

incidence of fatal coronary heart disease by 8%, but beta carotene had no effect on this end point. **CONCLUSION: Supplementation with a small dose of vitamin E has only marginal effect on the incidence of fatal coronary heart disease in male smokers with no history of myocardial infarction, but no influence on nonfatal myocardial infarction. Supplementation with beta carotene has no primary preventive effect on major coronary events**.

SUVIMAX (Vasquez et al, 1998) (#13,017 French adults); take either a daily capsule containing 120 milligrams of ascorbic acid, 30 milligrams of vitamin E, six milligrams of beta carotene, 100 micrograms of selenium, and 20 milligrams of zinc; or a placebo capsule; researchers found **no differences between the antioxidant and placebo group in terms of cancer incidence, or in cardiovascular disease incidence, or all-cause death.**

A Sub-Study of SUVIMAX (#1,162 subjects aged older than 50 years); results suggest **no beneficial effects of long-term daily low-dose supplementation of antioxidant vitamins and minerals on carotid atherosclerosis and arterial stiffness.**

The Nurses' Health Study and Folic Acid and Colon Cancer (Giovannucci et al, 1998) (#88,756 women taking vitamin C and B-carotene, for 8 years); **No benefit with respect to colon cancer after 4 years of use and had no significant risk reductions after 5 to 9 or 10 to 14 years of use. Long-term use of over 15 years of multivitamins may substantially reduce risk for colon cancer.** This **effect may be related to the folic acid** contained in multivitamins.

Dr. Andy Ness, of Bristol University, reported in the British Medical Journal in Dec. 2004, that there is the **possibility of increased risk of breast cancer in women taking folic acid supplements throughout pregnancy**. The researchers followed up **2,928 pregnant women** who had taken part in a supplemental trial in the 1960s. **The risk of death from breast cancer was much higher in women who had received high doses of the supplement than in those who had been given a placebo.** However, Godfrey Oakley and Jack Mandel, of Emory University, said other research studies indicate that more folic acid is likely to prevent breast cancer rather than cause it. **This is another glaring example of the contradictory and dangerous nature of the vitamin studies.**

Effect of B-group vitamins and antioxidant vitamins on hyperhomocysteinemia (Woodside, et al, 1998) (#101 men); 8-wk B-group vitamins with antioxidant vitamins, or placebo intervention. Homocysteine concentrations had significant decreases in both groups receiving B-group vitamins either with or without antioxidants. The effect of B-group vitamins alone over 8 wk was a reduction in homocysteine concentrations, whereas antioxidants alone produced a nonsignificant increase.

Relationships of serum carotenoids, retinol, alpha-tocopherol, and selenium with breast cancer risk: results from a prospective study in Columbia, Missouri (United States). (Dorgan et al, 1998) (#105 cases of histologically confirmed breast cancer) To evaluate relationships of serum carotenoids, alpha-tocopherol, selenium, and retinol with breast cancer prospectively, we conducted a case-control study nested in a cohort from the Breast Cancer Serum Bank in Columbia, Missouri (United States). Women free of cancer donated blood to this bank in 1977-87. **During up to 9.5 years of follow-up** (median = 2.7 years), **105 cases of histologically confirmed breast cancer** were diagnosed. For each case, two women alive and free of cancer at the age of the case's diagnosis and matched on age and date of blood collection were selected as controls. A nonsignificant gradient of decreasing risk of breast cancer with increasing serum beta-cryptoxanthin was apparent for all women. Serum lycopene also was associated inversely with risk, and among women who donated blood at least two years before diagnosis, a significant gradient of decreasing breast cancer risk with increasing lycopene concentration was evident. A marginally significant gradient of decreasing risk with increasing serum lutein/zeaxanthin also was apparent among these women. **They did not observe any evidence for protective effects of alpha- and beta-carotene, alpha-tocopherol, retinol, or selenium for breast cancer.** Results of this study suggest that the carotenoids beta-cryptoxanthin, lycopene, and lutein/zeaxanthin may protect against breast cancer.

Two prospective studies did not observe significant associations between blood retinol levels and subsequent risk of developing breast cancer, i.e., Hulten and Dorgan.

Incidence of cataract operations in Finnish male smokers unaffected by alpha tocopherol or beta carotene supplements. (Teikari et al, 1998) (#28,934 male smokers) Invesitgators examined the effect of alpha tocopherol and beta carotene supplementation on the incidence of age related cataract extraction. SETTING: The **Alpha-tocopherol Beta-carotene (ATBC) Study** was a randomised, double blind, placebo controlled, 2 x 2 factorial trial conducted in south western Finland. The cataract surgery study population of **28,934 male smokers 50-69 years** of age at the start. INTERVENTION: Random assignment to one of four regimens: alpha tocopherol 50 mg per day, beta carotene 20 mg per day, both alpha tocopherol and beta carotene, or placebo. Follow up continued for five to eight years (median 5.7 years) with a total of 159,199 person years. OUTCOME MEASURE: Cataract extraction, ascertained from the National Hospital Discharge Registry. RESULTS: 425 men had cataract surgery because of senile or presenile cataract during the follow up. Of these, 112 men were in the alpha tocopherol alone group, 112 men in the beta carotene alone group, 96 men in the alpha tocopherol and beta carotene group, and 105 men in the placebo group. When supplementation with alpha tocopherol and with beta carotene were introduced to a Cox proportional hazards model with baseline characteristics (age, education, history of diabetes, body mass index, alcohol consumption, number of cigarettes smoked daily, smoking duration, visual acuity, and total cholesterol), **neither alpha tocopherol nor beta carotene supplementation affected the incidence of cataract surgery.** CONCLUSION: **Supplementation with alpha tocopherol or beta carotene does not affect the incidence of cataract extractions among male smokers.**

Effects of α tocopherol and β carotene supplements on symptoms, progression, and prognosis of angina pectoris. (Rapola et al, 1998) (#1,795 Male smokers aged 50−69 years who had angina pectoris) They evaluated the effects of α tocopherol and β carotene supplements on recurrence and progression of angina symptoms, and incidence of major coronary events in men with angina pectoris. Design: Placebo controlled

211

clinical trial. The Finnish α tocopherol β carotene cancer prevention study primarily undertaken to examine the effects of α tocopherol and β carotene on cancer. Male smokers aged 50–69 years who had angina pectoris in the Rose chest pain questionnaire at baseline (n = 1795). Interventions: α tocopherol (vitamin E) 50 mg/day, β carotene 20 mg/day or both, or placebo in 2 × 2 factorial design. Main outcome measures: Recurrence of angina pectoris at annual follow up visits when the questionnaire was readministered; progression from mild to severe angina; incidence of major coronary events (non-fatal myocardial infarction and fatal coronary heart disease). **Results:** There were 2513 recurrences of angina pectoris during follow up (median 4 years). Compared to placebo, the odds ratios for recurrence in the active treatment groups were: α tocopherol only 1.06, α tocopherol and β carotene 1.02, β carotene only 1.06. **There were no significant differences in progression to severe angina among the groups given supplements or placebo.** Altogether 314 major coronary events were observed during follow up (median 5.5 years) and the risk for them did not differ significantly among the groups given supplements or placebo. **Conclusions: There was no evidence of beneficial effects for α tocopherol or β carotene supplements in male smokers with angina pectoris**, indicating no basis for therapeutic or preventive use of these agents in such patients.

The effects of antioxidant supplementation during Percoll preparation on human sperm DNA integrity (Hughes et al, 1998) (#150 patients)

The integrity of sperm DNA is crucial for the maintenance of genetic health. A major source of damage is reactive oxygen species (ROS) generation; therefore, antioxidants may afford protection to sperm DNA. The objectives of the study were, first, to measure the effects of antioxidant supplementation in vitro on endogenous DNA damage in spermatozoa using the single cell gel electrophoresis (comet) assay and, second, to assess the effect of antioxidant supplementation given prior to X-ray irradiation on induced DNA damage. Spermatozoa from **150 patients** were prepared by Percoll centrifugation in the presence of **ascorbic acid (300, 600 microM), alpha tocopherol (30, 60 microM), urate (200, 400 microM), or acetyl cysteine (5, 10 microM).** DNA damage was induced by 30 Gy X-irradiation. DNA strand breakage was measured using the comet assay. **Sperm DNA was protected from DNA damage by ascorbic acid (600 microM), alpha tocopherol (30 and 60 microM) and urate (400 microM).** These antioxidants provided protection from subsequent DNA damage by X-ray irradiation. *In contrast, acetyl cysteine or ascorbate and alpha tocopherol together induced further DNA damage to human sperm.* Supplementation in vitro with the antioxidants ascorbate, urate and alpha tocopherol separately has beneficial effects for sperm DNA integrity.

Dietary intake of antioxidant (pro)-vitamins, respiratory symptoms and pulmonary function: the MORGEN study

(Grievink et al. 1998) (#6,555 adults) A study was undertaken to investigate the relationships between the intake of the antioxidant (pro)-vitamins C, E and beta-carotene and the presence of respiratory symptoms and lung function. METHODS: Complete data were collected in a cross sectional study in a random sample of the Dutch population on **6555 adults** during 1994 and 1995. Antioxidant intake was assessed by a semi-quantitative food frequency questionnaire and respiratory symptoms (cough, phlegm, productive cough, wheeze, shortness of breath) were assessed by a self-administered questionnaire. Prevalence odds ratios for symptoms were calculated using logistic regression analysis. Linear regression analysis was used for forced expiratory volume in one second (FEVI) and forced vital capacity (FVC). The results are presented as a comparison between the 90th and 10th percentiles of antioxidant intake.

RESULTS: **Vitamin C intake was not associated with most symptoms but was inversely related with cough.** Subjects with a high intake of vitamin C had a 53 ml higher FEVI and 79 ml higher FVC than those with a low vitamin C intake. *Vitamin E intake showed no association with most symptoms and lung function, but had a positive association with productive cough. The intake of beta-carotene was not associated with most symptoms but had a positive association with wheeze.* However, subjects with a high intake of beta-carotene had a 60 ml higher FEVI and 75 ml higher FVC than those with a low intake of beta-carotene.

CONCLUSIONS: The results of this study suggest that **a high intake of vitamin C or beta-carotene is protective for FEVI and FVC compared with a low intake, but not for respiratory symptoms.**

Antioxidant nutrient intake and diabetic retinopathy: the San Luis Valley Diabetes Study. (Mayer-Davis et al. 1998) (#387 participants with type 2 diabetes) Diabetic retinopathy (DR) is a major cause of visual impairment and blindness in adults. Antioxidant nutrients, such as vitamins C and E and beta-carotene, may be protective of some eye disorders, such as cataract and age-related macular degeneration, but a relationship between these nutrients and DR has yet to be defined. The purpose of this study was to examine the relation between dietary and supplement intakes of vitamins C, E, and beta-carotene and the risk of DR. DESIGN: Both cross-sectional and longitudinal data were collected from participants in the San Luis Valley Diabetes Study, including non-Hispanic white and Hispanic adults in southern Colorado. PARTICIPANTS: A total of **387 participants with type 2 diabetes** completed at least 1 complete retinal examination and 24-hour dietary recall (including vitamin supplement use). MAIN OUTCOME MEASURES: Type 2 diabetes was defined according to World Health Organization criteria. DR was assessed by retinal photographs, using the Airlie House criteria to classify DR as none, background, preproliferative, or proliferative. Data for both eyes, from up to three clinic visits per participant, were used for analysis. Ordinal logistic regression analysis was used, taking advantage of multiple clinic visits by individual participants and observations from both eyes, to assess the risk for increased DR severity over time as a function of changes in intake of vitamin C, vitamin E, and beta-carotene. Six categories of intake for each nutrient (first to fourth quintiles and ninth and tenth deciles) were considered to ascertain any potential threshold effect. Analyses accounted for age, duration of diabetes, insulin use, ethnicity, glycated hemoglobin, hypertension, gender, and caloric intake. RESULTS: *An increase over time in vitamin C intake from the first to ninth deciles was associated with a risk for increased severity of Diabetic retinopathy (DR)*, although excess risk was not observed for the tenth decile or the second through fourth quintiles compared to the first quintile. *Increased intake of vitamin E was associated with increased severity of DR among those not taking insulin. Among those taking insulin, increased intake of beta-carotene was associated with a risk for severity of Diabetic retinopathy (DR).*

CONCLUSIONS: **No protective effect was observed between antioxidant nutrients and Diabetic retinopathy (DR).** *Depending on insulin use, there appeared to be a potential for deleterious effects of nutrient antioxidants.* Further research is needed to confirm associations of nutrient antioxidant intake and DR.

Prospective association between lipid soluble antioxidants and coronary heart disease in men. The Multiple Risk Factor Intervention Trial. (Evans et al. 1998) (#743) A nested case-control study was performed using participants enrolled in the **Multiple Risk Factor Intervention Trial (MRFIT)**. The cases involved nonfatal myocardial infarction or death from coronary heart disease. Serum samples (*n* = **734**) obtained at baseline and frozen for approximately 20 years were analyzed for the antioxidants, carotenoids, retinol, and α-, γ-, and total tocopherol. The concentrations of antioxidants were in the expected range and their association with low density lipoprotein (LDL) cholesterol reflected their absorption and transport mechanisms. Among nonsmokers, the odds ratios for quartile IV versus quartile I were 1.40 (0.40–4.89), for retinol, total carotenoids, and α-, γ-, and total tocopherol, respectively. The equivalent odds ratios (95% CI) for smokers were 0.90, and 0.52, respectively. **This analysis of antioxidant concentrations by quartiles indicated no significant association of antioxidant levels with the risk of coronary disease death or nonfatal myocardial infarction.**

Circa 1999

GISSI-Prevention Trial (GISSI-Prevenzione Investigators;1999) (#11,324 patients with recent MI); **No benefit from vitamin E**; 15% decrease in risk of death, nonfatal MI, and stroke from ω-3 PUFA.

With the release in 1999 of the Gruppo Italiano per lo Studio della Sopravvivenza nell'Infarto miocardico **(GISSI)** and the Heart Outcomes Prevention Evaluation **(HOPE)**, the role of antioxidants for secondary prevention was **again in doubt**. The **HOPE** study reported **no reduction in heart attack, stroke, or death in patients with heart disease or diabetes after use of vitamin E supplements for more than 4 years.** Similarly, the **GISSI trial, which followed**

11,324 patients with recent heart attacks, **showed no benefit from use of vitamin E supplements for up to 2 years.**

The lack of benefit with vitamin E, however, was corroborated by the results of the HOPE trial and therefore the refutation of the 'antioxidant hypothesis' was at hand. (Marchioli and Valagussa, 2000).

Women's Health Study (Lee et al., 1999) (#39,876 healthy women); 50 mg β-carotene (alternate days); **No effect on incidence of cancer, CVD, or total mortality; no benefit or harm from beta-carotene supplementation after a median of 4.1 years on the incidence of cancer and of cardiovascular disease.**

The beta-carotene part of the study was **stopped early,** in 1996, when other studies showed no protection and even a possible risk of cancer.

A study by Lee et al in the Nov. 2004 issue of American Journal of Clinical Nutrition, found *that **such supplements may actually promote the clogging of arteries.*** They evaluated cardiovascular disease in **1,923 post-menopausal women with diabetes,** part of the **Iowa Women's Health Study,** which collected data in 1986 about diets and vitamin C consumption in nearly **35,000 recruits.** The researchers found that **women with diabetes consuming at least 300 milligrams of vitamin C per day faced 2.3** *times the risk of death from stroke and 2 times the risk of dying from coronary artery disease* as did diabetic women who took in less of the vitamin C.

*A recent study showed that **mega-doses of vitamin C might actually speed up hardening of the arteries.*** Researchers from the University of Southern California studied **573** outwardly healthy middle-aged men and women. About 30% of them regularly took various vitamins. The study found no clear-cut sign that getting lots of vitamin C from food or a daily multivitamin does any harm. But *those taking vitamin C pills had accelerated thickening of the walls of the big arteries in their necks. The more they took, the faster the buildup.*

The Health Professionals Follow-Up Study (Ascherio et al. 1999) (#43,738 men); followed for 8 years. Vitamin E and vitamin C supple-

216

ments and specific carotenoids **did not substantially reduce risk for stroke in this cohort** with 8 year followup.

Familial hypercholesterolemia, intima-to-media thickness (FH IMT study) (Raal et al, 1999) (#15 with homozygous familial hypercholesterolemia); *homozygous familial hypercholesterolemia, intima-to-media thickness (FH IMT study) increased with vitamin E supplements (400 mg/day) for 2 years,* but decreased when subjects received statin therapy. **Vitamin E supplementation fails to slow (or inhibit) the progression of intima-to-media thickness in healthy men and women at low risk for cardiovascular disease.**

At about this time in 1999, researcher Rudolph Salganik, PhD, of the University of North Carolina research team stated, "The truth is we don't know how useful vitamins are for us." In the experiments by Salganik they found that mice that were deprived of all but the smallest traces of vitamins A and E had tumors that were 17 percent smaller than those with normal amounts of the vitamins in their diet. **The vitamin-deficient mice also had five times the number of dying cells in their tumors.** Their work suggested that using vitamins can backfire, actually helping the cancer spread more quickly. Salganik said, "If you suppress free radicals (with antioxidants), you suppress programmed cell death." (Salganik et al, 2000) (Albright et al, 2003).

Beta carotene supplementation in prevention of basal-cell and squamous-cell carcinomas of the skin (Green et al, 1999) (#1,383 participants) **There was no beneficial or harmful effect on the rates of either type of skin cancer, as a result of beta carotene supplementation.**

Vitamins A, C and E and the Risk of Breast Cancer: results from a case-control study in Greece. (Bohlke et al, 1999) (#820 patients with breast cancer plus 1,548 controls) Although several dietary compounds are hypothesized to have anticarcinogenic properties, the role of specific micronutrients in the development of breast cancer remains unclear. To address this issue, they assessed intake of retinol, β-carotene, vitamin C and vitamin E in relation to breast cancer risk in a case–control study in Greece. **Eight**

217

hundred and twenty women with histologically confirmed breast cancer were compared with 1548 control women. Dietary data were collected through a 115-item semiquantitative food frequency questionnaire. Data were modelled by logistic regression, with adjustment for total energy intake and established breast cancer risk factors, as well as mutual adjustment among the micronutrients. Among post-menopausal women, there was no association between any of the micronutrients evaluated and risk of breast cancer.

Among premenopausal women, β-carotene, vitamin C and vitamin E were each inversely associated with breast cancer risk, but after mutual adjustment among the three nutrients only β-carotene remained significant; the odds ratio (OR) for a one-quintile increase in β-carotene intake was 0.84. The inverse association observed with β-carotene intake, however, is slightly weaker than the association previously observed with vegetable intake in these data, raising the possibility that the observed β-carotene effect is accounted for by another component of vegetables.

Dietary antioxidants and risk of myocardial infarction in the elderly: the Rotterdam Study. (Klipstein-Grobusch et al, 1999) (#4,802 participants of the Rotterdam Study aged 55–95 y who were free of MI) Epidemiologic studies have shown dietary antioxidants to be inversely correlated with ischemic heart disease. They investigated whether dietary ß-carotene, vitamin C, and vitamin E were related to the risk of myocardial infarction (MI) in an elderly population. **Design:** The study sample consisted of **4,802 participants of the Rotterdam Study** aged 55–95 y who were free of MI at baseline and for whom dietary data assessed by a semiquantitative food frequency questionnaire were available. During a **4-y follow-up** period, 124 subjects had an MI. The association between energy-adjusted ß-carotene, vitamin C, and vitamin E intakes and risk of MI was examined by multivariate logistic regression. **Results:** Risk of MI for the highest compared with the lowest tertile of ß-carotene intake was 0.55, adjusted for age, sex, body mass index, pack-years, income, education, alcohol intake, energy-adjusted intakes of vitamin C and E, and use of antioxidative vitamin supplements. **When ß-carotene intakes from supplements were considered, the inverse relation with risk of MI was slightly more pronounced. Stratification by smoking status indicated that the association was most evident in current and former smokers.**

No association with risk of MI was observed for dietary vitamin C and vitamin E. **Conclusion:** The results of this **observational study** in the elderly population of the Rotterdam Study support the hypothesis that **high dietary ß-carotene intakes may protect against cardiovascular disease. However, they did not observe an association between vitamin C or vitamin E and MI.**

A prospective study of vitamin supplement intake and cataract extraction among US women. (Chasan-Taber et al, 1999) (#47,152 female nurses) We prospectively examined the association between vitamin supplement intake and the incidence of cataract extraction during 12 years of follow-up in a cohort of **47,152 female nurses**. Women were 45 years or older and free of diagnosed cancer in 1980; others were added as they reached 45 years of age, for a total of 73,956 women. During 720,082 years of follow-up, 1,377 senile cataracts were diagnosed and extracted. **Those who used multivitamins or separate supplements of vitamin C, E, or A did not have decreased risks of cataract as compared with nonusers even for use of 10 or more years.** After adjusting for cataract risk factors, including cigarette smoking, body mass index, and diabetes mellitus, users of vitamin C supplements for 10 or more years had a relative risk (RR) of 0.95. Associations were stronger among long-term vitamin C supplement users who were never-smokers and less than 60 years of age. These findings suggest that **there is little overall benefit of long-term use of vitamin supplements for risk of cataracts requiring extraction.**

Antioxidant treatment of patients with asthenozoospermia or moderate oligoasthenozoospermia with high-dose vitamin C and vitamin E: a randomized, placebo-controlled, double-blind study (Rolf et al, 1999) (#31 without genital infection but with asthenozoospermia) In a **randomized, placebo-controlled, double-blind study** we investigated whether high-dose oral treatment with vitamins C and E for 56 days was able to improve semen parameters of infertile men. Ejaculate parameters included semen volume, sperm concentration and motility, and sperm count and viability. **Thirty-one patients** without genital infection but with asthenozoospermia (< 50% motile spermatozoa) and normal or only moderately reduced sperm concentration (> 7 x 10(6) spermatozoa/ml) (according to WHO criteria) were examined. To investigate the influence of the epididymal storage period on semen parameters, the patients were asked to deliver two semen samples with abstinence times of 2 and 7 days both before and at the end of vitamin treatment. After randomization, the **patients received either 1000 mg vitamin C and 800 mg vitamin E (n = 15) or identical placebo capsules** (n = 16). **No changes in semen parameters were observed during treatment**, and no pregnancies were initiated during the treatment period. **Combined high-dose antioxidative treatment with vitamins C and E did not improve conventional semen parameters or the 24-h sperm survival rate.** Prolonged abstinence time increased ejaculate volume, sperm count, sperm concentration and the total number of motile spermatozoa.

Antioxidant supplementation in vitro does not improve human sperm motility (Donnelly et al, Fertil Steril. 1999) (#60 patients) Investigators determined the effects of supplementation of preparation media with ascorbate and alpha-tocopherol on subsequent sperm motility and reactive oxygen species production. DESIGN: **Prospective study** to analyze postpreparation human sperm motility parameters and reactive oxygen species production following antioxidant supplementation. SETTING: Andrology Laboratory, Royal Maternity Hospital, Belfast, Northern Ireland. PATIENT(S): **Sixty patients** attending the Andrology Laboratory for semen analysis. INTERVENTION(S): Normozoospermic and asthenozoospermic semen samples (n = 10 for each control and antioxidant group) were prepared by Percoll density centrifugation in **media supplemented with ascorbate or alpha-tocopherol to different concentrations within physiologic levels**. Controls were included that were not exposed to antioxidant. OUTCOME: Sperm motility parameters were assessed using computer-assisted semen analysis. The generation of reactive oxygen species was determined using luminol-dependent chemiluminescence. RESULT(S): **The production of reactive oxygen species by sperm was reduced by supplementation in vitro with ascorbate and alpha-tocopherol. However, *progressive motility, average path velocity, curvilinear velocity, straight-line velocity, and linearity were decreased significantly, with the greatest inhibition observed with the highest concentrations of antioxidants.*** CONCLUSION(S): **Supplementation of preparation media with ascorbate and alpha-tocopherol, either singly or in combination, is not beneficial to sperm motility. In short, *sperm motility was significantly decreased by antioxidants vitamin C and alpha-tocopherol.***

The effect of ascorbate and alpha-tocopherol supplementation in vitro on DNA integrity and hydrogen peroxide-induced DNA damage in human spermatozoa (Donnelly et al, Mutagenesis. 1999) (#Semen samples with normozoospermic and asthenozoospermic profiles (n = 15 for each control and antioxidant group) The aim of this study was to determine the effects of **supplementation with ascorbate and alpha-tocopherol, both singly and in combination**, during sperm preparation on subsequent sperm DNA integrity, induced DNA damage and reactive oxygen species (ROS) generation. Semen samples with normozoospermic and asthenozoospermic profiles (n = 15 for each control and antioxidant group) were prepared by Percoll density centrifugation where the medium had been supplemented with these antioxidants to a number of different concentrations, all within physiological levels. Controls were included which had no ascorbate or alpha-tocopherol added. DNA damage was induced using hydrogen peroxide (H_2O_2) and DNA integrity was determined using a modified alkaline single cell gel electrophoresis (Comet) assay, while ROS generation was measured using chemiluminescence. Addition of ascorbate to sperm preparation medium did not affect baseline DNA integrity but did provide sperm with complete protection against H_2O_2-induced DNA damage. Generation of H_2O_2-induced ROS was also significantly reduced after treatment with ascorbate, although baseline levels were unaffected by this antioxidant. Supplementation of sperm preparation medium with alpha-tocopherol did not influence baseline DNA integrity but provided sperm with dose-dependent protection against H_2O_2-induced DNA damage. Generation of H_2O_2-induced ROS was significantly reduced after treatment with alpha-tocopherol, although baseline ROS levels were unaffected by this antioxidant. **Addition of both ascorbate and alpha-tocopherol in combination to sperm preparation medium actually induced DNA damage and intensified the damage induced by H_2O_2,** however, **H_2O_2-induced ROS production was significantly reduced in a dose-dependent manner by supplementation with both vitamins.**

Circa 2000

Heart Outcome Prevention Evaluation Study (HOPE) (Yusuf et al, 2000) (#9,541 patients at high risk for cardiovascular events or diabetes) Dose: 400 IU (268 mg) a-tocopherol; patients at high risk for cardiovascular events or diabetes, **treatment with vitamin E for a mean of 4.5 years had no apparent effect on cardiovascular outcomes.** There were no significant differences in the numbers of deaths from cardiovascular causes, myocardial infarction, or stroke. There were no significant differences in the incidence of secondary cardiovascular outcomes or in death from any cause.

Meta-Analysis of Vitamin E in CVD, Ischemic Heart Disease (IHD) and Mortality (Dagenais et al. 2000) (#51,000 participants);

a meta-analysis of the four randomized trials done in Europe and America involving a total of **51,000** participants allocated to vitamin E or placebo for 1.4 to 6 years, **did not demonstrate a reduction in cardiovascular and IHD mortality and nonfatal myocardial infarction.** In 1999-2000, there were no data to support the use of these vitamins to reduce the risk of cardiovascular events.

This same data is discussed at the following reference: Dagenais GR, Marchioli R, Yusuf S, Tognoni G. Beta-carotene, vitamin C, and vitamin E and cardiovascular diseases. Curr Cardiol Rep. 2000 Jul;2(4):293-9.

In contrast, four large randomized trials did not reveal a reduction in cardiovascular events with beta-carotene use, and *may, in fact, increase IHD and total mortality in male smokers.*

Vitamin C and the Risk of Acute Myocardial Infarction. (Riemersma et al, 2000) (#180 males with a first AMI and 177 healthy volunteers)

Low-fat soluble-antioxidant status is associated with an increased risk of heart disease. The aim of this study was to examine whether low plasma concentrations of vitamin C confer an independent risk of acute myocardial infarction (AMI). **Design:** Male patients (n = 180) aged <65 y **with a first AMI and without an existing diagnosis of angina** (>6 mo) who were admitted within 12 h after onset of symptoms were compared with apparently healthy volunteers (n = 177). Plasma concentrations and dietary intakes of vitamin C were determined during hospitalization and 3 mo later. **Results:** Compared with the control subjects, **the patients had higher total cholesterol and lower HDL-cholesterol concentrations and more of them smoked.** The relative risk of AMI for the lowest compared with the highest quintile of plasma vitamin C during hospitalization was 8.37 after adjustment for classic risk factors. At 3 mo, mean (±SEM) plasma vitamin C concentrations in patients had increased significantly, from 19.6 ± 1.2 to 35.1 ± 1.9 µmol/L and no longer conferred a risk of AMI. Habitual dietary vitamin C intake of patients (before AMI) did not differ significantly from that of control subjects. The increase in plasma vitamin C after recovery from the infarction could not be explained by a similarly large increase in dietary vitamin C. **Conclusions: A low plasma concentration of vitamin C was not associated with an increased risk of AMI, irrespective of smoking status**. The apparent risk of AMI due to a low plasma vitamin C concentration was distorted by the acute phase response.

The effects of combined conventional treatment, oral antioxidants and essential fatty acids on sperm biology in subfertile men (Comhaire et al, 2000) (#27 infertile men) They evaluated the effects of combined conventional treatment, oral antioxidants **(N-acetylcysteine or vitamins A plus E)** and essential fatty acids (FA) on sperm biology in an open prospective study including 27 infertile men. The evaluation included sperm characteristics, seminal reactive oxygen species (ROS), FA of sperm membrane phospholipids, sperm oxidized DNA (8-OH-dG), and induced acrosome reaction (AR). **Treatment did not improve sperm motility and morphology, nor decrease the concentration of round cells and white blood cells in semen**. Sperm concentration increased in oligozoospermic men. Treatment significantly reduced ROS and 8-OH-dG. Treatment increased the AR, the proportion of polyunsaturated FA of the phospholipids, and sperm membrane fluidity. The overall pregnancy rate was 4.5% in 134 months. The per month pregnancy rate tended to be higher in partners of (ex)-smokers than in never-smokers.

Oral vitamin C and endothelial function in smokers: short-term improvement, but no sustained beneficial effect (Raitakari et al, 2000) (20 healthy young adult smokers) Investigators tested the hypothesis that antioxidant therapy would improve endothelial function in smokers. BACKGROUND: Several studies have documented a beneficial effect of short-term oral or parenteral vitamin C on endothelial physiology in subjects with early arterial dysfunction. Possible long-term effects of vitamin C on endothelial function, however, are not known. METHODS: they studied the effects of short- and long-term oral vitamin C therapy on endothelial function in **20 healthy young adult smokers** (age 36 +/- 6 years, 8 male subjects, 21 +/- 10 pack-years). Each subject was studied at baseline, 2 h after a single dose of 2 g vitamin C and 8 weeks after taking 1 g vitamin C daily, and after placebo, in a **randomized double-blind crossover study**. Blood samples were analyzed for plasma ascorbate levels and endothelial function was measured as flow-mediated dilation of the brachial artery, using high resolution ultrasound. Nitroglycerin-mediated dilation (endothelium-independent) was also measured at each visit. RESULTS: At baseline, plasma ascorbate level was low in the smokers, increased with vitamin C therapy after 2 h to 120 +/- 54 micromol/liter (p < 0.001) and remained elevated after eight weeks of supplementation at 92 +/- 32 micromol/liter. Flow-mediated dilation, however, increased at 2 h, but **there was no sustained beneficial effect after eight weeks**. Nitroglycerin-mediated dilation was unchanged throughout. CONCLUSION: **Oral vitamin C therapy improves endothelial dysfunction in the short term in healthy young smokers, but it has no beneficial long-term effect, despite sustained elevation of plasma ascorbate levels.**

Effects of vitamin E on chronic and acute endothelial dysfunction in smokers. (Neunteufl et al, 2000) (#22 healthy male smokers).

The aims of this study were to determine whether chronic or acute impairment of flow mediated vasodilation (FMD) in the brachial artery of smokers can be restored or preserved by the antioxidant vitamin E. BACKGROUND: Transient impairment of endothelial function after heavy cigarette smoking and chronic endothelial dysfunction in smokers result at least in part from increased oxidative stress. METHODS: We studied **22 healthy male smokers** (mean +/- SD, 23 +/- 9 cigarettes per day) randomly assigned to receive either 600 IU vitamin E per day (n = 11, age 28 +/- 6 years) or placebo (n = 11, age 27 +/- 6 years) for four weeks and 11 age-matched healthy male nonsmokers. Flow mediated vasodilation and endothelium-independent, nitroglycerin-induced dilation were assessed in the brachial artery using high resolution ultrasound (7.5 MHz) at baseline and after therapy. Subjects stopped smoking 2 h before the ultrasound examinations. At the end of the treatment period, a third scan was obtained 20 min after smoking a cigarette (0.6 mg nicotine, 7 mg tar) to estimate transient impairment of FMD. RESULTS: Flow mediated vasodilation at baseline was abnormal in the vitamin E and in the placebo group compared with nonsmoking controls. Using a two-way repeated measures analysis of variance (ANOVA) to examine the effects of vitamin E on

FMD, we found no effect for the grouping factor (p = 0.5834) in the ANOVA over time but a highly significant difference with respect to time (p = 0.0065). The interaction of the time factor and the grouping factor also proved to be significant (p = 0.0318). Flow mediated vasodilation values remained similar after treatment for four weeks in both groups but declined faster after smoking a cigarette in subjects taking placebo compared with those receiving vitamin E. The transient attenuation of FMD (calculated as the percent change in FMD) was related to the improvement of the antioxidant status, estimated as percent changes in thiobarbituric acid-reactive substances. Nitroglycerin-induced dilation did not differ between study groups at baseline or after therapy. CONCLUSIONS: **These results demonstrate that oral supplementation of vitamin E can attenuate transient impairment of endothelial function after heavy smoking due to an improvement of the oxidative status but cannot restore chronic endothelial dysfunction within four weeks in healthy male smokers.**

Smoking characteristics, antioxidant vitamins, and carotid artery wall thickness among life-long smokers. (de Waart et al. 2000) (#158 male life-long cardiovascular disease (CVD)-free smokers) They studied the associations between the common carotid-intima-media thickness (IMT), as a marker of atherosclerosis, and smoking characteristics and antioxidant vitamins among **158 male life-long cardiovascular disease (CVD)-free smokers.** An "increased" carotid IMT was defined as the upper 25%. The prevalence of increased IMT was 2.5 times higher among smokers inhaling smoke deeply into the lungs than among moderate and non-inhalers. This association decreased when adjusted for other CVD risk factors. **Smokers with an increased carotid IMT did not differ significantly in mean antioxidant vitamin intake and status with the remaining group.** However, classical CVD risk factors contributed importantly to increased carotid IMT. In our study, depth of inhalation was the only smoking characteristic associated with carotid IMT although attenuated after adjustment for traditional risk factors for CVD. Furthermore, **in these life-long smokers not using any vitamin supplements, no associations were found for antioxidant vitamins.**

Controlled trial of alpha-tocopherol and beta-carotene supplements on stroke incidence and mortality in male smokers. (Leppala et al. 2000) (#28,519 male cigarette smokers). Observational data suggest that diets rich in fruits and vegetables and with high serum levels of antioxidants are associated with decreased incidence and mortality of stroke. We studied the effects of alpha-tocopherol and beta-carotene supplementation. The incidence and mortality of stroke were examined in **28,519 male cigarette smokers aged 50 to 69 years without history of stroke who participated in the Alpha-Tocopherol, Beta-Carotene Cancer Prevention Study (ATBC Study).** The daily supplementation was 50 mg alpha-tocopherol, 20 mg beta-carotene, both, or placebo. The median follow-up was 6.0 years. A total of 1,057 men suffered from incident stroke: 85 men had subarachnoid hemorrhage; 112, intracerebral hemorrhage; 807, cerebral infarction; and 53, unspecified stroke. Deaths due to stroke within 3 months numbered 38, 50, 65, and 7, respectively (total 160). *alpha-Tocopherol supplementation increased the risk of subarachnoid hemorrhage 50%* but decreased that of cerebral infarction 14%, whereas *beta-carotene supplementation increased the risk of intracerebral hemorrhage 62%. alpha-Tocopherol supplementation also increased the risk of fatal subarachnoid hemorrhage 181%.* The overall net effects of either supplementation on the incidence and mortality from total stroke were nonsignificant. *alpha-Tocopherol supplementation increases the risk of fatal hemorrhagic strokes* but prevents cerebral infarction. The effects may be due to the antiplatelet actions of alpha-tocopherol. *beta-Carotene supplementation increases the risk of intracerebral hemorrhage*, but no obvious mechanism is available. (Leppala et al. 2000).

A randomized, 12-year primary-prevention trial of beta carotene supplementation for nonmelanoma skin cancer in the physician's health study. (Frieling et al, 2000) (#22,071) Although basic research provides plausible mechanisms for benefits of beta carotene supplementation on nonmelanoma skin cancer (NMSC) primarily consisting of basal cell carcinoma (BCC) and squamous cell carcinoma (SCC), observational studies are inconsistent. Randomized trial data are limited to 1 trial of secondary prevention that showed no effect of beta carotene on the incidence of NMSC after 5 years. OBJECTIVE: To test whether supplementation with beta carotene reduces the risk for development of a first NMSC, including BCC and

SCC. DESIGN: Randomized, double-blind, placebo-controlled trial with 12 years of beta carotene supplementation and follow-up. SETTING: Physicians' Health Study in the United States. PARTICIPANTS: Apparently healthy male physicians aged 40 to 84 years in 1982 (N = 22,071). INTERVENTION: Beta carotene, 50 mg, on alternate days. AIN OUTCOME MEASURE: Relative risk (RR) and 95% confidence interval (CI) for a first NMSC, BCC, and SCC. RESULTS: After adjusting for age and randomized aspirin assignment, there was no effect of beta carotene on the incidence of a first NMSC, BCC, or SCC. There was also no significant evidence of beneficial or harmful effects of beta carotene on NMSC by smoking status (current, past, or never). CONCLUSION: This large-scale, randomized, primary prevention trial among apparently healthy well-nourished men indicates that an average of 12 years of supplementation with beta carotene does not affect the development of a first NMSC, including BCC and SCC. (Frieling et al, 2000).

Serum carotenoids and atherosclerosis. The Rotterdam Study. (Klipstein-Grobusch et al, 2000) (#108 subjects with aortic atherosclerosis) High circulating levels of carotenoids have been thought to exhibit a protective function in the development of atherosclerosis. They investigated whether aortic atherosclerosis was associated with lower levels of the major serum carotenoids in alpha-carotene, beta-carotene, beta-cryptoxanthin, lutein, lycopene, and zeaxanthin-in a subsample of the elderly population of the Rotterdam Study. Aortic atherosclerosis was assessed by presence of calcified plaques of the abdominal aorta. The case-control analysis comprised **108 subjects with aortic atherosclerosis and controls**. In an age- and sex-adjusted logistic regression model, **serum lycopene was inversely associated with the risk of atherosclerosis**. The odds ratio for the highest compared to the lowest quartile of serum lycopene was 0.55. Multivariate adjustment did not appreciably alter these results. Stratification by smoking status indicated that the inverse association between lycopene and aortic calcification was most evident in current and former smokers. **No association with atherosclerosis was observed for quartiles of serum concentrations of alpha-carotene, beta-carotene, lutein, and zeaxanthin.** In conclusion, **this study provides evidence for a modest inverse association between levels of serum lycopene and presence of atherosclerosis, the association being most pronounced in current and former smokers**. Findings suggest that lycopene may play a protective role in the development of atherosclerosis.

227

Multivitamin use and mortality in a large prospective study.
(Watkins et al, 2000) (#1,063,023 adults) To determine the relation between multivitamin use and death from heart disease, cerebrovascular disease, and cancer, the authors examined a prospective cohort of 1,063,023 adult Americans in 1982-1989 and compared the mortality of users of multivitamins alone; vitamin A, C, or E alone; and multivitamin and vitamin A, C, or E in combination with that of vitamin nonusers by using multivariate Cox proportional hazard models. **Multivitamin users had heart disease and cerebrovascular disease mortality risks similar to those of nonusers, whereas combination users had mortality risks that were 15% lower than those of nonusers.** Multivitamin and combination use had minimal effect on cancer mortality overall, *although mortality from all cancers combined was increased among male current smokers who used multivitamins alone or in combination with vitamin A, C, or E,* but decreased in male combination users who had never or had formerly smoked. No such associations were seen in women. These observational data provide limited support for the hypothesis that **multivitamin use in combination with vitamin A, C, or E may reduce heart disease and cardiovascular disease mortality, but add to concerns raised by randomized studies that some vitamin supplements may adversely affect male smokers.**

Circa 2001

Dietary antioxidant vitamins, retinol, and breast cancer incidence in a cohort of Swedish women. (Michels et al, 2001) (#59,036 women free of cancer) Dietary antioxidant vitamins and retinol have been proposed to be protective against breast cancer on the basis of their ability to reduce oxidative DNA damage and their role in cell differentiation. Epidemiologic studies have not been convincing in supporting this hypothesis, but women with high exposure to free radicals and oxidative processes have not been specifically considered. They explored these issues in the Swedish Mammography Screening Cohort, a large population-based prospective cohort study in Sweden that comprised **59,036 women**, 40-76 years of age, who were free of cancer at baseline and who had answered a validated 67-item food frequency questionnaire. During 508,267 person-years of follow-up, 1,271 cases of invasive breast cancer were diagnosed. **There was no overall associa-**

tion between intake of ascorbic acid, beta-carotene, retinol or vitamin E and breast cancer incidence. High intake of ascorbic acid was inversely related to breast cancer incidence among overweight women and women with high consumption of linoleic acid (HR=0.72; 95% CI 0.52-1.02, for highest quintile of ascorbic acid intake and average consumption of more than 6 grams of linoleic acid per day). Among women with a body mass index of 25 or below, the hazard ratio for breast cancer incidence was 1.27, comparing the highest to the lowest quintile of ascorbic acid intake. Consumption of foods high in ascorbic acid may convey protection from breast cancer among women who are overweight and/or have a high intake of linoleic acid.

The Perth Carotid Ultrasound Disease Assessment Study (CUDAS) (McQuillan et al 2001) (#1,111 subjects); antioxidant vitamins (vitamins A, C and E, lycopene and alpha- and beta-carotene) were independently associated with common carotid artery intima-media (wall) thickness (IMT) or focal plaque, or both. This study provided limited support for the hypothesis that increased dietary intake of vitamin E and increased plasma lycopene may decrease the risk of atherosclerosis. **No benefit was demonstrated for supplemental antioxidant vitamin use.**

Previous observational reports of subclinical atherosclerosis evaluated by carotid IMT and antioxidant vitamins have also generally yielded quite confusing and often conflicting results. For example, a report from the **Atherosclerosis Risk In Communities (ARIC)** group found individuals with **the highest carotid IMT to have lower levels of plasma carotenoids but higher alpha-tocopherol and retinol levels compared to controls** (Iribarren et al, 1997).

Randomized Trial of Supplemental ß-Carotene to Prevent Second Head and Neck Cancer (Mayne et al, 2001) (#264 patients who had been curatively treated for a recent early-stage squamous cell carcinoma of the oral cavity, pharynx, or larynx.); randomized, placebo-controlled, double-blinded clinical trial; 50 mg of ß-carotene per day; After a median follow-up of 51 months, there **was no difference between the two groups** in the time to failure [second primary tumors plus local recurrences. **Supplemental ß-carotene had no significant effect on second head and neck cancer or lung cancer.** Whereas none of the effects were statistically significant, the *point estimates suggested a possible decrease in second head and neck cancer risk but a possible increase in lung cancer risk.*

Age-Related Eye Disease Study Research Group (AREDS)
(AREDS, 2001) (#4,757 participants); use of a high-dose formulation of vitamin C, vitamin E, and beta carotene in **4,757** participants, relatively well-nourished older adult cohort, had **no apparent effect on the 7-year risk of development or progression of age-related lens opacities or visual acuity loss.**

Meagher et al, **found no effect of vitamin E supplements on biochemical markers of lipid peroxidation in 30** healthy men and women, aged 18-60 years in an 8 wk. study. Garret A. FitzGerald and Emma Meagher, University of Pennsylvania Medical Center, said, "**There are no large-scale controlled trials indicating that healthy people derive any benefit from vitamin E supplements.** It would seem therapeutically judicious and economically prudent for such individuals to abstain from vitamin E consumption and await the evidence." (Meagher et al. 2001).

HDL Atherosclerosis Treatment study (HATS) (Brown et al, 2001) (#160 participants); an antioxidant cocktail (vitamin E, ß-carotene, vitamin C, and selenium) had a 0.7% progression in stenosis after 3 years, compared with 0.4% regression in the group on only simvastatin/niacin. Thus, *antioxidant supplements may have interfered with the efficacy of statin-plus-niacin therapy.* **No clinical or angiographically measurable benefit from anti-oxidants was found.** *When used in combination with simvastatin/niacin, antioxidants negated the benefit of the latter on plasma lipid profile and stenosis progression.*

Brown et al. conclude that antioxidant vitamins E and C and ß-carotene, alone or in combination, do not protect against cardiovascular disease. Their use for this purpose may create a diversion away from proven therapies. Because these vitamins blunt the protective HDL2 cholesterol response to HDL cholesterol–targeted therapy, they are potentially harmful in this setting. We conclude that they should rarely, if ever, be recommended for cardiovascular protection (Brown et al., 2002).

230

Thus, in agreement with many in the field, **it was concluded that the existing scientific database does not justify routine use of antioxidant supplements for the prevention and treatment of CVD** i.e., the Nutrition Committee of the American Heart Association Council on Nutrition, Physical Activity, and Metabolism. Antioxidant Vitamin Supplements and Cardiovascular Disease; the American College of Cardiology/American Heart Association 2002 Guideline Update; American College of Cardiology/American Heart Association Task Force on Practice Guidelines. Committee on the Management of Patients With Chronic Stable Angina. ACC/AHA 2002 guideline update for the management of patients with chronic stable angina.

Additionally, "Evidence-Based Guidelines for Cardiovascular Disease Prevention in Women" concluded that **antioxidant vitamin supplements should not be used to prevent CVD, pending the results of ongoing trials (Class III, Level A Evidence).**

Even in 2001, the thrust of the data was turning against the overly-hyped antioxidant supplements as evidenced in an article entitled, "Antioxidant supplements to prevent heart disease: Real hope or empty hype?" (Tran et al, 2001).

Vitamin C and Vitamin E Supplement Use and Colorectal Cancer Mortality in a large American Cancer Society cohort. (Cancer Prevention Study II cohort - CPS-II) (Jacobs, 2001) (#711,891 men and women in U.S.A.)

Some recent epidemiological studies have suggested that use of vitamin C or vitamin E supplements, both of which are important antioxidants, may substantially reduce the risk of colon or colorectal cancer. They examined the association between colorectal cancer mortality and use of individual vitamin C and E supplements in the American Cancer Society's Cancer Prevention Study II cohort. We used proportional hazards modeling to estimate rate ratios among **711,891 men and women** in the United States who completed a self-administered questionnaire at study enrollment in 1982, had no history of cancer, and were followed for mortality through 1996. During the 14 years of follow-up, 4404 deaths from colorectal cancer occurred. After adjustment for multiple colorectal cancer risk factors, regular use of vitamin C or E supplements, even long-term use, was not associated with colorectal cancer mortality. The combined-sex rate ratios were 0.89 for 10 or more years of vitamin C use and 1.08) for 10 or more years of vitamin E use. In subgroup analyses, use of vitamin C supplements for 10 or more years was associated with decreased risk of colorectal cancer mortality before age 65 years (rate ratio = 0.48; 95% CI, 0.28-0.81) and decreased risk of rectal cancer mortality at any age (rate ratio = 0.40; 95% CI, 0.20-0.80). **Our results do not support a substantial effect of vitamin C or E supplement use on overall colorectal cancer mortality.**

Risk of Ovarian Carcinoma and Consumption of Vitamins A, C and E and Specific Carotenoids: a prospective analysis.
(Fairfield et al, 2001) (#80,326 women)

Antioxidant vitamins may decrease risk of cancer by limiting oxidative DNA damage leading to cancer initiation. Few prospective studies have assessed relations between antioxidant vitamins and ovarian carcinoma. METHODS: The authors prospectively assessed consumption of vitamins A, C, and E and specific carotenoids, as well as fruit and vegetable intake, in relation to ovarian carcinoma risk among Women reported on **known and suspected ovarian carcinoma risk 80,326 participants in the Nurses' Health Study who had no history of cancer other than nonmelanoma skin carcinoma.** factors including reproductive factors, smoking, and use of vitamin supplements on biennial mailed questionnaires from 1976 to 1996. Food frequency questionnaires were included in 1980, 1984, 1986, and 1990. The authors confirmed 301 incident cases of invasive epithelial ovarian carcinoma during 16 years of dietary follow-up (1980-1996). Pooled logistic regression was used to control for age, oral contraceptive use, body mass index, smoking history, parity, and tubal ligation. RESULTS: The authors observed no association between ovarian carcinoma risk and antioxidant vitamin consumption from foods, or foods and supplements together. The multivariate relative risks (95% confidence intervals [CIs]) for ovarian carcinoma among women in the highest versus lowest quintile of intake were 1.04 for vitamin A from foods and supplements; 1.01 for vitamin C; 0.88 for vitamin E; and 1.10 for beta-carotene. **Among users of vitamin supplements, the authors found no evidence of an association between dose or duration of any specific vitamin and ovarian carcinoma risk,** although the authors had limited power to assess these relations. **No specific fruits or vegetables were associated significantly with ovarian carcinoma risk. The authors found** no association between ovarian carcinoma and consumption of total fruits or vegetables, or specific subgroups including cruciferous vegetables, green leafy vegetables, legumes, or citrus fruits. Women who consumed at least 2.5 total servings of fruits and vegetables as adolescents had a 46% reduction in ovarian carcinoma risk (relative risk, 0.54, 95% CI, 0.29-1.03; P value for trend 0.04). CONCLUSIONS: These data do not support an important relation between consumption of antioxidant vitamins from foods or supplements, or intake of fruits and vegetables, and incidence of ovarian carcinoma in this cohort. However, modest associations cannot be excluded, and the authors' finding of an inverse association for total fruit and vegetable intake during adolescence raises the possibility that the pertinent exposure period may be much earlier than formerly anticipated.

As of 2010, new studies may be exposing another "medical myth." Investigators, writing in the Journal of the National Cancer Institute, are now telling us that fruits and vegetables do not dramatically lower the risk of common diseases, including cancer. Since the early 1980s, the dietary guidelines for Americans, published jointly by the USDA and the DHHS, have reflected the accumulated scientific research concerning diet and health. Early 1980 guidelines pertaining to nutrition and cancer were to "eat a variety of foods" and "eat foods with adequate starch and fiber." In 1990, "eat a variety of foods" remained a guideline but "choose a diet with plenty of vegetables, fruits, and grain products" replaced the starch and fiber reference. This mirrored the growing data suggesting a lowered risk of cancer with increased vegetable and fruit consumption. But, in 1995, grain products were placed ahead of vegetables and fruit in the guidelines to better reflect the structure of the USDA food pyramid. Other recommendations included the 1989 National Research Council *Diet and Health* report supporting consumption of 5 fruit and vegetable servings per day and the 1991 National Cancer Institute–DHHS sponsorship of the 5-A-Day Program. Public health guidelines for food oriented toward high vegetable and fruit consumption continued up to the present. This scenario led to the rising popularity of vitamin supplements from the 1980s until today, but there have been huge problems with these trends. First, the vitamin supplements were shown to lack the effect of vitamins acquired through the diet and second, vitamins A (beta carotene) and E (alpha tocopherol) were shown to have particularly harmful potential. Yet, recommendations promoting vegetable and fruit consumption remained a center piece, until now. A

233

study of 500,000 Europeans joins a growing body of evidence undermining the high hopes that pushing "five-a-day" might slash Western cancer rates and it estimated that only around 2.5% of cancers could be averted by increasing fruit and vegetable intake. In short, **research has failed to substantiate the suggestion that as many as 50% of cancers could be prevented by boosting the public's consumption of fruit and vegetables.**

This latest study, which analyzed recruits from 10 countries to the highly-regarded **European Prospective Investigation into Cancer and Nutrition, confirms that the association between fruit and vegetable intake and reduced cancer risk is indeed weak**. The team, led by researchers from the Mount Sinai School of Medicine, in New York, calculated for lifestyle factors such as smoking and exercise before drawing their conclusions.

Carotenoids, Alpha-tocopherols, and Retinol in Plasma and Breast Cancer Risk in Northern Sweden. (Hulten et al, 2001)

(#201 cases and 290 referents) Using a nested case-referent design we evaluated the relationship between plasma levels of six carotenoids, alpha-tocopherol, and retinol, sampled before diagnosis, and later breast cancer risk. **Methods:** In total, 201 cases and 290 referents were selected from three population-based cohorts in northern Sweden, where all subjects donated blood samples at enrolment. All blood samples were stored at -80 degreesC. Cases and referents were matched for age, age of blood sample, and sampling centre. Breast cancer cases were identified through the regional and national cancer registries. **Results:** Plasma concentrations of carotenoids were positively intercorrelated. In analysis of three cohorts as a group **none of the carotenoids was found to be significantly related to the risk of developing breast cancer. Similarly, no significant associations between breast cancer risk and plasma levels of alpha -tocopherol or retinol were found.** However, **in postmenopausal women from a mammography cohort with a high number of prevalent cases, lycopene was significantly associated with a decreased risk of breast cancer.** A significant trend of an inverse association between lutein and breast cancer risk was seen in premenopausal women from two combined population-based cohorts with only incident cases. A non-significant reduced risk with higher plasma alpha -carotene was apparent throughout all the sub-analyses. **Conclusion: No significant associations were found between plasma** levels of carotenoids, alpha -tocopherol or retinol and breast cancer risk in analysis of three combined cohorts. However, results from stratified analysis by cohort membership and menopausal status suggest that lycopene and other plasma-carotenoids may reduce the risk of developing breast cancer and that menopausal status has an impact on the mechanisms involved.

Antioxidant vitamins and prevention of cardiovascular disease: epidemiological and clinical trial data. (Marchioli et al. 2001) (#not available) Naturally occurring antioxidants such as vitamin E, beta-carotene, and vitamin C can inhibit the oxidative modification of low density lipoproteins. This action could positively influence the atherosclerotic process and, as a consequence, the progression of coronary heart disease. A wealth of experimental studies provide a sound biological rationale for the mechanisms of action of antioxidants, whereas epidemiologic studies strongly sustain the "antioxidant hypothesis." **To date, however, clinical trials with beta-carotene supplements have been disappointing, and their use as a preventive intervention for cancer and coronary heart disease should be discouraged.** Only scanty data from clinical trials are available for vitamin C. As to vitamin E, discrepant results have been obtained by the Alpha-Tocopherol, Beta Carotene Cancer Prevention Study with a low-dose vitamin E supplementation (50 mg/d) and the Cambridge Heart Antioxidant Study (400-800 mg/d). The results of the GISSI-Prevenzione (300 mg/d) and HOPE (400 mg/d) trials suggest the absence of relevant clinical effects of vitamin E on the risk of cardiovascular events. Currently ongoing are several large-scale clinical trials that will help in clarifying the role of vitamin E in association with other antioxidants in the prevention of atherosclerotic coronary disease.

A controlled clinical trial of vitamin E supplementation in patients with congestive heart failure (Keith et al, 2001) (#56 with advanced heart failure). Oxidative stress is increased in patients with congestive heart failure and can contribute to the progressive deterioration observed in these patients. Increased oxidative stress is the result of either an increased production of free radicals or a depletion of endogenous antioxidants, such as vitamin E. OBJECTIVE: They aimed to determine whether vitamin E supplementation of patients with advanced heart failure would modify levels of oxidative stress, thereby preventing or delaying the deterioration associated

with free radical injury. DESIGN: **Fifty-six outpatients with advanced heart failure** (New York Heart Association functional class III or IV) were enrolled in **a double-blind randomized controlled trial for 12 wk.** At a baseline visit and at 2 follow-up visits, blood and breath samples were collected for the measurement of indexes of heart function and disease state, including malondialdehyde, isoprostanes, and breath pentane and ethane. Quality of life was also assessed at baseline and after 12 wk of treatment. RESULTS: Vitamin E treatment significantly increased plasma concentrations of alpha-tocopherol in the treatment group but **failed to significantly affect any other marker of oxidative stress or quality of life.** In addition, concentrations of atrial natriuretic peptide (a humoral marker of ventricular dysfunction), neurohormonal-cytokine markers of prognosis, tumor necrosis factor, epinephrine, and norepinephrine were unchanged with treatment and were not significantly different from those in the control group. CONCLUSION: **Supplementation with vitamin E did not result in any significant improvements in prognostic or functional indexes of heart failure or in the quality of life of patients with advanced heart failure.**

Circa 2002

The Vitamin E Atherosclerosis Prevention Study (VEAPS)

(Hodis et al, 2002) (#353 were randomized (176 placebo, 177 vitamin E) Epidemiological studies have demonstrated an inverse relationship between vitamin E intake and cardiovascular disease (CVD) risk. In contrast, randomized controlled trials have reported conflicting results as to whether vitamin E supplementation reduces atherosclerosis progression and CVD events. METHODS AND RESULTS: The study population consisted of men and women > or =40 years old with an LDL cholesterol level > or =3.37 mmol/L (130 mg/dL) and no clinical signs or symptoms of CVD. Eligible participants were randomized to DL-alpha-tocopherol 400 IU per day or placebo and followed every 3 months for an average of 3 years. A mixed effects model using all determinations of IMT was used to test the hypothesis of treatment differences in IMT change rates. Compared with placebo, alpha-tocopherol supplementation significantly raised plasma vitamin E levels ($P<0.0001$), reduced circulating oxidized LDL ($P=0.03$), and reduced LDL oxidative susceptibility ($P<0.01$). However, **vitamin E supplementation did not reduce the progression of IMT over a 3-year period compared with subjects randomized to placebo.** CONCLUSIONS: The results are consistent with previous randomized controlled trials and **extend the null results of vitamin E supplementation to the progression of IMT in healthy men and women at low risk for CVD.**

MRC/BHF (MRC/BHF, 2002) (#20,536); (600 mg vitamin E, 250 mg vitamin C and 20 mg beta-carotene daily) It has been suggested that increased intake of various antioxidant vitamins reduces the incidence rates of vascular disease, cancer, and other adverse outcomes. METHODS: **20,536 UK adults (aged 40-80) with coronary disease, other occlusive arterial disease, or diabetes** were randomly allocated to receive antioxidant vitamin supplementation (600 mg vitamin E, 250 mg vitamin C, and 20 mg beta-carotene daily) or matching placebo. Intention-to-treat comparisons of outcome were conducted between all vitamin-allocated and all placebo-allocated participants. Allocation to this vitamin regimen approximately doubled the plasma concentration of alpha-tocopherol, increased that of vitamin C by one-third, and quadrupled that of beta-carotene. FINDINGS: **There were no significant differences in all-cause mortality or in deaths due to vascular causes. Nor were there any significant differences in the numbers of participants having non-fatal myocardial infarction or coronary death, non-fatal or fatal stroke, or coronary or non-coronary revascularization.** For the first occurrence of any of these "major vascular events", there were no material differences either or in any of the various subcategories considered. There were no significant effects on cancer incidence or on hospitalization for any other non-vascular cause. INTERPRETATION: Among the high-risk individuals that were studied, these antioxidant vitamins appeared to be safe. But, **although this regimen increased blood vitamin concentrations substantially, it did not produce any significant reductions in the 5-year mortality from, or incidence of, any type of vascular disease, cancer, or other major outcome.**

Antioxidant Vitamins and US Physician CVD Mortality

(Muntwyler et al. 2002) (#83,639 male U.S.A. physicians); use of supplements (**vitamin E, vitamin C, and multivitamin supplements was reported by 29% of the participants (16,727 patients). US male physicians self-selected supplementation with vitamin E, vitamin C or multivitamins was not associated with a significant decrease in total CVD or CHD mortality.**

Women's Angiographic Vitamin and Estrogen (WAVE) Trial (Waters et al, 2002) (#423 postmenopausal women, with at least one 15% to 75% coronary stenosis); **neither HRT nor antioxidant vitamin supplements (vitamins C & E) provided any cardiovascular benefit**. Instead, *a potential for harm was suggested there is some* **evidence of potentially adverse effects of antioxidant supplements on CVD as assessed by angiographic end points**. In the Women's Angiographic Vitamin and Estrogen Study, *postmenopausal women with coronary disease on hormone replacement therapy given vitamin E plus vitamin C had an unexpected significantly higher all-cause mortality rate and a trend for an increased cardiovascular mortality rate* compared with the vitamin placebo women.

Mega-dose vitamins and minerals in the treatment of non-metastatic breast cancer: an historical cohort study (Lesperance et al, 2002) (#90 patients with non-metastatic breast cancer who received conventional treatment) a historical cohort of **90 patients with non-metastatic breast cancer who received conventional treatment (eg, surgery, chemotherapy, radiation therapy, and hormonal therapy) either alone or in combination with high doses of β-carotene, vitamin C, niacin, selenium, coenzyme Q10, and/or zinc. Breast cancer–specific survival (ie, patients censored only at death from breast cancer) and** *disease-free survival were shorter in the nutrient-supplemented group* **than in the non-supplemented group, but the differences were not statistically significant.**

Investigators stated that, "It is troubling that both (Lesperance et al, 2002 and Ferreira et al, 2004) reported results suggesting poorer survival with concurrent administration of antioxidants and cytotoxic therapy."

The Roche European American Cataract Trial (REACT) (Chylack et al. 2002) (#445 patients); followed for 3 years after treatment with oral Vitamins E, C and beta-carotene. After two years of treatment, there was **a small positive treatment effect in U.S.** patients. There was **no statistically significant benefit** of treatment in the U.K. group.

RMH Note: Long-term supplementation with alpha-tocopherol or beta-carotene was studied for an association with cataract prevalence and severity. An
end-of-trial random sample of **1,828** participants from the randomized, double-
blind, placebo-controlled clinical trial the alpha-tocopherol, beta-carotene cancer
prevention study. Supplementation with alpha-tocopherol or beta-carotene for 5
to 8 years does not influence the cataract prevalence among middle-aged, smoking men (Teikari et al, 1997).

Vitamin E supplementation and macular degeneration (Tay-
lor et al, 2002) (#1,193 subjects); Vitamin E 500 IU or placebo daily for
four years; Daily supplement with **vitamin E supplement does not prevent
the development or progression of early or later stages of age related
macular degeneration.**

Vitamin E on Cardiovascular and Microvascular Outcomes
in High-Risk Patients With Diabetes. Results of the HOPE
Study and MICRO-HOPE Substudy (Lonn et al. 2002) (#3,654
with diabetes); people with diabetes in a randomized clinical trial with a 2 x 2
factorial design, which evaluated the effects of vitamin E and of ramipril in patients
at high risk for CV events. The **daily administration of 400 IU vitamin E for
an average of 4.5 years to middle-aged and elderly people with diabetes
and CV disease** and/or additional coronary risk factor(s) has **no effect on CV
outcomes or nephropathy.**

Vitamin C and Vitamin E Supplement Use and Bladder
Cancer Mortality in a Large Cohort of US Men and Women
(Cancer Prevention Study II (CPS-II) (Jacobs et al., 2002) (#991,522
US adults in the Cancer Prevention Study II (CPS-II) cohort.); CPS-II
participants completed a self-administered questionnaire at enrollment in 1982
and were followed regarding mortality through 1998. Regular vitamin C supplement use (≥15 times per month) was not associated with bladder cancer mortality, regardless of duration. Regular vitamin E supplement use for ≥10 years was
associated with a reduced risk of bladder cancer mortality, but regular use of
shorter duration was not. **Subjects who regularly consumed a vitamin E
supplement for longer than 10 years had a reduced risk of death from
bladder cancer. No benefit was seen from vitamin C supplements.**

Supplemental Vitamin C & E and Multivitamin use and Stomach Cancer Mortality in U.S.A. (Cancer Prevention Study II cohort - CPS-II) (Jacobs et al. Jan. 2002) (#1,045,923 in U.S.A.) Supplementation with antioxidant vitamins has been associated with decreased risk of stomach cancer or regression of precancerous lesions in high-risk areas of China and Colombia. We examined the association between stomach cancer mortality and regular use (> or =15 times per month) of individual vitamin C supplements, individual vitamin E supplements, and multivitamins among **1,045,923 United States adults in the Cancer Prevention Study II (CPS-II) cohort.** CPS-II participants completed a questionnaire at enrollment in 1982 and were followed for mortality through 1998. During follow-up, there were 1,725 stomach cancer deaths (1,127 in men and 598 in women). After adjustment for multiple potential stomach cancer risk factors, **vitamin C use at enrollment was associated with reduced risk of stomach cancer mortality.** However, this reduction in risk was observed only among participants with short duration use at enrollment. **There was no association between stomach cancer mortality and regular use of vitamin E or multivitamins, regardless of duration of use.** Our results suggest that **the use of vitamin C, vitamin E, or multivitamin supplements may not substantially reduce risk of stomach cancer mortality in North American** populations in which stomach cancer rates are relatively low. Our results do not rule out effects of vitamin supplementation in areas in which stomach cancer rates are high and stomach cancer etiology may differ.

As in any observational study, the effects of potential confounding factors need to be considered, which is particularly true in analyses of vitamin supplement use because regular vitamin users are generally more likely to practice "health-conscious" behaviors. **At this time researchers cannot confidently recommend vitamin E supplements for the prevention of cancer because the evidence on this issue is inconsistent and limited.**

As of 2002, according to M.A. Moyad, "Some evidence for the use of these supplements exists, but **serious embellishment of study findings may be leading to an inappropriate use of these supplements** [selenium and vitamin E] **in a clinical setting.**" Moyad et al had published an article entitled, "Selenium and vitamin E supplements for prostate cancer: evidence or embellishment?" (Moyad et al, 2002).

Vitamin E and C Supplements and Risk of Dementia (Laurin et al, 2002) (#3,734 Japanese men) Honolulu-Asia Aging Study (HAAS), Intake of supplemental vitamins E and C was assessed from data collected in the 1988 mailed questionnaire and at the 1991 through 1993 asessment. **Men using both vitamin E and vitamin C (either for long or short term) were not at reduced risk for dementia or for the AD, AD with CVD, or VaD sub-types. No association was noted for supplements taken separately.** CONCLUSION: **we did not find a significant association of vitamin E or vitamin C supplement use and incident dementia**. Our results are based on incident cases and a longer period between the 1988 report of supplement use and assessment of dementia. Our data suggest that **supplemental intake of both vitamins E and C does not alter the risk for dementia.**

Retinol intake and bone mineral density in the elderly: the Rancho Bernardo Study. (Promislow et al, 2002) (#570 women and 388 men) **Retinol is involved in bone remodeling, and excessive intake has been linked to bone demineralization**, yet its role in osteoporosis has received little evaluation. We studied the associations of retinol intake with bone mineral density (BMD) and bone maintenance in an ambulatory community-dwelling cohort of 570 women and 388 men, aged 55-92 years at baseline. Regression analyses, adjusted for standard osteoporosis covariates, showed an inverse U-shaped association of retinol, assessed by food-frequency questionnaires in 1988-1992, with baseline BMD, BMD measured 4 years later, and BMD change. Supplemental retinol use, reported by 50% of women and 39% of men, was an effect modifier in women; **the associations of log retinol with BMD and BMD change were negative for supplement users and positive for nonusers at the hip, femoral neck, and spine. At the femoral neck, for every unit increase in log retinol intake, supplement users had 0.02 g/cm^2 ($p = 0.02$) lower BMD and 0.23% ($p = 0.05$) greater annual bone loss, and nonusers had 0.02 g/cm^2 ($p = 0.04$) greater BMD and 0.22% ($p = 0.19$) greater bone retention.** However, among supplement users, retinol from dietary and supplement sources had similar associations with BMD, suggesting total intake is more important than source. In both sexes, *increasing* *retinol became negatively associated with skeletal health at intakes not far beyond the recommended daily allowance (RDA), intakes reached predominately by supplement users.* This study suggests **there is a delicate balance between ensuring that the elderly consume sufficient vitamin A and simultaneously cautioning against excessive retinol supplementation.**

Vitamin A intake and hip fractures among postmenopausal women. (Feskanich et al, 2002) (#72,337 postmenopausal women)

Ingestion of toxic amounts of vitamin A affects bone remodeling and can have adverse skeletal effects in animals. The possibility has been raised that long-term high vitamin A intake could contribute to fracture risk in humans. **Objective** To assess the relationship between high vitamin A intake from foods and supplements and risk of hip fracture among postmenopausal women. **Design** Prospective analysis begun in 1980 with 18 years of follow-up within the Nurses' Health Study. **Setting** General community of registered nurses within 11 US states. **Participants** A total of 72,337 postmenopausal women aged 34 to 77 years. **Main Outcome Measures** Incident hip fractures resulting from low or moderate trauma, analyzed by quintiles of vitamin A intake and by use of multivitamins and vitamin A supplements, assessed at baseline and updated during follow-up. **Results** From 1980 to 1998, 603 incident hip fractures resulting from low or moderate trauma were identified. After controlling for confounding factors, *women in the highest quintile of total vitamin A intake (≥3000 µg/d of retinol equivalents [RE]) had a significantly elevated relative risk (RR) of hip fracture* compared with women in the lowest quintile of intake (<1250 µg/d of RE). *This increased risk was attributable primarily to retinol.* **The association of high retinol intake with hip fracture was attenuated among women using postmenopausal estrogens.** Beta carotene did not contribute significantly to fracture risk. Women currently taking a specific vitamin A supplement had a nonsignificant 40% increased risk of hip fracture compared with those not taking that supplement, and, among women not taking supplemental vitamin A, retinol from food was significantly associated with fracture risk. **Conclusions** *Long-term intake of a diet high in retinol may promote the development of osteoporotic hip fractures in women.* The amounts of retinol in fortified foods and vitamin supplements may need to be reassessed.

A prospective study on supplemental vitamin e intake and risk of colon cancer in women and men. (Wu et al, 2002) (#87,998 females from the Nurses' Health Study and 47, 344 males from the Health Professionals Follow-up Study) (#135,332 total participants) They conducted a prospective study on the association between supplemental vitamin E and colon cancer in **87,998 females from the Nurses' Health Study and 47, 344 males from the Health Professionals Follow-up Study.** There was some suggestion that men with supplemental vitamin E intake of 300 IU/day or more may be at lower risk for colon cancer when compared with never users [multivariate relative risk (RR), 300-500 IU/day versus never users, 0.73; >or=600 IU/day versus never users = 0.70, but CIs included 1. **In women, there was no evidence for an inverse association between vitamin E supplementation and risk of colon cancer. Our findings do not provide consistent support for an inverse association between supplemental vitamin E and colon cancer risk.** Considering the paucity of epidemiological data on this association, further studies of vitamin E and colon cancer are warranted.

The Collaborative Primary Prevention Project (PPP) (Chiabrando et al., 2002) (#144 participants with CHD risk factors); vitamin E (300 mg/day for about three years); reassess critically the role of vitamin E in CVD prevention; **Prolonged vitamin E supplementation did not reduce lipid peroxidation in subjects with major cardiovascular risk factors.**

Antioxidants to slow aging, facts and perspectives. (Bonnefoy et al. 2002) (#not applicable). FREE RADICALS AND THE THEORY OF AGING: Severe oxidative stress progressively leads to cell dysfunction and ultimately cell death. Oxidative stress is defined as an imbalance between pro-oxidants and/or free radicals on the one hand, and anti-oxidizing systems on the other. The oxygen required for living may indirectly be responsible for negative effects; these deleterious effects are due to the production of free radicals, which are toxic for the cells (superoxide anions, hydroxyl radicals, peroxyl radicals, hydrogen peroxide, hydroperoxides and peroxinitrite anions). Free radical attacks are responsible for cell damage and the targeted cells are represented by the cell membranes, which are particularly rich in unsaturated fatty acids, sensitive to oxidation reactions; DNA is also the target of severe

243

attacks by these reactive oxygen species (ROS). THE DEFENSE SYSTEMS: These are represented by the enzymes and free radical captors. The latter are readily oxidizable composites. **The free radical captor or neutralization systems of these ROS use a collection of mechanisms, vitamins (E and C), enzymes [superoxide dismutase (SOD), glutathion peroxidase (GPx) and others], and glutathion reductase (GSH), capable of neutralizing peroxinitrite.** The efficacy of this system is dependent on the genome for the enzymatic defence systems, and on nutrition for the vitamins. Some strategies aimed at reducing oxidative stress-related alterations have been performed in animals. However, **only a few can be used and are efficient in humans,** such as avoidance of unfavourable environmental conditions (radiation, dietary carcinogens, smoking...) and antioxidant dietary supplementation. DIETARY SUPPLEMENTATION: Epidemiological data suggests that antioxidants may have a beneficial effect on many age-related diseases: atherosclerosis, cancer, some neurodegenerative and ocular diseases. However, **the widespread use of supplements is hampered by several factors: the lack of prospective and controlled studies; insufficient knowledge on the pro-oxidant, oxidant and antioxidant properties of the various supplements; growing evidence that free radicals are not only by-products, but also play an important role in cell signal transduction, apoptosis and infection control.** RECOMMENDATIONS: Although current data indicate that antioxidants cannot prolong maximal life span, the beneficial impact of antioxidants on various age-related degenerative diseases may forecast an improvement in life span and enhance quality of life. **The current lack of sufficient data does not permit the systematic recommendation of anti-oxidants.** Nevertheless, antioxidant-rich diets with fruit and vegetables should be recommended.

Selenium and vitamin E supplements for prostate cancer: evidence or embellishment? (Moyad et al. 2002) (# not available)

Selenium and vitamin E are probably 2 of the most popular dietary supplements considered for use in the reduction of prostate cancer risk. This enthusiasm is reflected in the initiation of the **Selenium and Vitamin E Chemoprevention Trial (SELECT).** Is there sufficient evidence to support the use of these supplements in a large-scale prospective trial for patients who want to reduce the risk of prostate cancer? Results from numerous laboratory and observational studies support the use of these supplements, and data from recent prospective trials also add partial support. However, a closer analysis of the data reveals some interesting and unique associations. **Selenium supplements provided a benefit only for those individuals who had lower levels of baseline plasma selenium.** *Other subjects, with normal or higher levels, did not benefit and may have an increased risk for prostate cancer.* The concept that supplements reduce prostate cancer risk only in those at a higher risk and/or those with lower plasma levels of these compounds is supported by trials examining beta-carotene supplements. Smokers may be the only individuals who benefit, as has also been shown with vitamin E supplementation. In 4 recent prospective studies, vitamin E was found to reduce the risk of prostate cancer in past/recent and current smokers and those with low levels of this vitamin. *Vitamin E supplements in higher doses (> or =100 IU) were also associated with a higher risk of aggressive or fatal prostate cancer in nonsmokers from a past prospective study.* The dose of vitamin E in the SELECT trial (400 IU/day) is 8 times higher than what has been suggested to be effective (50 IU/day) by the largest randomized prospective trial in which the incidence rate of prostate cancer was used as an endpoint. Recent research also suggests that dietary vitamin E may be associated with a lower risk of prostate cancer than the vitamin E supplement. Additionally, **recent results from all past cardiovascular prospective, randomized trials suggest that vitamin E shows little benefit for cardiovascular disease risk, especially at the dose being used in the SELECT trial.** Other intriguing positive findings from past prospective studies of supplements suggest that aspirin and other nonsteroidal anti-inflammatory drugs have a role in reducing the risk of prostate cancer or other types of cancer (eg, colon cancer). It may be time to conduct a large costly trial to reconsider the use of selenium and vitamin E supplements for the reduction of prostate cancer risk. Some evidence for the use of these supplements exists, but **serious embellishment of study findings may be leading to an inappropriate use of these supplements in a clinical setting.**

Vitamin E in cardiovascular disease: has the die been cast? (Yusoff. K. 2002) (#not applicable)

Cardiovascular disease, in particular coronary artery disease (CAD), remains the most important cause of morbidity and mortality in developed countries and, in the near future, more so in the developing world. Atherosclerotic plaque formation is the underlying basis for CAD. Growth of the plaque leads to coronary stenosis, causing a progressive decrease in blood flow that results in angina pectoris. Acute myocardial infarction and unstable angina were recently recognised as related to plaque rupture, not progressive coronary stenosis. Acute thrombus formation causes an abrupt coronary occlusion. The characteristics of the fibrin cap, contents of the plaque, rheological factors and active inflammation within the plaque contribute to plaque rupture. Oxidative processes are important in plaque formation. Oxidized low density lipoproteins (LDL) but not unoxidized LDL is engulfed by resident intimal macrophages, transforming them into foam cells which develop into fatty streaks, the precursors of the atherosclerotic plaque. Inflammation is important both in plaque formation and rupture. Animal studies have shown that antioxidants reduce plaque formation and lead to plaque stabilization. In humans, high intakes of antioxidants are associated with lower incidence of CAD, despite high serum cholesterol levels. This observation suggests a role for inflammation in CAD and that reducing inflammation using antioxidants may ameliorate these processes. Men and women with high intakes of vitamin E were found to have less CAD. Vitamin E supplementation was associated with a significant reduction in myocardial infarction and cardiovascular events in the incidence of recurrent myocardial infarction. **In the hierarchy of evidence in evidence-based medicine, data from large placebo-controlled clinical trials is considered necessary. Results from various mega-trials have not shown benefits (nor adverse effects) conferred by vitamin E supplementation, suggesting that vitamin E has no role in the treatment of CAD. These results do not seem to confirm, at the clinical level, the effect of antioxidants against active inflammation during plaque rupture.** However, a closer examination of these studies showed a number of limitations, rendering them inconclusive in addressing the role of vitamin E in CAD prevention and treatment. Further studies that specifically address the issue of vitamin E in the pathogenesis of atherosclerosis and in the treatment of CAD need be performed. These studies should use the more potent antioxidant property of alpha-tocotrienol vitamin E.

Prospective study of carotenoids, tocopherols, and retinoid concentrations and the risk of breast cancer. (Sato et al, 2002)

(#590) Previous prospective studies have raised the possibility that the antioxidant properties of carotenoids and vitamin E (alpha-tocopherol) and the role of vitamin A (retinol) in cellular differentiation may be associated with a reduced risk of subsequent breast cancer. To investigate the association between serum and plasma **concentrations of retinol, retinyl palmitate, alpha-carotene, beta-carotene, beta-cryptoxanthin, lutein, lycopene, total-carotenoids, alpha-tocopherol, and gamma-tocopherol with subsequent development of breast cancer,** a nested case control study was conducted among female residents of Washington County, Maryland, who had donated blood for a serum bank in 1974 or 1989. Cases (n = 295) and controls (n = 295) were matched on age, race, menopausal status, and date of blood donation, and the analyses were stratified by cohort participation. Median concentrations of beta-carotene, lycopene, and total carotene were significantly lower in cases compared with controls in the 1974 cohort (13.1, 12.5, and 7.9% difference; P = 0.01, 0.04, and 0.04, respectively) and for lutein in the 1989 cohort (6.7% difference; P = 0.02). The risk of developing breast cancer in the highest fifth was approximately half of that of women in the lowest fifth for beta-carotene [odds ratio (OR) = 0.41; 95% confidence interval (CI) 0.22-0.79; P trend = 0.007], lycopene (OR = 0.55; 95% CI 0.29-1.06; P trend = 0.04), and total carotene (OR = 0.55; 95% CI 0.29-1.03; P trend = 0.02) in the 1974 cohort. **There was generally a protective association for other micronutrients in both cohorts, although none reached statistical significance.** The results suggest that carotenoids may protect against the development of breast cancer. (Sato et al, 2002). 000

Lack of effect of oral vitamin C on blood pressure, oxidative stress and endothelial function in Type II diabetes (Darko et al, 2002) (#35) Type II diabetes is characterized by increased oxidative stress, endothelial dysfunction and hypertension. We investigated whether short-term treatment with oral vitamin C reduces oxidative stress and improves endothelial function and blood pressure in subjects with Type II diabetes. Subjects (n =35) received vitamin C (1.5 g daily in three doses) or matching placebo for 3 weeks in a randomized, double-blind, parallel-group design. Plasma concentrations of 8-epi-prostaglandin F(2alpha) (8-epi-PGF(2alpha)), a non-enzymically derived oxidation product of arachidonic acid, were used as a marker of oxidative stress. Endothelial function was assessed by measuring forearm blood flow responses to brachial artery infusion of the endothelium-dependent vasodilator acetylcholine (with nitroprusside as an endothelium-independent control) and by the pulse wave responses to systemic albuterol (endothelium-dependent vasodilator) and glyceryl trinitrate (endothelium-independent vasodilator).

Plasma concentrations of vitamin C increased from 58+/-6 to 122+/-10 micromol/l after vitamin C, but 8-epi-PGF(2alpha) levels (baseline, 95+/-4 pg/l; after treatment, 99+/-5 pg/l), blood pressure (baseline, 141+/-5/80+/-2 mmHg; after treatment, 141+/-5/81+/-3 mmHg) and endothelial function, as assessed by the systemic vasodilator response to albuterol and by the **forearm blood flow response to acetylcholine, were not significantly different from baseline or placebo.** Thus **treatment with vitamin C (1.5 g daily) for 3 weeks does not significantly improve oxidative stress, blood pressure or endothelial function in patients with Type II diabetes.** (Darko et al, 2002)

Antioxidant vitamins and risk of cardiovascular disease. Review of large-scale randomised trials (Clarke, Armitage, 2002) **(#70,000 people from 3 large-scale trials in healthy populations and on vitamin E supplementation involving 29,000 patients at high-risk of cardiovascular disease from 5 large-scale trials)** People who consume a diet rich in fruit and vegetables have lower risks of cancer, cardiovascular disease and all-cause mortality. Many prospective cohort studies have reported inverse associations between dietary intake or blood levels of beta-carotene and risks of cancer. Several large-scale trials were set up to assess whether beta-carotene supplementation might reduce the risk of cancer. Subsequently, evidence emerged from basic research which indicated that oxidative modification of low-density lipoprotein cholesterol increases its atherogenicity. The evidence from basic research, and epidemiological evidence for a possible protective effect of antioxidant vitamins for cardiovascular disease was strongest for vitamin E. More recently, further trials were set up to examine if supplementation with anti-oxidant vitamins might also reduce the risk of cardiovascular disease. This review summarises the available randomised evidence from published trials of beta-carotene supplementation involving **70,000 people from 3 large-scale trials in healthy populations and on vitamin E supplementation involving 29,000 patients at high-risk of cardiovascular disease from 5 large-scale trials. The results of these trials have been disappointing and failed to confirm any protective effect of these vitamins for either cancer or for cardiovascular disease.** (Clarke, Armitage, 2002)

Circa 2003

Vitamins E & A fail to reduce incidence or mortality of lung cancer: Cochrane Database Syst Rev. 2003. (Caraballoso et al., 2003) (#109,394 participants); **The electronic databases MED-LINE (1966-july 2001), EMBASE (1974-july 2001) and the Cochrane Controlled Trial Register (CENTRAL, Issue 3/2001) and bibliographies were searched. Duration of treatment varied from 2 to 12 years and follow-up was from two to five years.** When beta-carotene was combined with retinol, data from a single study showed that there was a statistically significant, *increased risk of lung cancer incidence and mortality* in people with risk factors for lung cancer who took both vitamins. **There is currently no evidence to support recommending vitamins such as alpha-tocopherol, beta-carotene or retinol, alone or in combination, to prevent lung cancer. A harmful effect was found for beta-carotene with retinol at pharmacological doses in people with risk factors for lung cancer.**

Use of antioxidant vitamins for the prevention of cardiovascular disease: meta-analysis of randomized trials. (Vivekananthan et al., 2003) (The vitamin E trials involved a total of #81,788 patients, and the beta-carotene trials involved #138,113); **Vitamin E did not provide any benefit in lowering mortality compared to control treatments, and it did not significantly decrease the risk of cardiovascular death or stroke** ("cerebrovascular accident"). **The lack of any beneficial effect was seen consistently regardless of the doses of vitamins used and the diversity of the patient populations.** Therefore, **the CCF (Cleveland Clinic) researchers conclude that this study does "not support the routine use of vitamin E."**

Beta carotene led to small but statistically significant increase in all-cause mortality and a slight increase in cardiovascular death.

The researchers call their findings *"especially concerning"* because beta carotene doses are commonly included in over-the-counter vitamin supplements and multivitamin supplements that have been advocated for widespread use.

**The study says that using vitamin supplements that contain beta caro-
tene should be "actively discouraged" because of the increase in the
risk of death.** They also *recommend discontinuing study of beta carotene
supplements because of their risk.*

Despite the negative findings of most of the clinical trials, **manufacturers
continue to promote antioxidants as though they have been proven
beneficial "wonder drugs."** Many also hype mixtures of beta-carotene and
other carotenoids, which, they suggest, may provide the same benefits as fruits
and vegetables, which they know is not true. **The FDA will not permit any
of these substances to be labeled or marketed with claims that they
can prevent disease.** The increased death rate from lung cancer in smokers
who took beta carotene is evidence enough that high doses of vitamins and
minerals are not necessarily harmless and may be fatal.

Two studies examined effects of *dietary* antioxidant intake on incidence of Alzheimer's disease. One found protective effects of E and C (Engelhart et al, 2002); the
other only for E (Morris et al, 2002). **Neither found a benefit from vitamin
supplements.** A third study found **no effect** of dietary carotenes, vitamin C,
or vitamin E, nor effects of vitamin C or E supplements (Luchsinger et al. (2003).

**Antioxidant Vitamins Effect on Alzheimer's Disease: Wash-
ington Heights-Inwood Columbia Aging Project** (Luchsinger
et al, 2003) (#980 elderly subjects) relationship between AD and the intake
of carotenes, vitamin C, and vitamin E in 980 elderly subjects in the Washington Heights-Inwood Columbia Aging Project who were free of dementia at baseline and were followed for a mean time of 4 years. CONCLUSION: **Neither
dietary, supplemental, nor total intake of carotenes and vitamins C and
E was associated with a decreased risk of AD in this study.**

**Neoplastic and Antineoplastic Effects of Beta Carotene on
Colorectal Adenoma Recurrence: Results of a Randomized
Trial** (Baron et al, 2003) (#864 subjects who had had an adenoma
removed and were polyp-free); the effect of beta carotene supplementation
on colorectal adenoma recurrence among subjects in a multicenter double-blind,
placebo-controlled clinical trial of antioxidants for the prevention of colorectal

adenomas. Results: **Among subjects who neither smoked cigarettes nor drank alcohol, beta-carotene was associated with a marked decrease in the risk of one or more recurrent adenomas, but beta carotene supplementation conferred a modest increase in the risk of recurrence among those who smoked.** *For participants who smoked cigarettes and also drank more than one alcoholic drink per day, beta carotene doubled the risk of adenoma recurrence.*

Routine Vitamin Supplementation To Prevent Cardiovascular Disease: A Summary of the Evidence for the U.S. Preventive Services Task Force (Morris and Carson, 2003); Data Sources:

Cochrane Controlled Trials Registry and MEDLINE (1966 to September 2001), reference lists, and experts. **Conclusions:** Some good-quality cohort studies have reported an association between the use of vitamin supplements and lower risk for cardiovascular disease. **Randomized, controlled trials of specific supplements, however, have failed to demonstrate a consistent or significant effect of any single vitamin or combination of vitamins on incidence of or death from cardiovascular disease.**

In 2003, the **U.S. Preventive Services Task Force** (an independent panel sponsored by the Agency for Healthcare Research and Quality) concluded that **"the evidence is insufficient to recommend for or against the use of supplements of vitamins A, C, or E; multivitamins with folic acid; or antioxidant combinations for the prevention of cancer or cardiovascular disease."**

An **American Heart Association Science Advisory statement** (Circulation 110, 637-641 (2004)) concluded, **"At this time, the scientific data do not justify the use of antioxidant vitamin supplements for CVD risk reduction."**

Midlife Dietary Intake of Antioxidants and Risk of Late-Life Incident Dementia: The Honolulu-Asia Aging Study (Laurin et al, 2003) (#2,459 men) Data were obtained from the Honolulu-Asia Aging Study, a prospective community-based study of Japanese-American men. Analysis included 2,459 men with complete dietary data who were dementia-free at the

first assessment. CONCLUSION: **Intakes of beta-carotene, flavonoids, and vitamins E and C were not associated with the risk of dementia or its subtypes.** This analysis suggests that **midlife dietary intake of antioxidants does not modify the risk of late-life dementia or its most prevalent subtypes.**

Serum retinol levels and the risk of fracture. (Michealsson et al, 2003) (#2,322 men)

Although studies in animals and epidemiologic studies have indicated that a high vitamin A intake is associated with increased bone fragility, no biologic marker of vitamin A status has thus far been used to assess the risk of fractures in humans. **Methods**: We enrolled 2,322 men, 49 to 51 years of age, in a population-based, longitudinal cohort study. Serum retinol and beta carotene were analyzed in samples obtained at enrollment. Fractures were documented in 266 men during 30 years of follow-up. Cox regression analysis was used to determine the risk of fracture according to the serum retinol level. **Results: *The risk of fracture was highest among men with the highest levels of serum retinol.*** Multivariate analysis of the risk of fracture in the highest quintile for serum retinol (>75.62 µg per deciliter [2.64 µmol per liter]) as compared with the middle quintile (62.16 to 67.60 µg per deciliter [2.17 to 2.36 µmol per liter]) showed that the rate ratio was 1.64 (95 percent confidence interval, 1.12 to 2.41) for any fracture and 2.47 (95 percent confidence interval, 1.15 to 5.28) for hip fracture. *The risk of fracture was further increased within the highest quintile for serum retinol. Men with retinol levels in the 99th percentile (>103.12 µg per deciliter [3.60 µmol per liter]) had an overall risk of fracture that exceeded the risk among men with lower levels by a factor of seven* (P<0.001). The level of serum beta carotene was not associated with the risk of fracture. **Conclusions**: Our findings, which are consistent with the results of studies in animals, as well as in vitro and epidemiologic dietary studies, suggest that **current levels of vitamin A supplementation and food fortification in many Western countries may need to be reassessed.**

Impact of simvastatin, niacin, and/or antioxidants on cholesterol metabolism in CAD patients with low HDL. (Matthan et al, 2003) (#123 HATS participants)

The HDL Atherosclerosis Treatment Study (HATS) demonstrated a clinical benefit in coronary artery disease patients with low HDL cholesterol (HDL-C) levels treated with simvastatin and niacin (S-N) or S-N plus antioxidants (S-N+A) compared with antioxidants alone or placebo. Angiographically documented stenosis regressed in the S-N group but progressed in all other groups. To assess the mechanism(s) responsible for these observations, surrogate markers of cholesterol absorption and synthesis were measured in a subset of 123 **HATS participants** at 24 months (on treatment) and at 38 months (off treatment). Treatment with S-N reduced desmosterol and lathosterol levels (cholesterol synthesis indicators) 46% and 36%, respectively, and elevated campesterol and ß-sitosterol levels (cholesterol absorption indicators) 70% and 59%, respectively, relative to placebo and antioxidant but not S-N+A. **Treatment with antioxidants alone had no significant effect**. Combining S-N with antioxidants reduced desmosterol and lathosterol by 37% and 31%, and elevated campesterol and ß-sitosterol levels by 54% and 46%, but differences did not attain significance.

A randomized trial of beta carotene and age-related cataract in US physicians.

(Christen et al, 2003) (#22,071 Male US physicians aged 40 to 84 years) OBJECTIVE: To examine the development of age-related cataract in a trial of beta carotene supplementation in men. DESIGN: **Randomized, double-masked, placebo-controlled trial**. METHODS: **Male US physicians** aged 40 to 84 years (n = 22 071) were randomly assigned to receive either beta carotene (50 mg on alternate days) or placebo for 12 years. MAIN OUTCOME MEASURES: Age-related cataract and extraction of age-related cataract, defined as an incident, age-related lens opacity, responsible for a reduction in best-corrected visual acuity to 20/30 or worse, based on self-report confirmed by medical record review. RESULTS: There was no difference between the beta carotene and placebo groups in the overall incidence of cataract or cataract extraction. In subgroup analyses, the effect of beta carotene supplementation appeared to be modified by smoking status at baseline (P =.02). Among current smokers, there were 108 cases of cataract in the beta carotene group and 133 in the placebo group. Among current nonsmokers, there was no significant difference in the number of cases in the 2 treatment groups. The results for cataract extraction appeared to be similarly modified by baseline smoking status (P =.05). CONCLUSIONS: **Randomized trial data from a large population of healthy men indicate no overall benefit or harm of 12 years of beta carotene supplementation on cataract or cataract extraction**. However, among current smokers at baseline, beta carotene appeared to attenuate their excess risk of cataract by about one fourth.

Plasma carotenoids and tocopherols and risk of myocardial infarction in a low-risk population of US male physicians. (Hak et al. 2003) (#531 physicians diagnosed with MI) One of the two available studies even suggests a higher risk for MI with higher gamma-tocopherol concentrations. Results from the prospective nested case-control Physicians' Health Study published by Hak et al. showed in a multi-variate analysis that *men with high plasma γ-tocopherol levels tended to have an increased risk of nonfatal and fatal MI.*

- Health's Professional Study, which included 39,910 U.S. male physicians. (*Circulation.* 2003;108:802.) Increased intake of carotenoids and vitamin E may protect against myocardial infarction (MI). However, prospective data on blood levels of carotenoids other than beta-carotene and vitamin E (tocopherol) and risk of MI are sparse. *Methods and Results—* We conducted a prospective, nested case-control analysis among male physicians without prior history of cardiovascular disease who were followed for up to 13 years in the Physicians' Health Study. Samples from **531 physicians diagnosed with MI** were analyzed together with samples from paired control subjects, matched for age and smoking, for 5 major carotenoids (alpha- and beta-carotene, beta-cryptoxanthin, lutein, and lycopene), retinol, and alpha- and gamma-tocopherol. **Overall, we found no evidence for a protective effect against MI for higher baseline plasma levels of retinol or any of the carotenoids measured.** Among current and former smokers but not among never-smokers, higher baseline plasma levels of beta-carotene tended to be associated with lower. **Men with higher plasma levels of gamma-tocopherol tended to have an increased risk of MI.** *Conclusions—* **These prospective data do not support an overall protective relation between plasma carotenoids or tocopherols and future MI risk among men without a history of prior cardiovascular disease.**

Selenium supplementation and secondary prevention of nonmelanoma skin cancer in a randomized trial. (Duffield-Lillico, 2003) (#1,312). The Nutritional Prevention of Cancer Trial was a double-blind, randomized, placebo-controlled clinical trial designed to test whether selenium as selenized yeast (200 microg daily) could prevent nonmelanoma skin cancer among 1312 patients from the Eastern United States who had previously had this disease. Results from September 15, 1983, through December 31, 1993, showed no association between treatment and the incidence of basal and squamous cell carcinomas of the skin. This report summarizes the entire blinded treatment period, which ended on January 31, 1996. The association between treatment and time to first nonmelanoma skin cancer diagnosis and between treatment and time to multiple skin tumors overall and within subgroups, defined by baseline characteristics, was evaluated. Although results through the entire blinded period continued to show that selenium supplementation was not statistically significantly associated with the risk of basal cell carcinoma (hazard ratio [HR] = 1.09, 95% confidence interval [CI] = 0.94 to 1.26), *selenium supplementation was associated with statistically significantly elevated risk of squamous cell carcinoma* and of total nonmelanoma skin cancer. **Results from the Nutritional Prevention of Cancer Trial conducted among individuals at high risk of nonmelanoma skin cancer continue to demonstrate that selenium supplementation is ineffective at preventing basal cell carcinoma and that *it increases the risk of squamous cell carcinoma and total nonmelanoma skin cancer.*** (Duffield-Lillico, 2003). ===

Acetylcysteine, coronary procedure and prevention of contrast-induced worsening of renal function: which benefit for which patient? (Kefer et al. 2003) (#108) This study was designed to

determine whether acetylcysteine could provide a protective effect on renal function in a population of patients with normal renal function or mild to moderate chronic renal failure, usually referred for a coronary procedure. BACKGROUND: Contrast-induced nephropathy is a well-recognized complication of coronary angiography. Recent studies suggest that saline hydration and acetylcysteine reduce the incidence of contrast-induced worsening of renal function in patients with pre-existing chronic renal failure who are undergoing computed tomography examinations. METHODS: **One hundred eight patients were blindly and randomly assigned to receive either acetylcysteine or placebo** before and after administration of contrast agent in association with a moderate hydration protocol. Serum creatinine and urea nitrogen were measured before and 24 hours after coronary procedure. RESULTS: The mean serum creatinine concentration remained unchanged 24 hours after contrast agent administration in both groups: from 1.04 +/- 0.26 to 1.03 +/- 0.29 mg/dl in the acetylcysteine group and from 1.16 +/- 1.1 to 1.06 +/- 0.41 mg/dl in the control group. We divided the population into 3 subgroups according to their creatinine clearance: no significant change of serum creatinine concentration was observed in patients with normal renal function nor in patients with pre-existing mild to moderate chronic renal failure in both groups. **There was no significant difference for the incidence of contrast-induced nephropathy between both groups**. CONCLUSIONS: **Our data do not support the systematic use of acetylcysteine before a coronary procedure in patients with normal renal function or mild to moderate chronic renal failure, to prevent contrast-induced nephropathy**. Division of Cardiology, University of Louvain, Brussels, Belgium.

Oral acetylcysteine does not protect renal function from moderate to high doses of intravenous radiographic contrast (Boccalandro et al. 2003) (#106 consecutive patients)

The use of radiographic contrast during cardiac catheterization can cause acute renal failure with an increase in morbidity and mortality. Prophylactic acetylcysteine plus intravenous hydration have been shown to prevent contrast-induced nephropathy (CIN) in patients with chronic renal failure undergoing computed tomography scan, who receive low doses of intravenous contrast. Whether the use of prophylactic acetylcysteine can decrease the incidence of CIN when larger doses of contrast are used remains to be determined. We sought to evaluate whether the prophylactic administration of acetylcysteine plus intravenous hydration is superior to intravenous hydration alone in prevention of CIN in patients with chronic renal failure undergoing cardiac catheterization and receiving moderate to high doses of intravenous contrast (> 1 cc/kg). **Seventy-three consecutive patients with renal insufficiency who received intravenous hydration and 600 mg of acetylcysteine** twice a day 24 hr before and the day of the cardiac catheterization were compared with **106 consecutive patients** who received hydration alone. Baseline and 48-hr serum creatinine concentrations were compared between the two groups before and after cardiac catheterization. **Multivariate and univariate analysis** were performed to assess the effects of acetylcysteine and other clinical variables in the change of serum creatinine after the procedure. Both groups had comparable clinical characteristics and received similar volumes of intravenous hydration. The volume of contrast used was similar for the two groups. A mean change in serum creatinine of 0.17 +/- 0.54 mg/dl for the acetylcysteine group vs. 0.19 +/- 0.40 mg/dl for the control group was observed at 48 hr. The incidence CIN was 13% in the acetylcysteine vs. 12% in the control group. **Acetylcysteine, whether analyzed with multivariate or univariate analysis, failed to demonstrate a significant effect in the change of serum creatinine after cardiac catheterization.** In patients with chronic renal insufficiency, acetylcysteine in a dose of 600 mg twice a day before and after cardiac catheterization, along with intravenous fluids, is as effective as fluids alone in the prevention of CIN when moderate to high doses of contrast are used. University of Texas Medical School and Memorial Hermann Hospital, Houston, Texas.

Circa 2004

There has been evidence of adverse effects of antioxidant supplements on cardio-vascular disease (CVD) as assessed by angiographic end points. In the Women's Angiographic Vitamin and Estrogen Study (2002), postmenopausal women with coronary disease on hormone replacement therapy given vitamin E plus vitamin C had an unexpected significantly higher all-cause mortality rate and a trend for an increased cardiovascular mortality rate compared with the vitamin placebo women. Similarily, in the HDL-Atherosclerosis Treatment Study (2001), subjects with angiographically demonstrated coronary artery disease on simvastatin/nia-cin and an antioxidant cocktail (vitamin E, ß-carotene, vitamin C, and selenium) had a 0.7% progression in stenosis after 3 years, compared with 0.4% regression in the group on only simvastatin/niacin. Thus, antioxidant supplements may have interfered with the efficacy of statin-plus-niacin therapy. Further evaluation of data showed that the addition of the antioxidant vitamins blunted the expected rise in the protective HDL-2 cholesterol and apolipoprotein AI subfractions of HDL. Overall, the studies showing either positive or adverse effects (especially for vitamins E, vitamins E and C, and the antioxidant cocktails) are much smaller studies than the larger clinical trials that consistently have not shown any benefi-cial effects of antioxidant supplements on several CVD end points. The existing scientific database, as of 2004, does not justify routine use of antioxidant supple-ments for the prevention and treatment of CVD.

Supplemental vitamin C increase cardiovascular disease risk in women with diabetes (Iowa Women's Health study) (Lee et al, 2004) (#1,923 postmenopausal women who reported being diabetic) When dietary and supplemental vitamin C were analyzed separately, only supplemental vitamin C showed a positive association with mortality endpoints. CONCLUSION: *A high vitamin C intake from supplements is associated with an increased risk of cardiovascular disease mortality in postmenopausal women with diabetes.*

Cochrane Database Syst Rev. 2004: Vitamins E & A fail to reduce incidence or mortality of gastrointestinal cancer. (Cochrane Database Syst Rev.) (Bjelakovic et al, 2004) (#170,525 participants); **14 randomized trials (170,525 participants), assessing beta-carotene (9 trials), vitamin A (4 trials), vitamin C (4 trials), vitamin E (5 trials), and selenium (6 trials).** Neither the fixed effect nor random effects meta-analyses showed significant effects of supplementation with antioxidants on the incidences of gastrointestinal cancers.

Among the seven high-quality trials reporting on mortality (131,727 participants), the fixed effect unlike the random effects meta-analysis showed that *antioxidant supplements significantly increased mortality. Beta-carotene and vitamin A and beta-carotene and vitamin E significantly increased mortality,* while beta-carotene alone only tended to do so. Selenium showed significant beneficial effect on gastrointestinal cancer incidences. *When the selenium trials were excluded, both analyses showed a statistically significant increase in mortality, which was particularly strong in patients taking beta carotene and vitamin A.*

No single antioxidant or combination of antioxidants significantly reduced the incidence of esophageal, gastric, colorectal, pancreatic, or hepatic cancer.

CONCLUSIONS: They could not find evidence that antioxidant supplements prevent gastrointestinal cancers. On the contrary, they seem to increase overall mortality.

ATBC 6-year followup study (2004) (Thornwall et al., 2004) (#29,133 male smokers); evaluated the 6-year post-trial effects of α-tocopherol and ß-carotene supplementation on coronary heart disease (CHD) in the α-tocopherol, ß-carotene cancer prevention (ATBC) study. *ß-Carotene seemed to increase the post-trial risk of first-ever non-fatal MI* but there is no plausible mechanism to support it. Among men with pre-trial myocardial infarction, **no effects were observed** in post-trial risk of major coronary event.

HOPE study of vitamin E on renal insufficiency (2004) (Mann et al, 2004) (#993 people with a serum creatinine > or =1.4 to 2.3 mg/dL. And renal insufficiency); In people with mild-to-moderate renal insufficiency at high cardiovascular risk, vitamin E at a dose of 400 IU/day had **no apparent effect on cardiovascular outcomes**.

Randomized trials of vitamin E in the treatment and prevention of cardiovascular disease (2004) (Eidelman et al., 2004) (7 large-scale randomized trials) **Six of the 7 trials showed no significant effect of vitamin E on cardiovascular disease. In an overview, vitamin E had neither a statistically significant nor a clinically important effect on any important cardiovascular event or its components: nonfatal myocardial infarction or cardiovascular death.**

A point of emphasis is that, **"The use of agents of proven lack of benefit, especially those easily available over the counter, may contribute to underuse of agents of proven benefit and failure to adopt healthy lifestyles."**

Effect of supplemental vitamin E for the prevention and treatment of cardiovascular disease (Shekelle et al, 2004) (#Eighty-four eligible trials) 84 trials were identified. For the outcomes of all-cause mortality, cardiovascular mortality, fatal or nonfatal myocardial infarction, and blood lipids, **neither supplements of vitamin E alone nor vitamin E given with other agents yielded a statistically significant beneficial or adverse pooled relative risk for all-cause mortality, cardiovascular mortality, and nonfatal myocardial infarction,** respectively. CONCLUSIONS: **There is**

good evidence that vitamin E supplementation does not beneficially or adversely affect cardiovascular outcomes.

SU.VI.MAX Study (2004) (Hercberg et al, 2004) (#A total of 13,017 French adults (7,876 women aged 35-60 years and 5141 men aged 45-60 years); All participants took a single daily capsule of a combination of 120 mg of ascorbic acid, 30 mg of vitamin E, 6 mg of beta carotene, 100 mug of selenium, and 20 mg of zinc, or a placebo. Median follow-up time was 7.5 years. RESULTS: **No major differences were detected between the groups in total cancer incidence, ischemic cardiovascular disease incidence or all-cause mortality.** However, unexplainably, after 7.5 years, low-dose anti-oxidant supplementation lowered total cancer incidence and all-cause mortality in men but not in women.

Meta-analysis: high-dosage vitamin E supplementation may increase all-cause mortality (Miller et al., 2004) (#135,967 subjects); **Meta-analysis, including more than 135,000 subjects, concluded that** *high doses of vitamin E increased mortality.* **Researchers said the current U.S. dietary guidelines do not recommend vitamin E supplementation.** As a result, you will never see beta carotene supplements recommended again," Miller said.

Further, in January 2005, Gullar, Hanley and Miller et al. published a dose–response meta-analysis showing that high-dosage (≥ 400 IU/d) vitamin E supplementation was associated with a small but statistically significant increased risk for mortality (relative risk, 1.04 [95% CI, 1.01 to 1.07]) (1). The precise dosage of vitamin E at which the relative risk for mortality exceeded 1 and the magnitude of the risk increase were uncertain. However, the analysis showed that high-dosage vitamin E supplementation was likely to be harmful (Gullar et al, 2005).

Also, a 2004 study found vitamin E did not reduce onset of Alzheimer's.

Five studies included **102,735 patients** taking various doses of **vitamin C but showed no effect on cardiovascular disease mortality** (Morris and Carson, 2003).

260

Therefore, vitamin E (alpha-tocopherol) alone in doses of 400 units is of questionable value, and larger doses may cause intracranial hemorrhage or interact negatively with lipid-lowering drugs. Vitamin E should not be used in patients who have bleeding disorders or patients on anticoagulants or acetylsalicylic acid (ASA).

The role of vitamin E in the prevention of coronary events and stroke. Meta-analysis of randomized controlled trials

(Alkhenizan and Al-Omran, 2004) (#80,645 subjects) a meta-analysis, using the Cochrane Group Methodology, of all available randomized controlled trials (RCTs) to evaluate the role of vitamin E in the prevention of CVD. Nine studies met inclusion criteria. **Vitamin E supplementation was not associated with a reduction in total mortality or total CVD mortality**, but it was associated with a small statistically significant reduction in non-fatal myocardial infarction in patients with pre-existing coronary artery disease.

Oats, Antioxidants and Endothelial Function in Overweight, Dyslipidemic Adults (Katz et al, 2004) (#30) (16 males ≥age 35;

14 postmenopausal females) were assigned, in random order, to oats (60 g oatmeal), vitamin E (400 IU) plus vitamin C (500 mg), the combination of oats and vitamins, or placebo; brachial artery reactivity scans (BARS) following a single dose of each treatment, and again following 6 weeks of daily ingestion, with 2-week washout periods. CONCLUSION: **The direction of effect was negative for vitamins C and E and the oat/vitamin combination with both acute and sustained treatment.**

Vitamin C worsens coronary atherosclerosis in those with two copies of the haptoglobin 2 gene. (Levy et al, 2004) (#299

postmenopausal women) Antioxidant trials have not demonstrated efficacy in slowing cardiovascular disease but could not rule out benefit for specific patient subgroups. Antioxidant therapy reduces LDL oxidizability in haptoglobin 1 allele homozygotes (Hp 1-1), but not in individuals with the haptoglobin 2 allele (Hp 2-1 or Hp 2-2). They therefore hypothesized that haptoglobin type would be predictive of the effect of vitamin therapy on coronary atherosclerosis as assessed by angiography.

They tested this hypothesis in the Women's Angiographic Vitamin and Estrogen (WAVE) trial, a prospective angiographic study of vitamins C and E with or without hormone replacement therapy (HRT) in postmenopausal women. Haptoglobin type was determined in **299 women** who underwent baseline and follow-up angiography. The annualized change in the minimum luminal diameter (MLD) was examined in analyses stratified by vitamin use, haptoglobin type, and diabetes status.

RESULTS—They found a significant benefit on the change in MLD with vitamin therapy as compared with placebo in Hp 1-1 subjects (0.079 ± 0.040 mm, P = 0.049). This benefit was more marked in diabetic subjects (0.149 ± 0.064 mm, P = 0.021). On the other hand, there was a trend toward a more rapid decrease in MLD with vitamin therapy in Hp 2-2 subjects, which was more marked in diabetic subjects (0.128 ± 0.057 mm, P = 0.027). HRT had no effect on these outcomes.

CONCLUSIONS—The relative benefit or harm of vitamin therapy on the progression of coronary artery stenoses in women in the WAVE study was dependent on haptoglobin type. This influence of haptoglobin type seemed to be stronger in women with diabetes. When the results of one randomized controlled trial were reanalyzed based on haptoglobin genotype, **antioxidant therapy (1,000 mg/day of vitamin C + 800 IU/day of vitamin E) was associated with improvement of coronary atherosclerosis in diabetic women with two copies of the haptoglobin 1 gene** *but worsening of coronary atherosclerosis in those with two copies of the haptoglobin 2 gene.*

Vitamin C for preventing and treating the common cold. Cochrane Database Syst Rev. 2004;(4):CD000980.

(Douglas et al, 2004) (#11,350 study participants) The role of vitamin C (ascorbic acid) in the prevention and treatment of the common cold has been a subject of controversy for 60 years, but is widely sold and used as both a preventive and therapeutic agent. OBJECTIVES: To discover whether oral doses of 0.2 g or more daily of vitamin C reduces the incidence, duration or severity of the common cold when used either as continuous prophylaxis or after the onset of symptoms. SEARCH STRATEGY: We searched the Cochrane Central Register of Controlled Trials (CENTRAL) (The Cochrane Library Issue 4, 2006); MEDLINE

(1966 to December 2006); and EMBASE (1990 to December 2006). SELECTION CRITERIA: Papers were excluded if a dose less than 0.2 g per day of vitamin C was used, or if there was no placebo comparison. DATA COLLECTION AND ANALYSIS: Two review authors independently extracted data and assessed trial quality. 'Incidence' of colds during prophylaxis was assessed as the proportion of participants experiencing one or more colds during the study period. 'Duration' was the mean days of illness of cold episodes. MAIN RESULTS: Thirty trial comparisons involving **11,350 study participants** contributed to the meta-analysis on the relative risk (RR) of developing a cold whilst taking prophylactic vitamin C. The pooled RR was 0.96. A subgroup of six trials involving a total of 642 marathon runners, skiers, and soldiers on sub-arctic exercises reported a pooled RR of 0.50. Thirty comparisons involving 9676 respiratory episodes contributed to a meta-analysis on common cold duration during prophylaxis. A consistent benefit was observed, representing a reduction in cold duration of 8% for adults and 13.6% for children. Seven trial comparisons involving 3294 respiratory episodes contributed to the meta-analysis of cold duration during therapy with vitamin C initiated after the onset of symptoms. **No significant differences from placebo were seen.** Four trial comparisons involving 2,753 respiratory episodes contributed to the meta-analysis of cold severity during therapy and **no significant differences from placebo were seen.** CONCLUSIONS: **The failure of vitamin C supplementation to reduce the incidence of colds in the normal population indicates that routine mega-dose prophylaxis is not rationally justified for community use.** But evidence suggests that it could be justified in people exposed to brief periods of severe physical exercise or cold environments. **A meta-analysis of 30 placebo-controlled prevention trials found that vitamin C supplementation in doses up to 2 grams/day did not decrease the incidence of colds.**

Vitamin E supplementation and cataract: randomized controlled trial. (McNeil et al, 2004) (#1,193 eligible subjects with early or no cataract) Investigators determined whether treatment with vitamin E (500 IU daily) reduces either the incidence or rate of progression of age-related cataracts. DESIGN: A prospective, randomized, double-masked, placebo-controlled clinical trial entitled the Vitamin E, Cataract and Age-Related Maculopathy Trial. PARTICIPANTS: Of 1,906 screened volunteers, **1,193 eligible subjects with early or no cataract**, aged 55 to 80 years, were enrolled and followed up for 4 years.

INTERVENTION: Subjects were assigned randomly to receive either **500 IU of natural vitamin E** in soybean oil encapsulated in gelatin or a placebo with an identical appearance. MAIN OUTCOME MEASURES: The incidence and progression rates of age-related cataract were assessed annually with both clinical lens opacity gradings and computerized analysis of Scheimpflug and retroillumination digital lens images obtained with a Nidek EAS-1000 lens camera. The analysis was undertaken using data from the eye with the more advanced opacity for each type of cataract separately and for any cataract changes in each individual. RESULTS: Overall, **87% of the study population completed the 4 years of follow-up**, with 74% of the vitamin E group and 76% of the placebo group continuing on their randomized treatment allocation throughout this time. For cortical cataract, the 4-year cumulative incidence rate was 4.5% among those randomized to vitamin E and 4.8% among those randomized to placebo. For nuclear cataract, the corresponding rates were 12.9% and 12.1%. For posterior subcapsular cataract, the rates were 1.7% and 3.5%, whereas for any of these forms of cataract, they were 17.1% and 16.7%, respectively. Progression of cortical cataract was seen in 16.7% of the vitamin E group and 18.4% of the placebo group. Corresponding rates for nuclear cataract were 11.4% and 11.9%, whereas those of any cataract were 16.5% and 16.7%, respectively. **There was no difference in the rate of cataract extraction between the 2 groups**. Lens characteristics of the participants withdrawn from the randomized medications were not different from those who continued. CONCLUSIONS: **Vitamin E given for 4 years at a dose of 500 IU daily did not reduce the incidence of or progression of nuclear, cortical, or posterior subcapsular cataracts. These findings do not support the use of vitamin E to prevent the development or to slow the progression of age-related cataracts.**

Antioxidant vitamins and coronary heart disease risk: a pooled analysis of 9 cohorts. (Knekt et al, 2004) (#293,172 subjects free of CHD at baseline) Epidemiologic studies have suggested a lower risk of coronary heart disease (CHD) at higher intakes of fruit, vegetables, and whole grain. Whether this association is due to antioxidant vitamins or some other factors remains unclear. They studied the relation between the intake of antioxidant vitamins and CHD risk. **Design:** A **cohort study pooling 9 prospective studies** that included information on intakes of **vitamin E, carotenoids, and vitamin C** and that met specific criteria was carried out. During a

10-y follow-up, 4647 major incident CHD events occurred in **293,172 subjects** who were free of CHD at baseline. **Results: Dietary intake of antioxidant vitamins was only weakly related to a reduced CHD risk after adjustment for potential nondietary and dietary confounding factors.** Compared with subjects in the lowest dietary intake quintiles for vitamins E and C, those in the highest intake quintiles had relative risks of CHD incidence of 0.84 and 1.23, respectively, and the relative risks for subjects in the highest intake quintiles for the various carotenoids varied from 0.90 to 0.99. **Subjects with higher supplemental vitamin C intake had a lower CHD incidence.** Compared with subjects who did not take supplemental vitamin C, those who took >700 mg supplemental vitamin C/d had a relative risk of CHD incidence of 0.75. **Supplemental vitamin E intake was not significantly related to reduced CHD risk. Conclusions:** The results suggest a reduced incidence of major CHD events at high supplemental vitamin C intakes. **The risk reductions at high vitamin E or carotenoid intakes appear small.**

A review of the epidemiological evidence for the 'antioxidant hypothesis' by the British Nutrition Foundation (the Food Standards Agency). (Stanner et al, 2004) (British Nutrition Foundation independent review) The British Nutrition Foundation was recently commissioned by the Food Standards Agency to conduct a review of the government's research programme on Antioxidants in Food. Part of this work involved an independent review of the scientific literature on the role of antioxidants in chronic disease prevention, which is presented in this paper. BACKGROUND: There is consistent evidence that diets rich in fruit and vegetables and other plant foods are associated with moderately lower overall mortality rates and lower death rates from cardiovascular disease and some types of cancer. The 'antioxidant hypothesis' proposes that vitamin C, vitamin E, carotenoids and other antioxidant nutrients afford protection against chronic diseases by decreasing oxidative damage. RESULTS: Although scientific rationale and observational studies have been convincing, **randomized primary and secondary intervention trials have failed to show any consistent benefit from the use of antioxidant supplements on cardiovascular disease or cancer risk, with some trials even suggesting possible harm in certain subgroups.** These trials have usually involved the administration of single antioxidant nutrients given at relatively high doses. The results of trials investigating the effect of

265

a balanced combination of antioxidants at levels achievable by diet are awaited. CONCLUSION: **The suggestion that antioxidant supplements can prevent chronic diseases has not been proved or consistently supported by the findings of published intervention trials.** Further evidence regarding the efficacy, safety and appropriate dosage of antioxidants in relation to chronic disease is needed. The most prudent public health advice remains to increase the consumption of plant foods, as such dietary patterns are associated with reduced risk of chronic disease.

Impact of antioxidants, zinc, and copper on cognition in the elderly: a randomized, controlled trial. (Yaffe et al, 2004) (#2,166 elderly persons) Participants in the Age-Related Eye Disease Study

were randomly assigned to receive daily antioxidants (vitamin C, 500 mg; vitamin E, 400 IU; beta carotene, 15 mg), zinc and copper (zinc, 80 mg; cupric oxide, 2 mg), antioxidants plus zinc and copper, or placebo. A cognitive battery was administered to **2,166 elderly persons** after a median of 6.9 years of treatment. Treatment groups did not differ on any of the six cognitive tests (p > 0.05 for all). **These results do not support a beneficial or harmful effect of antioxidants or zinc and copper on cognition in older adults.**

No long-term effect of combined vitamins E and C on coronary and peripheral endothelial function (Kinlay et al. 2004) (#30).

They tested whether long-term administration of antioxidant vitamins C and E improves coronary and brachial artery endothelial function in patients with coronary artery disease (CAD). BACKGROUND: Endothelial function is a sensitive indicator of vascular health. Oxidant stress and oxidized low-density lipoprotein (LDL) impair endothelial function by reducing nitric oxide bioavailability in the artery wall. METHODS: They **randomly assigned 30 subjects with CAD to combined vitamin E (800 IU per day) and C (1000 mg per day)** or to placebos in a double-blind trial. Coronary artery endothelial function was measured as the change in coronary artery diameter to acetylcholine infusions (n = 18 patients), and brachial artery endothelial function was assessed by flow-mediated dilation (n = 25 patients) at baseline and six months. Plasma markers of oxidant stress (oxidized LDL and autoantibodies) were also measured.

RESULTS: Plasma alpha-tocopherol and ascorbic acid increased with active therapy. Compared to placebo, there was no improvement in coronary and brachial endothelial vasomotor function over six months. **Although vitamins C and E tended to reduce F2-isoprostanes, they failed to alter oxidized LDL or autoantibodies to oxidized LDL.** CONCLUSIONS: **Long-term oral vitamins C and E do not improve key mechanisms in the biology of atherosclerosis or endothelial dysfunction, or reduce LDL oxidation in vivo.**

Age-related cataract in a randomized trial of beta-carotene in women. (Christen et al. 2004) (#39,876). Investigators examined the development of age-related cataract in a trial of beta-carotene supplementation in women.

METHODS: The **Women's Health Study** is a randomized, double-masked, placebo-controlled trial originally designed to test the balance of benefits and risks of beta-carotene (50 mg on alternate days), vitamin E, and aspirin in the primary prevention of cancer and cardiovascular disease among **39,876 female health professionals** aged 45 years or older. The beta-carotene component of the trial was terminated early after a median treatment duration of 2.1 years. Main outcome measures were visually-significant cataract and cataract extraction, based on self-report confirmed by medical record review.

RESULTS: There were 129 cataracts in the beta-carotene group and 133 in the placebo group. For cataract extraction, there were 94 cases in the beta-carotene group and 89 cases in the placebo group. Subgroup analyses suggested a possible beneficial effect of beta-carotene in smokers. **CONCLUSIONS: These randomized trial data from a large population of apparently healthy female health professionals indicate that two years of beta-carotene treatment has no large beneficial or harmful effect on the development of cataract during the treatment period.**

Fruit, vegetable, and antioxidant intake and all-cause, cancer, and cardiovascular disease mortality in a community-dwelling population in Washington County, Maryland (CLUE) (Genkinger et al. 2004) (#6,151). Higher intake of fruits, vegetables, and antioxidants may help protect against oxidative damage, thus

lowering cancer and cardiovascular disease risk. This Washington County, Maryland, prospective study examined the association of fruit, vegetable, and antioxidant intake with all-cause, cancer, and cardiovascular disease death. CLUE participants who donated a blood sample in 1974 and 1989 and completed a food frequency questionnaire in 1989 (N = 6,151) were included in the analysis. Participants were followed to date of death or January 1, 2002. Compared with those in the bottom fifth, participants in the highest fifth of fruit and vegetable intake had a lower risk of all-cause, cancer, and cardiovascular disease mortality. Higher intake of cruciferous vegetables was associated with lower risk of all-cause mortality. **No statistically significant associations were observed between dietary vitamin C, vitamin E, and beta-carotene intake and mortality.** Overall, **greater intake of fruits and vegetables was associated with lower risk of all-cause, cancer, and cardiovascular disease death**. These findings support the general health recommendation to consume multiple servings of fruits and vegetables (5-9/day).

Vitamin E and beta-carotene supplementation and hospital-treated pneumonia incidence in male smokers. (Hemila et al, 2004) (#29,133 men aged 50 to 69 years, who smoked at least five cigarettes per day)

Vitamin E and beta-carotene affect various measures of immune function and accordingly might influence the predisposition of humans to infections. However, only few controlled trials have tested this hypothesis. OBJECTIVE: To examine whether vitamin E or beta-carotene supplementation affects the risk of pneumonia in a controlled trial. DESIGN: The Alpha-Tocopherol Beta-Carotene Cancer Prevention (ATBC) study, a randomized, double-blind, placebo-controlled trial that examined the effects of vitamin E, 50 mg/d, and beta-carotene, 20 mg/d, on lung cancer using a 2 x 2 factorial design. The trial was conducted in the general community in southwestern Finland in 1985 to 1993; the intervention lasted for 6.1 years (median). The hypothesis being tested in the present study was formulated after the trial was closed. PARTICIPANTS: **ATBC study cohort of 29,133 men aged 50 to 69 years, who smoked at least five cigarettes per day**, at baseline. OUTCOME: The first occurrence of hospital-treated pneumonia was retrieved from the national hospital discharge register (898 cases). RESULTS: **Vitamin E supplementation had no overall effect on the incidence of pneumonia nor had beta-car-**

otene supplementation. Nevertheless, the age of smoking initiation was a highly significant modifying factor. **Among subjects who had initiated smoking at a later age** (> or =21 years; n = 7,469 with 196 pneumonia cases), **vitamin E supplementation decreased the risk of pneumonia**, whereas **beta-carotene supplementation increased the risk.** CONCLUSIONS: Data from this large controlled trial suggest that **vitamin E and beta-carotene supplementation have no overall effect on the risk of hospital-treated pneumonia in older male smokers**, but our subgroup finding that vitamin E seemed to benefit subjects who initiated smoking at a later age warrants further investigation.

Early Infant Multivitamin Supplementation Is Associated With Increased Risk for Food Allergy and Asthma (Milner et al. 2004) (#over 8,000) Dietary vitamins have potent immunomodulating effects in vitro. Individual vitamins have been shown to skew T cells toward either T-helper 1 or T-helper 2 phenotypic classes, suggesting that they may participate in inflammatory or allergic disease. With the exception of antioxidant protection, there has been little study on the effect of early vitamin supplementation on the subsequent risk for asthma and allergic disease. The objective of this study was to determine whether early vitamin supplementation during infancy affects the risk for asthma and allergic disease during early childhood. Methods: Cohort data were analyzed from the National Center for Health Statistics 1988 National Maternal-Infant Health Survey, which followed pregnant women and their newborns, and the 1991 Longitudinal Follow-up of the same patients, which measured health and disease outcomes. Patients were stratified by race and breastfeeding status. Factors that are known to be associated with alteration of risk for asthma or food allergies were identified using univariate logistic regression. Those factors were then analyzed in multivariate logistic regression models. Early vitamin supplementation was defined as vitamin use within the first 6 months. Results. There were >8000 total patients in the study. The overall incidence of asthma was 10.5% and of food allergy was 4.9%. In univariate analysis, male gender, smoker in the household, child care, prematurity (<37 weeks), being black, no history of breastfeeding, lower income, and lower education were associated with higher risk for asthma. Child care, higher levels of education, income, and history of breastfeeding were associated with a higher risk for food allergies. *In multivariate logistic analyses, a history of vitamin use within the first 6*

months of life was associated with a higher risk for asthma in black infants. Early vitamin use was also associated with a higher risk for food allergies in the exclusively formula-fed population. *Vitamin use at 3 years of age was associated with increased risk for food allergies* but not asthma in both breastfed and exclusively formula-fed infants. Conclusions. **Early vitamin supplementation is associated with increased risk for asthma in black children and food allergies in exclusively formula-fed children.** Additional study is warranted to examine which components most strongly contribute to this risk.

Dietary carotenoids and risk of lung cancer in a pooled analysis of seven cohort studies. (Mannisto et al, 2004) (#399,765 participants)

Intervention trials with supplemental beta-carotene have observed either no effect or a harmful effect on lung cancer risk. Because food composition databases for specific carotenoids have only become available recently, epidemiological evidence relating usual dietary levels of these carotenoids with lung cancer risk is limited. We analyzed the association between lung cancer risk and intakes of specific carotenoids using the primary data from seven cohort studies in North America and Europe. Carotenoid intakes were estimated from dietary questionnaires administered at baseline in each study. We calculated study-specific multivariate relative risks (RRs) and combined these using a random-effects model. The multivariate models included smoking history and other potential risk factors. During follow-up of up to 7-16 years across studies, 3,155 incident lung cancer cases were diagnosed among **399,765 participants. beta-Carotene intake was not associated with lung cancer risk** (pooled multivariate RR = 0.98; 95% confidence interval, 0.87-1.11; highest versus lowest quintile). **The RRs for alpha-carotene, lutein/zeaxanthin, and lycopene were also close to unity. beta-Cryptoxanthin intake was inversely associated with lung cancer risk** (RR = 0.76; 95% confidence interval, 0.67-0.86; highest versus lowest quintile). **These results did not change after adjustment for intakes of vitamin C (with or without supplements), folate (with or without supplements), and other carotenoids and multivitamin use.** The associations generally were similar among never, past, or current smokers and by histological type. Although smoking is the strongest risk factor for lung cancer, greater intake of foods high in beta-cryptoxanthin, such as citrus fruit, may modestly lower the risk. (Mannisto et al, 2004). 000

Antioxidants and cardiovascular disease: Still a topic of interest. (Nojiri et al, 2004)

Cardiovascular disease constitutes a major public health concern in industrialized nations. Over recent decades, a large body of evidence has accumulated indicating that free radicals play a critical role in cellular processes implicated in atherosclerosis. Herein, we present a mechanism of oxidative stress, focusing mainly on the development of an oxidized low density lipoprotein, and the results of a clinical trial of antioxidant therapy and epidemiological studies on the relationships between nutrient antioxidants, such as vitamin E, vitamin C, β-carotene, coenzyme Q, flavonoids and L-arginine, and coronary events. These studies indicated that a diet high in antioxidants is associated with a reduced risk of cardiovascular disease, but **did not confirm a strong causality link**. With regard to vitamin E, observational studies suggested that the daily use of at least 400 International Units of vitamin E is associated with beneficial effects on coronary events. However, it is apparently too early to define the clinical benefits of vitamin E for cardiovascular disease. **From the results of several randomized interventional trials, it appears that no single antioxidant given to subjects at high doses has substantial benefits**, and the question of whether nutrient antioxidants truly protect against cardiovascular disease remains open. (Nojiri et al, 2004) 000

Effects of Long-Term Daily Low-Dose Supplementation With Antioxidant Vitamins and Minerals on Structure and Function of Large Arteries. (Zureik et al, 2004) (#1,162) Lim-

ited data exist from randomized trials evaluating, noninvasively, the impact of antioxidant supplementation on vascular structure and function. *Methods and Results—* This is **a substudy of the SU.VI.MAX Study**, which is **a randomized, double-blind, placebo-controlled, cardiovascular and cancer primary prevention trial**. Eligible participants (free of symptomatic chronic diseases and apparently healthy) were randomly allocated to daily receive either **a combination of antioxidants (120 mg vitamin C, 30 mg vitamin E, 6 mg beta carotene, 100 μg selenium, and 20 mg zinc) or placebo** and followed-up over an average of **7.2±0.3 years**. At the end-trial examination, the carotid ultrasound examination and carotid–femoral pulse-wave velocity (PWV) measurement were performed blindly in **1,162 subjects** aged older than 50 years and

271

living in the Paris area. The percentage of subjects with carotid plaques was higher in the intervention group compared with the placebo group. Common carotid intima-media thickness (mean±SD) was not different between the 2 groups. Mean PWV tended to be lower (indicating less stiff aortic arteries) in the intervention group but the difference did not reach statistical significance (P=0.13). **Conclusion— These results suggest no beneficial effects of long-term daily low-dose supplementation of antioxidant vitamins and minerals on carotid atherosclerosis and arterial stiffness.** (Zureik et al, 2004) 000

Circa 2005

Use of multivitamins and prostate cancer mortality in a large cohort of US men. (Stevens et al, 2005) (#475,726 men who were cancer-free) Investigators assessed the association between the use of multivitamins and prostate cancer mortality. METHODS: A total of 5585 deaths from prostate cancer were identified during 18 years of follow-up of 475,726 men who were cancer-free and provided complete information on multivitamin use at enrollment in the Cancer Prevention Study II (CPS-II) cohort in 1982. Cox proportional hazards modeling was used to measure the association between multivitamin use at baseline and death from prostate cancer and to adjust for potential confounders. RESULTS: **The death rate from prostate cancer was marginally higher among men who took multivitamins regularly (> or =15 times/month) compared to non-users; this risk was statistically significant only for those multivitamin users who used no additional (vitamin A, C, or E) supplements.** In addition, risk was greatest during the initial four years of follow-up. CONCLUSIONS: *Regular multivitamin use was associated with a small increase in prostate cancer death rates* in our study, and this association was limited to a subgroup of users.

Vitamin A Supplementation for Reducing the Risk of Mother-to-child Transmission of HIV Infection: Cochrane systematic review 2005. (Wiysonge et al, 2005) (#3,033 females) Mother-to-child transmission (MTCT) of HIV is the dominant mode of acquisition of HIV infection for children, currently resulting in more than 2000 new pediatric HIV infections each day worldwide. Observational studies have found significant associations between low serum vitamin A levels and increased risk of MTCT of

HIV. We systematically reviewed currently available randomized controlled trials (RCTs) to evaluate the efficacy of vitamin A supplementation in preventing MTCT of HIV and other adverse pregnancy outcomes.

RCTs published between January 1980 and September 2005 were identified by searching the Cochrane Controlled Trials Register, PubMed, EMBASE, AIDSearch and conference proceedings, and contacting researchers. At least two authors independently assessed trial eligibility, and quality, and extracted data. We conducted meta-analysis using a fixed effects method, assessed heterogeneity between study results using the χ^2 test of homogeneity, and used Higgins' I^2 to quantify the heterogeneity.

Four trials, involving **3,033 participants**, met the inclusion criteria. There was no evidence of an effect of vitamin A supplementation on MTCT of HIV infection. However, there was evidence of heterogeneity between the three trials with information on MTCT of HIV. **While the trials conducted in South Africa and Malawi did not find evidence that the effect of Vitamin A supplementation was different from that of placebo, *the trial in Tanzania found evidence that vitamin A supplementation increased the risk of MTCT of HIV compared with placebo and multivitamins (excluding vitamin A).***

However, **synthesis of the currently available data does not show evidence of an effect of vitamin A supplementation on the risk of MTCT of HIV**, though there is an indication that vitamin A supplementation improves birth weight. The data suggest that the association between low serum vitamin A levels and increased risk of MTCT of HIV, seen in observational studies, could have alternative explanations, for example, low serum vitamin A levels may be a marker of advanced HIV infection and not causally related to MTCT of HIV.

The Alzheimer's Disease Cooperative Study (ADCS) Group
(Petersen et al, 2005) (#769 subjects); **Neither vitamin E nor donepezil delays progression from amnestic mild cognitive impairment to Alzheimer's disease** in **769** subjects over 3 years.

Vitamin E Supplementation in Alzheimer's Disease, Parkinson's Disease, Tardive Dyskinesia, and Cataract: Part 2 (Pham

273

et al, 2005); Using the MeSH terms alpha-tocopherol, tocopherols, vitamin E, Parkinson disease, tardive dyskinesia, Alzheimer disease, cataract, and clinical trials, a literature review was conducted to identify peer-reviewed articles in MEDLINE (1966-July 2005). RESULT: The clinical studies demonstrated contradicting results regarding the benefits of vitamin E in Parkinson's disease, tardive dyskinesia, and cataract. **There is enough evidence from large, well-designed studies to discourage the use of vitamin E in Parkinson's disease, cataract, and Alzheimer's disease.**

Dementia and Alzheimer's Disease in Community-Dwelling Elders Taking Vitamin C and/or Vitamin E: (Fillenbaum et al, 2005) (#626 elderly) a subgroup from the **Duke Established Populations for Epidemiologic Studies of the Elderly**; a longitudinal study. Information gathered during in-home interviews included sociodemographic characteristics, health status, health service use, and vitamin use. **Neither use of any vitamins C and/or E (used by 8% of subjects at baseline) nor high-dose use reduced the time to dementia or AD.** CONCLUSION: In this community in the southeastern US where vitamin supplement use is low, **use of vitamins C and/or E did not delay the incidence of dementia or AD.**

HOPE-TOO Extension (Lonn et al, 2005) (#3,994 original study enrollees) **The Heart Outcomes Prevention Evaluation (HOPE) investigators report an extension of the 9,541-patient HOPE Vitamin E trial. In the 2.5-year extension of HOPE (HOPE-TOO)** 174 of the original 267 centers continued an extended follow-up. From these centers, **3,994** of the 7,030 original study enrollees who were still alive elected to continue the randomized vitamin E/placebo drug assignment. **After a mean of 7.2 years of follow-up, vitamin E did not significantly reduce the relative risk (RR) of total cancer incidence, of cancer death, or a composite of cardiovascular events including cardiovascular death, nonfatal myocardial infarction, and stroke, or of individual components of this composite end point. These findings of lack of benefit from vitamin E (natural source, 400 IU** α-tocopheryl acetate) **during the extended study are consistent with the original HOPE report and with recent meta-analyses of** Vivekananthan and Miller. *Another subgroup finding in HOPE-TOO was a vitamin E–associated increased risk of heart failure incidence that appeared*

in a secondary end point analysis in the 4.5-year report and persisted in the 7-year extended follow-up, as did the risk of hospitalization for heart failure.

An increased risk of heart failure was associated with vitamin E supplementation in multiple analyses, including a 19% increased risk of all heart failure events and a 40% increase in the risk of hospital admission due to heart failure. CONCLUSION: **In patients with vascular disease or diabetes mellitus, long-term vitamin E supplementation does not prevent cancer or major cardiovascular events and may increase the risk for heart failure.**

Patients in *the vitamin E group had a higher risk of heart failure and hospitalization for heart failure*. Similarly, among patients enrolled at the centers participating in the HOPE-TOO trial, **there were no differences in cancer incidence, cancer deaths, and major cardiovascular events, but *higher rates of heart failure and hospitalizations for heart failure*.**

The National Cancer Institute (NCI) issued a statement following the HOPE-TOO study as follows: **"NCI has never recommended that people take vitamin E outside a clinical trial for the prevention of cancer."**

In 2005, **the evidence for a relationship between vitamin E and heart disease and selenium and cancer was reviewed by the U.S. FDA. It was determined that there was insufficient evidence to permit a qualified health claim for vitamin E and cancer,** whereas there was some evidence for permitting a qualified health claim for selenium and cancer. The FDA also concluded that the primary prevention studies did not provide evidence for the relationship between vitamin E and reduced risk of CVD (Trombo, 2005).

The British Medical Journal reported on 8/05/05 that **multivitamins and minerals do not prevent respirator, stomach, skin or other infections in the elderly.** An estimated 10% of elderly over 70 years are thought to have vitamin and mineral deficiencies, which may lead to poor immune responses and increased infections. Dr. Avenell studied 900 elderly and **found no benefit in preventing infections with the use of multivitamins and mineral supplements.**

Recent attempts to validate the teachings of the free radical theory, in the prevention of cardiovascular disease and cancer have failed, as demonstrated with vitamin E and beta carotene. Data on hundreds of thousands of patients have resulted in antioxidant failure and ineffectiveness in the prevention of disease. Moreover, previously unrecognized risks caused by nutrient toxicity and nutrient interactions have surfaced during intervention studies. (Lichtenstein and Russell, 2005).

Women's Health Study (WHS) (Lee et al, 2005) (#39,876 apparently healthy US women); "Vitamin E in the Primary Prevention of Cardiovascular Disease and Cancer" **600 IU of natural-source vitamin E taken every other day provided no overall benefit for major cardiovascular events or cancer, did not affect total mortality, and decreased cardiovascular mortality in healthy women.** These **data do not support recommending vitamin E supplementation for cardiovascular disease or cancer prevention among healthy women.**

Randomized controlled trials like the **WHS** are considered the gold standard in medical research and provide the most reliable results.

A randomized trial of antioxidant vitamins to prevent second primary cancers in head and neck cancer patients (Bairati et al, 2005 Apr 6) (#540 patients with stage I or II head and neck cancer treated by radiation therapy) 540 patients with stage I or II head and neck cancer treated by radiation therapy. Supplementation with alpha-tocopherol (400 IU/day) and beta-carotene (30 mg/day). In the course of the trial, **beta-carotene supplementation was discontinued after 156 patients had enrolled because of ethical concerns.** Compared with patients receiving placebo, *patients receiving alpha-tocopherol supplements had a higher rate of second primary cancers during the supplementation period* but a lower rate after supplementation was discontinued. Similarly, *the rate of having a recurrence or second primary cancer was higher during* but lower after supplementation with alpha-tocopherol. The proportion of participants free of second primary cancer overall after 8 years of follow-up was similar in both arms. CONCLUSIONS: *alpha-Tocopherol supplementation produced unexpected* *adverse effects on the occurrence of second primary cancers and on cancer-free survival.*

Note: Patients taking an antioxidant were 1.65 times more likely to suffer a return of their original cancer during the three years they were on the supplement. The risk was highest among those taking only vitamin E (1.86 times higher). Five years after they stopped taking the supplement, their recurrence risk had fallen to the same level as those in the placebo group. Although suggestive of harm, these results were not statistically significant.

Randomized trial of antioxidant vitamins to prevent acute adverse effects of radiation therapy in head and neck cancer patients (Bairati et al, 2005 Aug 20) (#540 patients with stage I or II head and neck cancer treated by radiation therapy)

A randomized trial was conducted to determine whether supplementation with antioxidant vitamins could reduce the occurrence and severity of acute adverse effects of radiation therapy and improve quality of life without compromising treatment efficacy. During the course of the trial, *supplementation with beta-carotene was discontinued because of ethical concerns. Quality of life was not improved by the supplementation. The rate of local recurrence of the head and neck tumor tended to be higher in the supplement arm of the trial.* CONCLUSION: Supplementation with high doses of alpha-tocopherol and beta-carotene during radiation therapy could reduce the severity of treatment adverse effects. However, **this trial suggests that use of high doses of antioxidants as adjuvant therapy might compromise radiation treatment efficacy.**

Note: Researchers were concerned to find that the rate of local recurrence (that is, a return of the original cancer) was 54 percent higher among patients on the combination pill than those on placebo. There was a smaller but still worrisome increase among those on vitamin E only.

NCI COMMENT: "This is a large, well-done study with good compliance from the participants," said Eva Szabo, M.D., of the National Cancer Institute's Division of Cancer Prevention. "The results demonstrate that the use of vitamin E supplementation is not beneficial to patients with stage I or II head and neck cancer, either as a chemoprevention agent or to enhance quality of life during radiation therapy."

At about this point in 2005, the followers of the free radical theory were scrambling to explain the repeated RCT failures. **Despite overwhelming evidence on the damaging consequences of oxidative stress and its role in experimental diabetes, large scale clinical trials with classic antioxidants failed to demonstrate any benefit for diabetic patients** (Johansen et al, 2005).

Effect of intensive lipid lowering, with or without antioxidant vitamins, compared with moderate lipid lowering on myocardial ischemia (Stone et al, 2005) (#300 patients with stable coronary disease); **randomized, double-blind, placebo-controlled trial; antioxidant vitamins C (1000 mg/d) and E (800 mg/d); Angina frequency decreased in each group. There was no incremental effect of supplemental vitamins C and E on any ischemia outcome. CONCLUSIONS: Intensive lipid lowering with atorvastatin to an LDL level of 80 mg/dL, with or without antioxidant vitamins, does not provide any further benefits in ambulatory ischemia, exercise time to onset of ischemia, and angina frequency than moderate lipid lowering with diet and low-dose lovastatin to an LDL level of <120 mg/dL.**

Vitamin C and vitamin E for Alzheimer's disease. (Boothby and Doering, 2005) To evaluate the literature on supplemental vitamin C and vitamin E therapy in the prevention and treatment of Alzheimer's disease (AD). DATA SOURCES: Literature retrieval was accessed through MEDLINE (1966-March 2005) using the key words antioxidants, vitamin C, vitamin E, Alzheimer's disease, and dementia. International Pharmaceutical Abstracts (1970-March 2005), Current Contents (1996-March 2005), Cochrane Database of Systematic Reviews (1994-March 2005), and Ebsco's Academic Search Elite (1975-March 2005) were searched with the same key words. STUDY SELECTION AND DATA EXTRACTION: Articles related to the objective that were identified through PubMed were included. DATA SYNTHESIS: Oral supplementation of vitamin C (ascorbic acid) and vitamin E (D-alfa-tocopherol acetate) alone and in combination have been shown to decrease oxidative DNA damage in animal studies in vivo, in vitro, and in situ. Recent results of a prospective observational study (n = 4,740) suggest that **the combined use of vitamin E 400 IU daily and vitamin C 500 mg daily for at least 3 years was associated with the**

reduction of AD prevalence and incidence. **Contradicting this is a previous prospective observational study (n = 980) evaluating the relationship between 4 years of vitamin C and E intake and the incidence of AD, which detected no difference in the incidence of AD during the 4-year follow-up. Recent meta-analysis results suggest that doses of vitamin E > or =400 IU daily for more than one year are associated with increased all-cause mortality. Mega-trial results suggest that vitamin E doses > or =400 IU daily for 6.9 years in patients with preexisting vascular disease or diabetes mellitus increase the incidence of heart failure, with no other outcome benefits noted.** CONCLUSIONS: In the absence of prospective, randomized, controlled clinical trials documenting benefits that outweigh recently documented morbidity and mortality risks, **vitamin E supplements should not be recommended for primary or secondary prevention of AD.** Although the risks of taking high doses of vitamin C are lower than those with vitamin E, the lack of consistent efficacy data for vitamin C in preventing or treating AD should discourage its routine use for this purpose.

An Interesting Case of "Supplement Nephropathy"

A 48-year-old Japanese woman previously in good health was found to have severe proximal tubular dysfunction with a high serum level of ascorbic acid (57.3 microg/ml, reference range: 1.9 - 15.0 microg/ml). Renal biopsy specimen showed marked tubulointerstitial damage, i.e. tubular atrophy, dilatation of tubular lumen with flattened tubular epithelial cells, vacuolization of proximal and distal tubular epithelial cells, and severe interstitial fibrosis with mild infiltration of mononuclear cells. Calcified lesions, which caused tubular obstruction or stenosis, were also seen in interstitial area adjacent to degenerated proximal tubuli. Hypokalemic nephropathy, probably due to long-term use of laxatives, was clearly shown. However, **calcified lesions seemed to be caused by inappropriate excessive daily ingestion of ascorbic acid (6,000 mg/day),** calcium lactate, and vitamin D because of the patient's misunderstanding that these supplements could keep her in a good health. This condition may be clinically called "supplement nephropathy" (Ohtake et al, 2005).

279

Effects of vitamins C and E on oxidative stress markers and endothelial function in patients with systemic lupus erythematosus: a double blind, placebo controlled pilot study. (Tam et al. 2005) (#39 patients with SLE). Effects of vitamins C and E on oxidative stress markers and endothelial function in patients with systemic lupus erythematosus: a double blind, placebo controlled pilot study. Patients with systemic lupus erythematosus **(SLE) experience excess morbidity and mortality due to coronary artery disease** (CAD) that cannot be fully explained by the classical CAD risk factors. Among emerging CAD risk factors, oxidative stress is currently being emphasized. They evaluated the effects of long term antioxidant vitamins on markers of oxidative stress and antioxidant defense and endothelial function in **39 patients with SLE.** METHODS: Patients were **randomized** to receive either placebo or vitamins (**500 mg vitamin C and 800 IU vitamin E daily**) for 12 weeks. Markers of oxidative stress included malondialdehyde (MDA) and allantoin. Antioxidants measured included erythrocyte superoxide dismutase and glutathione peroxidase, plasma total antioxidant power (as FRAP value), and ascorbic acid and vitamin E concentrations. Endothelial function was assessed by flow-mediated dilatation (FMD) of the brachial artery and plasma concentration of von Willebrand factor (vWF) and plasminogen activator inhibitor type 1 (PAI-1). Primary outcome of the study included the change in lipid peroxidation as revealed by MDA levels. Secondary outcomes included changes in allantoin and antioxidant levels and change in endothelial function.

RESULTS: After treatment, plasma ascorbic acid and alpha-tocopherol concentrations were significantly increased only in the vitamin-treated group, associated with a significant decrease in plasma MDA. Other oxidative stress markers and antioxidant levels remained unchanged in both groups. FMD and vWF and PAI-1 levels remained unchanged in both groups. CONCLUSION: **Combined administration of vitamins C and E was associated with decreased lipid peroxidation, but did not affect endothelial function in patients with SLE after 3 months of therapy.**

Antioxidants for preventing pre-eclampsia. (Rumbold et al, Apr 18, 2005. CD004072) (#35,812 women and 37,353 pregnancies) No difference was seen between women taking any vitamins compared with controls for total fetal loss, early or late miscarriage or stillbirth and most of the other primary outcomes. *Women supplemented with vitamin C alone or combined with other supplements were at increased risk of giving birth preterm.* **Taking vitamin supplements, alone or in combination with other vitamins, prior to pregnancy or in early pregnancy, does not prevent women experiencing miscarriage or stillbirth.**

Vitamin E supplementation in pregnancy. (Rumbold et al, Apr 18, 2005. CD004072) (#566 women).

No difference was found between women supplemented with vitamin E in combination with other supplements during pregnancy compared with placebo for the risk of stillbirth, neonatal death, perinatal death, preterm birth, intrauterine growth restriction or birthweight, using fixed-effect models. **The data are too few to say if vitamin E supplementation either alone or in combination with other supplements is beneficial during pregnancy.**

Antioxidant vitamin supplementation in the prevention of cardiovascular disease (they are not recommended). (Yuen et al. 2005).

Oxidative stress, in particular oxidative modification of LDL-cholesterol, appears to be of great importance in the pathogenesis of atherosclerosis. Various observational epidemiological studies have suggested that antioxidant vitamin intake is associated with reduced cardiovascular morbidity and mortality. Also, experimental studies in animals have demonstrated that antioxidant vitamins slow the progression of atherosclerosis. However, **prospective controlled clinical trials have failed to demonstrate a benefit of antioxidant vitamin supplementation in primary or secondary prevention of cardiovascular disease.** Thus, **the use of antioxidants and vitamin supplements as a preventive or therapeutic intervention can not be recommended.**

Effect of multivitamin and multimineral supplements on morbidity from infections in older people (MAVIS trial) (Avenell et al. 2005) (910 men and women 65 or over)

To examine whether supplementation with multivitamins and multiminerals influences self reported days of infection, use of health services, and quality of life in people aged 65 or over. **Design** Randomized, placebo controlled trial, with blinding of participants, outcome assessors, and investigators. **Setting** Communities associated with six general practices in Grampian, Scotland. **Participants 910 men and women** aged 65 or over who did not take vitamins or minerals. **Interventions** Daily multivitamin and multimineral supplementation or placebo for one year. **Main outcome measures** Primary outcomes were contacts with primary care for infections, self reported days of infection, and quality of life. Secondary outcomes included antibiotic prescriptions, hospital admissions, adverse events, and compliance. **Results** Supplementation did not significantly affect contacts with primary care and days of infection per person. Quality of life was not affected by supplementation. No statistically significant findings were found for secondary outcomes or subgroups. **Conclusion. Multivitamins offered no protection from infection for patients 65 and older. Routine multivitamin and multimineral supplementation of older people living at home does not affect self reported infection related morbidity.**

Vitamin K3 triggers human leukemia cell death through hydrogen peroxide generation and histone hyperacetylation =

(Lin, Kang, Zheng, 2005) (#) Vitamin K3 (VK3) is a well-known anticancer agent, but its mechanism remains elusive. In the present study, VK3 was found to simultaneously induce cell death, reactive oxygen species (ROS) generation, including superoxide anion (O_2^{*-}) and hydrogen peroxide (H_2O_2) generation, and histone hyperacetylation in **human leukemia HL-60 cells** in a concentration- and time-dependent manner. Catalase (CAT), an antioxidant enzyme that specifically scavenges H_2O_2, could significantly diminish both histone acetylation increase and cell death caused by VK3, whereas superoxide dismutase (SOD), an enzyme that specifically eliminates O_2^{*-}, showed no effect on both of these, leading to the conclusion that H_2O_2 generation, but not O_2^{*-} generation, contributes to VK3-induced histone hyperacetylation and cell death. This conclusion was confirmed by the finding that enhancement of VK3-induced H_2O_2 generation by vitamin C (VC) could significantly promote both the histone hyperacetylation and cell death. Further studies suggested that histone hyperacetylation played an important role in VK3-induced cell death, since sodium butyrate, a histone deacetylase (HDAC) inhibitor, showed no effect on ROS generation, but obviously potentiated VK3-induced histone hyperacetylation and cell death. Collectively, **these results demonstrate a novel mechanism for the anticancer activity of VK3, i.e.,**

VK3 induced tumor cell death through H2O2 generation, which then further induced histone hyperacetylation. (Lin et al, 2005).

Fruits and vegetables and ovarian cancer risk in a pooled analysis of 12 cohort studies. (Koushik et al, 2005) (#560,441 women) Because fruits and vegetables are rich in bioactive compounds with potential cancer-preventive actions, increased consumption may reduce the risk of ovarian cancer. **Evidence on the association between fruit and vegetable intake and ovarian cancer risk has not been consistent.** We analyzed and pooled the primary data from 12 prospective studies in North America and Europe. Fruit and vegetable intake was measured at baseline in each study using a validated food-frequency questionnaire. To summarize the association between fruit and vegetable intake and ovarian cancer, study-specific relative risks (RR) were estimated using the Cox proportional hazards model, and then combined using a random-effects model. Among **560,441 women**, 2,130 cases of invasive epithelial ovarian cancer occurred during a maximum follow-up of 7 to 22 years across studies. **Total fruit intake was not associated with ovarian cancer risk**-the pooled multivariate RR for the highest versus the lowest quartile of intake was 1.06 [95% confidence interval (95% CI), 0.92-1.21; P value, test for trend = 0.73; P value, test for between-studies heterogeneity = 0.74]. Similarly, results for total vegetable intake indicated no significant association (pooled multivariate RR, 0.90; 95% CI, 0.78-1.04, for the highest versus the lowest quartile; P value, test for trend = 0.06; P value, test for between-studies heterogeneity = 0.31). Intakes of botanically defined fruit and vegetable groups and individual fruits and vegetables were also not associated with ovarian cancer risk. Associations for total fruits and vegetables were similar for different histologic types. These results suggest that **fruit and vegetable consumption in adulthood has no important association with the risk of ovarian cancer.** (Koushik et al, 2005). 000

Perioperative N-acetylcysteine to prevent renal dysfunction in high-risk patients undergoing cabg surgery: a randomized controlled trial (Burns KE, et al. 2005) (#295 patients required elective or urgent CABG) Renal dysfunction is a complication of coronary artery bypass graft (CABG) surgery performed with cardiopulmonary bypass (CPB) that is associated with increased morbidity and mortality.

N-acetylcysteine, an antioxidant and vasodilator, counteracts renal ischemia and hypoxia. OBJECTIVE: To determine whether perioperative intravenous (IV) N-acetylcysteine preserves renal function in high-risk patients undergoing CABG surgery with CPB compared with placebo. DESIGN, SETTING, AND PATIENTS: Randomized, quadruple blind, placebo-controlled trial (October 2003-September 2004) in operating rooms and general intensive care units (ICUs) of 2 Ontario tertiary care centers. The **295 patients required elective or urgent CABG** and had at least 1 of the following: preexisting renal dysfunction, at least 70 years old, diabetes mellitus, impaired left ventricular function, or undergoing concomitant valve or redo surgery. INTERVENTIONS: Patients received 4 (2 intraoperative and 2 postoperative) doses of IV N-acetylcysteine (600 mg) (n = 148) or placebo (n = 147) over 24 hours. MAIN OUTCOME MEASURES: The primary outcome was the proportion of patients developing postoperative renal dysfunction, defined by an increase in serum creatinine level greater than 0.5 mg/dL (44 micromol/L) or a 25% increase from baseline within the first 5 postoperative days. Secondary outcomes included postoperative interventions and complications, the requirement for renal replacement therapy (RRT), adverse events, hospital mortality, and ICU and hospital length of stay. RESULTS: There was no difference in the proportion of patients with postoperative renal dysfunction in the N-acetylcysteine and placebo groups, respectively. We noted nonsignificant differences in postoperative interventions and complications, the need for RRT and serious adverse events, hospital mortality, and ICU and hospital length of stay between the N-acetylcysteine and placebo groups. A post hoc subgroup analysis of patients (baseline creatinine level >1.4 mg/dL [120 micromol/L]) showed a nonsignificant trend toward fewer patients experiencing postoperative renal dysfunction in the N-acetylcysteine group compared with the placebo group. CONCLUSIONS: **N-acetylcysteine did not prevent postoperative renal dysfunction, interventions, complications, or mortality in high-risk patients undergoing CABG surgery with CPB.** Further research is required to identify CABG patients at risk for postoperative renal events, valid markers of renal dysfunction, and to establish renal thresholds associated with important clinical outcomes. Division of Critical Care Medicine, University of Western Ontario, London, Ontario, Canada.

Circa 2006

Vitamin-mineral supplementation and the progression of atherosclerosis (Bleys et al, 2006) (searched the MEDLINE, EMBASE, and CENTRAL databases); a meta-analysis of randomized controlled trials; meta-analysis showed **no evidence of a protective effect of antioxidants** (vitamins E and C, ß-carotene, or selenium) **or B vitamin supplements on the progression of atherosclerosis.**

Also in 2006, there was "The Efficacy and Safety of Multivitamin and Mineral Supplement Use To Prevent Cancer and Chronic Disease in Adults: A Systematic Review for a National Institutes of Health State-of-the-Science Conference by Huang et al. They concluded that the **evidence was insufficient to prove the presence or absence of benefits from use of multivitamin and mineral supplements to prevent cancer and chronic disease** (Huang et al, 2006).

Multivitamin/mineral supplements and prevention of chronic disease. (Huang et al, 2006 May) OBJECTIVES: To review and synthesize published literature on the efficacy of multivitamin/mineral supplements and certain single nutrient supplements in the primary prevention of chronic disease in the general adult population, and on the safety of multivitamin/mineral supplements and certain single nutrient supplements, likely to be included in multivitamin/mineral supplements, in the general population of adults and children. DATA SOURCES: All articles published through February 28, 2006, on MEDLINE, EMBASE, and the Cochrane databases. REVIEW METHODS: Each article underwent double reviews on title, abstract, and inclusion eligibility. Two reviewers performed data abstraction and quality assessment. **Differences in opinion were resolved through consensus adjudication.** RESULTS: Few trials have addressed the efficacy of multivitamin/mineral supplement use in chronic disease prevention in the general population of the United States. One trial on poorly nourished Chinese showed supplementation with combined Beta-carotene, vitamin E and selenium reduced gastric cancer incidence and mortality, and overall cancer mortality. In a French trial, combined vitamin C, vitamin E, Beta-carotene, selenium, and zinc reduced cancer risk in men but not in women. No cardiovascular benefit was evident in both trials. **Multivitamin/mineral supplement use had no benefit for preventing cataract.** Zinc/antioxidants had benefits for preventing advanced age-related macular degeneration in persons at high risk for the disease. **With few exceptions, neither Beta-carotene nor vitamin E had benefits for preventing cancer, cardiovascular disease, cataract, and age-related macular degeneration.** *Beta-carotene supplementation increased lung cancer risk in smokers and persons exposed to asbestos.* **Folic acid alone or combined with vitamin B12 and/or vitamin B6 had no significant effects on cognitive function.** Selenium may confer benefit for cancer prevention but not cardiovascular disease prevention. Calcium may prevent bone mineral density loss in postmenopausal women, and may reduce vertebral fractures, but not non-vertebral fractures. The evidence suggests dose-dependent benefits of vitamin D with/without calcium for retaining bone mineral density and preventing hip fracture, non-vertebral fracture and falls. **We found no consistent pattern of increased adverse effects of multivitamin/mineral supplements except for skin yellowing by Beta-carotene.** CONCLUSIONS: Multivitamin/mineral supplement use may prevent cancer in individuals with poor or suboptimal nutritional status. The heterogeneity in the study populations limits generalization to United States population. **Multivitamin/mineral supplements conferred no benefit in preventing cardiovascular disease or cataract**, and may prevent advanced age-related macular degeneration only in high-risk individuals. The overall quality and quantity of the literature on the safety of multivitamin/mineral supplements is limited.

The Efficacy and Safety of Multivitamin and Mineral Supplement Use To Prevent Cancer and Chronic Disease in Adults: A Systematic Review for a National Institutes of Health State-of-the-Science Conference. (Huang et al, 2006 Sept)

BACKGROUND: Multivitamin and mineral supplements are the most commonly used dietary supplements in the United States. PURPOSE: To synthesize studies on the efficacy and safety of multivitamin/mineral supplement use in primary prevention of cancer and chronic disease in the general population. DATA SOURCES: English-language literature search of the MEDLINE, EMBASE, and Cochrane databases through February 2006 and hand-searching of pertinent journals and articles. STUDY SELECTION: **Randomized, controlled trials in adults were reviewed to assess efficacy, and randomized, controlled trials and observational studies in adults or children were reviewed to**

assess safety. DATA EXTRACTION: Paired reviewers extracted data and independently assessed study quality. DATA SYNTHESIS: 12 articles from 5 randomized, controlled trials that assessed efficacy and 8 articles from 4 randomized, controlled trials and 3 case reports on adverse effects were identified. Study quality was rated fair for the studies on cancer, cardiovascular disease, cataracts, or age-related macular degeneration and poor for the studies on hypertension. **In a poorly nourished Chinese population, combined supplementation with beta-carotene, alpha-tocopherol, and selenium reduced the incidence of and mortality rate from gastric cancer and the overall mortality rate from cancer by 13% to 21%. In a French trial, combined supplementation with vitamin C, vitamin E, beta-carotene, selenium, and zinc reduced the rate of cancer by 31% in men but not in women. Multivitamin and mineral supplements had no significant effect on cardiovascular disease or cataracts**, except that combined beta-carotene, selenium, alpha-tocopherol, retinol, and zinc supplementation reduced the mortality rate from stroke by 29% in the Linxian study and that a combination of 7 vitamins and minerals stabilized visual acuity loss in a small trial. **Combined zinc and antioxidants slowed the progression of advanced age-related macular degeneration in high-risk persons.** No consistent adverse effects of multivitamin and mineral supplements were evident. LIMITATIONS: Only randomized, controlled trials were considered for efficacy assessment. Special nutritional needs, such as use of folic acid by pregnant women to prevent birth defects, were not addressed. Findings may not apply to use of commercial multivitamin supplements by the general U.S. population. CONCLUSIONS: **Evidence is insufficient to prove the presence or absence of benefits from use of multivitamin and mineral supplements to prevent cancer and chronic disease.**

Antioxidants Vitamin C and Vitamin E for the Prevention and Treatment of Cancer (Coulter et al, 2006) (Thirty-eight studies); **The systematic review of the literature does not support the hypothesis that the use of supplements of vitamin C or vitamin E in the doses tested helps prevent and/or treat cancer in the populations tested. The findings from randomized clinical trials were generally negative.**

Vitamin C levels in Type 2 diabetes and low vitamin C levels does not improve endothelial dysfunction or insulin resistance (Chen et al, 2006) (#32 type 2 diabetics); randomized, double-blind, placebo-controlled study of vitamin C (800 mg/day for 4 wk; CONCLUSION: **high-dose oral vitamin C therapy**, resulting in incomplete replenishment of vitamin C levels, is **ineffective at improving endothelial dysfunction and insulin resistance in Type 2 diabetes.**

Meta-analysis: antioxidant supplements for primary and secondary prevention of colorectal adenoma (2006) (Bjelakovic et al., 2006) (#17,620 participants); eight randomized trials (**17,620 participants**). **Neither fixed-effect nor random-effect model meta-analyses showed statistically significant effects of supplementation with beta-carotene, vitamins A, C, E and selenium alone or in combination.** They found no convincing evidence that antioxidant supplements have significant beneficial effect on primary or secondary prevention of colorectal adenoma.

Australian Collaborative Trial of Supplements (ACTS) (Rumbold et al, 2006) (#1,877 pregnant women); Supplementation with vitamin C (1,000 mg) and vitamin E (400 IU) **during pregnancy does not reduce the risk of pre-eclampsia in nulliparous women, the risk of intrauterine growth restriction, or the risk of death or other serious outcomes in their infants. Certain adverse maternal outcomes were more common in the antioxidant group than in the placebo group, but these findings could be explained by chance alone.**

This is **the second major report in a month to find no apparent benefit for the supplements in preventing pre-eclampsia.** These results followed closely similar findings reported in the March 30, 2006 issue of The Lancet by researchers at Kings College London. In that **randomized study of high-risk women, not only did these vitamins fail to reduce the pre-eclampsia risk,** but *the rate of low-birth-weight babies was higher and the rate for gestational hypertension was higher for women in the vitamin group.*

There were no significant differences between the groups for death or serious outcomes among the infants. The vitamins were not associated

with any primary benefits for the infants. **Although vitamin therapy reduced the risk of respiratory distress syndrome in the infants, the downside of the therapy for the women was an increased risk of hospitalization for hypertension and the use of anti-hypertensive medication.**

Of concern, there was a downside for the women taking vitamins. *Women in the vitamin group had an increased risk of being hospitalized antenatally for hypertension and having to take antihypertensive medication.* In addition, *a subgroup of women in the vitamin group had a higher frequency of abnormal liver-function tests.*

"Until more data are available, given the scant evidence of benefit and the potential for harm, supplemental antioxidant therapy for the prevention of pre-eclampsia should be limited to women enrolled in randomized trials and should not be prescribed as part of routine practice."

The Antioxidants in Prevention of Cataracts Study (APC Study): effects of antioxidant supplements on cataract progression in South India. (Gritz et al, 2006) (#798 subjects) (supplementation with beta carotene, vitamins C and E) To determine if antioxidant supplements (beta carotene and vitamins C and E) can decrease the progression of cataract in rural South India. METHODS: The **Antioxidants in Prevention of Cataracts (APC) Study** was a 5 year, randomized, triple masked, placebo controlled, field based clinical trial to assess the ability of interventional antioxidant supplements to slow cataract progression. The primary outcome variable was change in nuclear opalescence over time. Secondary outcome variables were cortical and posterior subcapsular opacities and nuclear color changes; best corrected visual acuity change; myopic shift; and failure of treatment. Annual examinations were performed for each subject by three examiners, in a masked fashion. Multivariate modeling using a general estimating equation was used for analysis of results, correcting for multiple measurements over time. RESULTS: Initial enrolment was **798 subjects.** Treatment groups were comparable at baseline. There was high compliance with follow up and study medications. **There was progression in cataracts. There was no significant difference between placebo and active treatment groups**

for either the primary or secondary outcome variables. CONCLU-
SION: **Antioxidant supplementation with beta carotene, vitamins C
and E did not affect cataract progression in a population with a high
prevalence of cataract whose diet is generally deficient in antioxidants.**

Vitamins in Pre-eclampsia (VIP) Trial Consortium (Poston
et al., 2006) (#2,410 women at increased risk for preeclampsia, analayzed 2,395) 1000 mg of vitamin C and 400 IU vitamin E; Concomitant supplementation with vitamin C and vitamin E does not prevent pre-eclampsia in
women at risk, but does *increase the rate of babies born with a low birth
weight.* "As such, use of these high-dose antioxidants is not justified in pregnancy." Thus, vitamin C and vitamin E supplementation are not helpful for woman
with preeclampsia and may be harmful to the fetus

Drs Lindheimer and Sibai write. **"It appears surprising that administration
of vitamins in amounts that did not exceed maximum daily tolerable
allowances should be associated with adverse effects**, but the VIP data are
not the first time that vitamin or nutrient supplements administered to populations already consuming considerable amounts of the supplement's content in
their diets have had adverse effects.

**SU.VI.MAX (2006) Antioxidants do not affect fasting blood
glucose** (Czernichow et al, 2006) (#3,146 subjects); a daily capsule containing 120 mg vitamin C, 30 mg vitamin E, 6 mg ß-carotene, 100 µg Se, and 20 mg
Zn or a placebo. Observational data suggest a protective effect of several antioxidants on fasting plasma glucose (FPG) and type 2 diabetes. However, randomized
trials have yielded inconsistent results. **Supplementation with antioxidants
at nutritional doses for 7.5 y had no effect on FPG** in men or women who
followed a balanced diet. An inverse association of baseline ß-carotene dietary
intake and plasma concentrations with FPG was found, probably because ß-carotene is an indirect marker of fruit and vegetable intakes.

Full title for Czernichow et al. 2006 article: Antioxidant supplementation does
not affect fasting plasma glucose in the Supplementation with Antioxidant Vitamins and Minerals (SU.VI.MAX) study in France: association with dietary intake
and plasma concentrations.

290

Vitamin E and Risk of Type 2 Diabetes in the Women's Health Study (Liu et al., 2006) (#38,716 apparently healthy U.S. women); efficacy of vitamin E supplements for primary prevention of type 2 diabetes among apparently healthy women. **In this large trial with 10-year follow-up, alternate-day doses of 600 IU vitamin E provided no significant benefit for type 2 diabetes in initially healthy women.**

Because of repeated failures of antioxidant vitamins, the Harvard School of Public Health posted the following: **"The evidence accumulated so far isn't promising. Randomized trials of vitamin C, vitamin E, and beta-carotene haven't revealed much in the way of protection from heart disease, cancer, or aging-related eye diseases." (Harvard School of Public Health, website accessed 2/09/06).**

Vitamin E supplementation and cognitive function in women: The Women's Health Study (2006) (Kang et al., 2006) (#39,876 healthy US women); **There were no differences in global score between the vitamin E and placebo groups at the first assessment (5.6 years** after randomization) **or at the last assessment (9.6 years** of treatment). CONCLUSION: **Long-term use of vitamin E supplements did not provide cognitive benefits among generally healthy older women.**

Supplemental and dietary vitamin E, beta-carotene, and vitamin C intakes and prostate cancer risk (PLCO Trial) (Kirsh et al, 2006) (#29,361 men during up to 8 years of follow-up); the screening arm of the Prostate, Lung, Colorectal, and Ovarian Cancer Screening Trial; Overall, **there was no association between prostate cancer risk and dietary or supplemental intake of vitamin E, beta-carotene, or vitamin C. CONCLUSIONS: Our results do not provide strong support for population-wide implementation of high-dose antioxidant supplementation for the prevention of prostate cancer.** However, vitamin E supplementation in male smokers and beta-carotene supplementation in men with low dietary beta-carotene intakes were associated with reduced risk of this disease.

Intakes of vitamins A, C and E and folate and multivitamins and lung cancer: a pooled analysis of 8 prospective studies. (Cho et al, 2006) (#430,281 persons over a maximum of 6-16 years in the studies) Intakes of vitamins A, C and E and folate have been hypothesized to reduce lung cancer risk. We examined these associations in a pooled analysis of the primary data from 8 prospective studies from North America and Europe. Baseline vitamin intake was assessed using a validated food-frequency questionnaire, in each study. We calculated study-specific associations and pooled them using a random-effects model. During follow-up of 430,281 persons over a maximum of 6-16 years in the studies, 3,206 incident lung cancer cases were documented. Vitamin intakes were inversely associated with lung cancer risk in age-adjusted analyses; the associations were greatly attenuated after adjusting for smoking and other risk factors for lung cancer. The pooled multivariate relative risks, comparing the highest vs. lowest quintile of intake from food-only, were 0.96 for vitamin A, 0.80 for vitamin C, 0.86 for vitamin E and 0.88 for folate. The association with vitamin C was not independent of our previously reported inverse association with beta-cryptoxanthin. Further, **vitamin intakes from foods plus supplements were not associated with a reduced risk of lung cancer in multivariate analyses, and use of multivitamins and specific vitamin supplements was not significantly associated with lung cancer risk.** The results generally did not differ across studies or by sex, smoking habits and lung cancer cell type. In conclusion, **these data do not support the hypothesis that intakes of vitamins A, C and E and folate reduce lung cancer risk.**

In discussing the role of micronutrients, Shenkin said, "Micronutrients play a central role in metabolism and in the maintenance of tissue function, but effects in preventing or treating disease which is not due to micronutrient deficiency cannot be expected from increasing the intake. **Provision of excess supplements to individuals who do not need them may be harmful.** Clinical benefit is most likely in those individuals who are severely depleted and at risk of complications, and is unlikely if this is not the case. Much more research is needed to characterize better markers of micronutrient status both in terms of metabolic effects and antioxidant effects (Shenkin, 2006).

The Melbourne Atherosclerosis Vitamin E Trial (MAVET): a study of high dose vitamin E in smokers. (Magliano et al, 2006)

(#409 male and female smokers) Their aim was to evaluate whether vitamin E (500 IU) slowed the progression of carotid atherosclerosis in a population of chronic smokers over 4 years as measured by ultrasound determination of carotid intima-media thickness (IMT) and systemic arterial compliance (SAC). Methods: The Melbourne Atherosclerosis Vitamin E Trial (MAVET) was **a randomized, double-blind, placebo-controlled trial** in which **409 male and female smokers** aged 55 years and over were randomized to receive 500 IU per day of natural vitamin E or placebo. The primary endpoint was progression of carotid atherosclerosis determined by intima-media thickness of the right common carotid artery. Secondary outcomes were change in systemic arterial compliance and low-density lipoprotein (LDL) oxidative susceptibility over time. Results: *The mean increase in intima-media thickness over time in the vitamin E group was 0.0041 mm/year faster than placebo* (95% confidence interval -0.0021 to 0.0102 mm/year, $P=0.20$). Similarly, a non-significant difference between vitamin E and placebo was found for rate of change in systemic arterial compliance ($P=0.11$). Vitamin E supplementation did, however, significantly reduce LDL oxidative susceptibility ($P<0.001$). Conclusion: **Vitamin E supplementation is ineffective in reducing the progression of carotid atherosclerosis as measured by intima-media thickness in chronic smokers. This finding extends our knowledge of lack of effectiveness of vitamin E supplementation in populations with high oxidant stress.**

An interesting overview was offered by Siekmeier and Marz in 2006 in the following abstract: **Can antioxidants prevent atherosclerosis?**

In vitro studies have shown that antioxidants (e. g. beta-carotene, vitamin C and vitamin E) can interfere with some pathomechanisms of atherosclerosis and therefore might have a protective effect. From the investigated antioxidants vitamin E showed the best effect. Some animal and epidemiological studies confirmed such a protective effect in vivo especially after administration of high doses of vitamin E. However, **most of the placebo-controlled studies for primary or secondary prevention failed to show a protective effect even after administration of high doses. In addition, other studies demonstrated a risk for adverse effects due to antioxidant supplementation (beta-carotene and vitamin E).** Our review summarizes the principle of antioxidant supplementation and a number of relevant epidemiological and clinical studies for prevention of atherosclerosis. The obtained results suggest that **supplementation of antioxidants cannot be recommended for the normal population** (Siekmeier and Marz, 2006).

In 2007, they said," The protective effect of antioxidants on atherosclerotic pathomechanisms has been confirmed in vitro, but only in some animal studies. Various epidemiological and observational studies have produced conflicting results on the protective effect of antioxidants. **Most studies of primary or secondary prevention failed to show a protective effect of antioxidants against atherosclerosis** (Siekmeier and Marz, 2007).

Dietary supplementation with different vitamin C doses: no effect on oxidative DNA damage in healthy people. (Herbert et al. 2006) (#160 volunteers). Antioxidants are believed to prevent many types of disease. Some previous studies suggest that dietary supplementation with vitamin C results in a decrease in the level of one of the markers of oxidative damage-8-oxoguanine in the DNA of peripheral blood mononuclear cells (PBMC). AIM OF TRIAL: To investigate the effect of different dose levels of dietary supplementation with vitamin C on oxidative DNA damage. METHODS: A **randomized double-blind placebo-controlled trial** was carried out using three different levels (80, 200 and 400 mg) of dietary vitamin C supplementation in a healthy population of **160 volunteers**; supplementation was for a period of 15 weeks followed by a 10 week washout period. Peripheral blood samples were obtained every 5 weeks from baseline to 25 weeks. RESULTS: An increase in PBMC vitamin C levels was not observed following supplementation in healthy volunteers. **There was no effect found on 8-oxoguanine measured using HPLC with electrochemical detection for any of the three supplemented groups compared to placebo.** 8-oxoadenine levels were below the limit of detection of the HPLC system used here. CONCLUSIONS: **Supplementation with vitamin C had little effect on cellular levels in this group of healthy individuals**, suggesting their diets were replete in vitamin C. The dose range of vitamin C used did not affect oxidative damage in PBMC DNA.

Carotenoids and cardiovascular health American Journal of Clinical Nutrition. (Voutilainen et al. 2006) (#not applicable)
Cardiovascular disease (CVD) is the main cause of death in Western countries.

Nutrition has a significant role in the prevention of many chronic diseases such as CVD, cancers, and degenerative brain diseases. The major risk and protective factors in the diet are well recognized, but interesting new candidates continue to appear. It is well known that a greater intake of fruit and vegetables can help prevent heart diseases and mortality. **Because fruit, berries, and vegetables are chemically complex foods, it is difficult to pinpoint any single nutrient that contributes the most to the cardioprotective effects.** Several potential components that are found in fruit, berries, and vegetables are probably involved in the protective effects against CVD. Potential beneficial substances include antioxidant vitamins, folate, fiber, and potassium. Antioxidant compounds found in fruit and vegetables, such as vitamin C, carotenoids, and flavonoids, may influence the risk of CVD by preventing the oxidation of cholesterol in arteries. In this review, the role of main dietary carotenoids, ie, lycopene, beta-carotene, alpha-carotene, beta-cryptoxanthin, lutein, and zeaxanthin, in the prevention of heart diseases is discussed. Although it is clear that a higher intake of fruit and vegetables can help prevent the morbidity and mortality associated with heart diseases, **more information is needed to ascertain the association between the intake of single nutrients, such as carotenoids, and the risk of CVD. Currently, the consumption of carotenoids in pharmaceutical forms for the treatment or prevention of heart diseases cannot be recommended.**

Vitamins C and E and the risks of preeclampsia and perinatal complications (Rumbold et al, 2006) (#1,877) Supplementation with antioxidant vitamins has been proposed to reduce the risk of preeclampsia and perinatal complications, but the effects of this intervention are uncertain. METHODS: They conducted a multicenter, **randomized trial of nulliparous women between 14 and 22 weeks of gestation**. Women were assigned to daily supplementation with 1000 mg of vitamin C and 400 IU of vitamin E or placebo (microcrystalline cellulose) until delivery. Primary outcomes were the risks of maternal preeclampsia, death or serious outcomes in the infants (on the basis of definitions used by the Australian and New Zealand Neonatal Network), and delivering an infant whose birth weight was below the 10th percentile for gestational age. RESULTS: Of the **1,877 women enrolled in the study, 935 were randomly assigned to the vitamin group and 942 to the placebo group (total - 1,877)**. Baseline characteristics of the two groups were similar. **There were no significant differences between the vitamin and**

placebo groups in the risk of preeclampsia, death or serious outcomes in the infant, or having an infant with a birth weight below the 10th percentile for gestational age.

CONCLUSIONS: Supplementation with vitamins C and E during pregnancy does not reduce the risk of preeclampsia in nulliparous women, the risk of intrauterine growth restriction, or the risk of death or other serious outcomes in their infants.

Smoking, alcohol drinking, green tea consumption and the risk of esophageal cancer in Japanese men. (Ishikawa et al, 2006) (#9,008 men in Cohort 1 and 17,715 men in Cohort 2) Although smoking and alcohol drinking are established risk factors of esophageal cancer, their public health impact is unclear. Furthermore, the effect of green tea is controversial. **METHODS:** The present study was based on a pooled analysis of two prospective cohort studies. A self-administered questionnaire about health habits was distributed to **9,008 men in Cohort 1 and 17,715 men in Cohort 2**, aged 40 years or older, with no previous history of cancer. We identified 38 and 40 patient cases with esophageal cancer among the subjects in Cohort 1 (9.0 years of follow-up) and Cohort 2 (7.6 years of follow-up), respectively. Cox proportional hazards regression was used to estimate hazard ratios (HRs) of the risk of esophageal cancer incidence.

RESULTS: *Cigarette smoking, alcohol drinking and green tea consumption were significantly associated with an increased risk of esophageal cancer.* Compared with men who had never smoked, never drunk alcohol or green tea, the pooled multivariate HRs for men who were currently smoking > or =20 cigarettes/day, drinking alcohol daily, or drinking > or =5 cups green tea/day, respectively. *The population attributable fractions of esophageal cancer incidence that was attributable to smoking, alcohol drinking and green tea consumption were 72.0%, 48.6%, and 22.1%, respectively.*

CONCLUSIONS: Among the variables studied, *smoking has the largest public health impact on esophageal cancer incidence in Japanese men, followed by alcohol drinking and green tea drinking* (Ishikawa et al, 2006).

Intake of major carotenoids and the risk of epithelial ovarian cancer in a pooled analysis of 10 cohort studies. (Koushik et al, 2006) (#521,911 women)

Carotenoids, found in fruits and vegetables, have the potential to protect against cancer because of their properties, including their functions as precursors to vitamin A and as antioxidants. We examined the associations between intakes **of alpha-carotene, beta-carotene, beta-cryptoxanthin, lutein/zeaxanthin and lycopene and the risk of invasive epithelial ovarian cancer.** The primary data from 10 prospective cohort studies in North America and Europe were analyzed and then pooled. Carotenoid intakes were estimated from a validated food frequency questionnaire administered at baseline in each study. Study-specific relative risks (RR) were estimated using the Cox proportional hazards model and then combined using a random-effects model. Among 521,911 women, 2,012 cases of ovarian cancer occurred during a follow-up of 7-22 years across studies. **The major carotenoids were not significantly associated with the risk of ovarian cancer.** The pooled multivariate RRs (95% confidence intervals) were 1.00 (0.95-1.05) for a 600 microg/day increase in alpha-carotene intake, 0.96 for a 2,500 microg/day increase in beta-carotene intake, 0.99 for a 100 microg/day increase in beta-cryptoxanthin intake, 0.98 for a 2,500 microg/day increase in lutein/zeaxanthin intake and 1.01 for a 4,000 microg/day increase in lycopene intake. These associations did not appreciably differ by study (p-values, tests for between-studies heterogeneity >0.17). Also, the observed associations did not vary substantially by subgroups of the population or by histological type of ovarian cancer. **These results suggest that consumption of the major carotenoids during adulthood does not play a major role in the incidence of ovarian cancer.** (Koushik et al, 2006). 000

Circa 2007

Mortality in Randomized Trials of Antioxidant Supplements for Primary and Secondary Prevention; Systematic Review and Meta-analysis (Bjelakovic et al, 2007) (#232,606 participants);
68 randomized trials with 232,606 participants (385 publications. In 47 low-bias trials with 180,938 participants, the antioxidant supplements significantly increased mortality. In low-bias risk trials, after exclusion of selenium trials, **beta**

297

carotene, vitamin A, and vitamin E, singly or combined, significantly increased mortality. Vitamin C and selenium had no significant effect on mortality. Bjelakovic's analysis found no evidence that taking beta-carotene, vitamin A or vitamin E extends life span. Conservatively, *the supplements increase the likelihood of dying by about 5 percent. When looked at separately, they found that Vitamin A increased death risk by 16 per cent, beta carotene by 7 per cent and Vitamin E by 4 per cent.*

Vitamin C gave contradictory results, but when given singly or in combination with other vitamins in good-quality trials, increased the death rate by 6 per cent. Selenium was the only supplement to emerge with any credit. It appears to cut death rates by 10 per cent when given on its own or with other supplements in high-quality trials, but the result is **not statistically significant**.

In 2007, Dr. Kristine Yaffe of the San Francisco Veterans Affairs Medical Center and University of California at San Francisco stated, **"For the clinician, there is no convincing justification to recommend the use of antioxidant dietary supplements to maintain cognitive performance in cognitively normal adults or in those with mild cognitive impairment." (Nov. 2007 Archives of Internal Medicine)**

Multivitamin Use and Risk of Prostate Cancer in the National Institutes of Health–AARP Diet and Health Study (Lawson et al, 2007) (#295,344 men); investigated the association between multivitamin use and prostate cancer risk; *use of multivitamins more than seven times per week, when compared with never use, was associated with a doubling in the risk of fatal prostate cancer* (The study of Lawson et al. is observational, and therefore confounding by indication and other confounding cannot be excluded. But the sample studied is very large, which reduces random errors, and the study seems well conducted.) **According to Victoria Stevens of the American Cancer Society, "There certainly is no evidence in healthy, relatively well-nourished people that vitamins or anti-oxidants protect against chronic diseases."**

The results are in accord with the results of systematic reviews and meta-analyses of randomized clinical trials (Vivekananthan et al., 2003)

(Bjelakovic et al, Cochrane Database Syst Rev. 2004) (Stevens et al, 2005) (Bjela-kovic et al, 2007) (Caraballoso et al., 2003).

This study of Lawson et al is so disturbing that I have expanded it in the section, "Where are the bodies?"

A Randomized Factorial Trial of Vitamins C and E and Beta Carotene in the Secondary Prevention of Cardiovascular Events in Women: Results From the Women's Antioxidant Cardiovascular Study (Cook et al. 2007) (#8,171 female health professionals at increased risk); ascorbic acid (500 mg/d), vitamin E (600 IU every other day), and beta carotene (50 mg every other day) on the combined outcome of myocardial infarction, stroke, coronary revascularization, or CVD. **There was no overall effect of ascorbic acid, vitamin E, or beta carotene on the primary combined end point or on the individual secondary outcomes of myocardial infarction, stroke, coronary revascularization, or CVD death. There were no overall effects of ascorbic acid, vitamin E, or beta carotene on cardiovascular events among women at high risk for CVD.**

The Women's Antioxidant Cardiovascular Study (WACS) examined the effects of vitamins C and E and beta carotene, as well as their combinations, in a randomized factorial trial among women at increased risk of vascular events. There were no significant effects on the primary end point of total cardiovascular disease, suggesting that **widespread use of these agents for cardiovascular disease prevention does not appear warranted.**

Use of Supplements of Multivitamins, Vitamin C, and Vitamin E in Relation to Mortality (Pocobelli et al, 2007) (#77,719 subjects aged 50–76 years) Washington State residents aged 50–76 years who completed a mailed self-administered questionnaire in 2000–2002. **Multivitamin use was not related to total mortality.** However, **vitamin C and vitamin E use were associated with small decreases in risk.** In cause-specific analyses, use of multivitamins and use of vitamin E were associated with decreased risks of CVD mortality. In contrast, **vitamin C use was not associated with CVD mortality. Multivitamin and vitamin E use were not**

associated with cancer mortality. Some of the associations we observed were small and may have been due to unmeasured healthy behaviors that were more common in supplement users.

Health Professionals Follow-up Study (2007): Effect of vitamins C, E, A and carotenoids and the occurrence of oral pre-malignant lesions (Maserejian et al, 2007) (#42,340 men enrolled in the Health Professionals Follow-up Study) (#207 found with oral premalignant lesions); researchers found no clear relationship with beta-carotene, lycopene, or lutein/zeaxanthin. *A trend for increased risk of oral pre-malignant lesions was observed with vitamin E, especially among current smokers and with vitamin E supplements. Beta-carotene also increased the risk among current smokers.* However, **dietary vitamin C was significantly associated with a reduced risk of oral premalignant lesions:** those with the highest intake had a 50 percent reduction in risk compared to those with the lowest intake.

Antioxidant meta-analysis for the treatment of macular degeneration (2007) (Chong et al, 2007) (#149,203 subjects); Investigators looked at 11 studies which included **149,203 people** - seven were prospective studies and three were randomized controlled trials and found that vitamins A and E, zinc, lutein, zeaxanthin, α- carotene, β-carotene, β-cryptoxanthin and lycopene have **either a very slight or no effect** on primary prevention of early age-related macular degeneration. The researchers concluded that **there is not enough evidence to support the role of dietary antioxidants, including the use of dietary antioxidant supplements, for the primary prevention of early AMD.**

Effect of *RRR*-α-tocopherol supplementation on carotid atherosclerosis in patients with stable coronary artery disease (CAD) (Devaraj et al, 2007) (#90 patients with CAD); **No significant difference was observed in the mean change in total carotid IMT in the placebo and alpha-tocopherol groups.** In addition, **no significant difference in cardiovascular events was observed. Conclusions:** High-dose *RRR*-α-tocopherol supplementation in patients with CAD was safe and signifi-

cantly reduced plasma biomarkers of oxidative stress and inflammation but **had no significant effect on carotid IMT during 2 years**.

Overview of the Women's Antioxidant Cardiovascular Study (WACS) (2007) (Zaharris et al, 2007) (#8,171 women); ascorbic acid (500 mg/d), vitamin E (600 IU every other day), and beta carotene (50 mg every other day); **There was no overall effect of ascorbic acid, vitamin E, or beta carotene on the primary combined end point or on the individual secondary outcomes of myocardial infarction, stroke, coronary revascularization, or CVD death.** But those randomized to both active ascorbic acid and vitamin E experienced fewer strokes. **Conclusion: There were no overall effects of ascorbic acid, vitamin E, or beta carotene on cardiovascular events among women at high risk for CVD.**

Serum alpha-tocopherol, concurrent and past vitamin E intake, and mild cognitive impairment (Dunn et al, 2007) (#526 subjects) 526 participants in a single-site ancillary study to the Women's Health Initiative, the Cognitive Change in Women study; In bivariate analyses, neither past dietary vitamin E intake (<8 mg/day vs more) nor current vitamin E supplement use was associated with impairment. CONCLUSION: **There was weak or no evidence of a protective effect of previous vitamin E intake on cognitive function.**

The role of vitamin E in the prevention of cancer: a meta-analysis of randomized controlled trials. (Alkhenizan and Hafez, 2007) (#167,025 subjects) There are conflicting results in published randomized controlled trials (RCTs) on the role of vitamin E in the prevention of cancer. We conducted a meta-analysis of RCTs to evaluate the role of vitamin E in the prevention of cancer in adults. METHODS: We included RCTs in which the outcomes of the intake of vitamin E supplement alone or with other supplements were compared to a control group. The primary outcomes were total mortality, cancer mortality, total incidence of cancer, and incidence of lung, stomach, esophageal, pancreatic, prostate, breast and thyroid cancers. All identified trials were reviewed independently by the two reviewers to determine whether trials should be included or excluded. The quality of all included studies was scored independently by the two reviewers. RESULTS: Twelve studies, which included **167,025**

participants, met the inclusion criteria. There were no statistically significant differences in total mortality among the different groups of patients included in this meta-analysis. **Vitamin E was associated with a significant reduction in the incidence of prostate cancer, but it did not reduce the incidence of any other types of cancer.** CONCLUSIONS: **Vitamin E supplementation was not associated with a reduction in total mortality, cancer incidence, or cancer mortality,** but it was associated with a statistically significant reduction in the incidence of prostate cancer. Vitamin E can be used in the prevention of prostate cancer in men who are at high risk of prostate cancer. **This recently published meta-analysis of 12 randomized controlled trials concluded that vitamin E supplementation was not associated with overall cancer incidence, cancer mortality, or total mortality.**

Chemoprevention of Primary Liver Cancer: A Randomized, Double-Blind Trial in Linxian, China. (Qu et al, 2007) (#29,450 initially healthy adults) Primary liver cancer is a common malignancy with a dismal prognosis. New primary prevention strategies are needed to reduce mortality from this disease. They examined the effects of supplementation with four different combinations of vitamins and minerals on primary liver cancer mortality among **29,450** initially healthy adults from Linxian, China. Methods: Participants were randomly assigned to take either a vitamin–mineral combination ("factor") or a placebo daily for 5.25 years (March 1986–May 1991). Four factors (at doses one to two times the US Recommended Daily Allowance)—retinol and zinc (factor A); riboflavin and niacin (factor B); ascorbic acid and molybdenum (factor C); and beta-carotene, alpha-tocopherol, and selenium (factor D)—were tested in a partial factorial design. The study outcome was primary liver cancer death occurring from 1986 through 2001. Adjusted Cox proportional hazards models were used to calculate hazard ratios (HRs) and 95% confidence intervals (CIs) of liver cancer death with and without each factor. All P values are two-sided. Results: A total of 151 liver cancer deaths occurred during the analysis period. **No statistically significant differences in liver cancer mortality were found comparing the presence and absence of any of the four intervention factors.** However, **both factor A and factor B reduced liver cancer mortality in individuals younger than 55 years at randomization but not in older individuals.** Factor C reduced liver cancer death, albeit with only borderline statistical significance in males but not in females. Cumulative risks of liver cancer death were 6.0 per 1000 in the placebo arm, 5.4 per 1000 in the arms with two factors, and 2.4 per 1000 in the arm with all four factors. Conclusion: **None of the factors tested reduced overall liver cancer mortality.** However, three factors reduced liver cancer mortality in certain subgroups.

As of 2007, no study had yet provided conclusive evidence of the beneficial effect of antioxidant supplementation in critically ill patients. The clinical evidence provided so far showed that there are several factors which might determine the efficacy of antioxidant supplementation in critically ill patients. There may be a need for large multi-center prospective randomized control trials to assess the effects of different types and doses of antioxidant supplementation in selected groups of patients with different types of critical illness. In critical illness, overwhelming inflammatory mediator response to infective or non-infective stimuli results in excessive production of free radicals (Mishra, 2007).

Risk of Mortality with Vitamin E Supplements: The Cache County Study. (Hayden et al, 2007) A recent meta-analysis reported increased mortality in clinical trial participants randomized to high-dose vitamin E. We sought to determine whether these mortality risks with vitamin E reflect adverse consequences of its use in the presence of cardiovascular disease. Methods: In a defined population aged 65 years or older, baseline interviews captured self- or proxy-reported history of cardiovascular illness. A medicine cabinet inventory verified nutritional supplement and medication use. Three sources identified subsequent deaths. Cox proportional hazards methods examined the association between vitamin E use and mortality. Results: After adjustment for age and sex, there was no association in this population between vitamin E use and mortality. Predictably, deaths were more frequent with a history of diabetes, stroke, coronary artery bypass graft surgery, or myocardial infarction, and with the use of warfarin, nitrates, or diuretics. None of these conditions or treatments altered the null main effect with vitamin E, but *mortality was increased in vitamin E users who had a history of stroke, coronary bypass graft surgery, or myocardial infarction and, independently, in those taking nitrates, warfarin, or diuretics.*

Although not definitive, a consistent trend toward reduced mortality was seen in vitamin E users without these conditions or treatments. Conclusions: In this

303

population-based study, **vitamin E use was unrelated to mortality, but this apparently null finding seems to represent a combination of increased mortality in those with severe cardiovascular disease and a possible protective effect in those without.**

Multivitamin-multimineral supplements and eye disease: age-related macular degeneration and cataract. (Seddon, 2007)

This **Dietary Ancillary Study of the Eye Disease Case-Control Study (EDCCS)** was designed to evaluate the relation between nutrition and AMD. Conclusion: **a multivitamin-multimineral supplement with a combination of vitamin C, vitamin E, β-carotene, and zinc (with cupric oxide) is recommended for age-related macular degeneration (AMD) but not cataract.** Observational studies for cataract provide only weak support for multivitamins or other vitamin supplements. The results of observational studies suggest that a healthy lifestyle with a diet containing foods rich in antioxidants, especially lutein and zeaxanthin, and n–3 fatty acids appears beneficial for AMD and possibly cataract.

As regarded multivitamins, investigators were becoming aware that **the influence that some dietary supplements, especially multivitamin-multimineral supplements, may have on the occurrence of chronic diseases was largely unknown, but there was a potential for adverse effects associated with their use** (Mulholland and Benford, 2007).

Antioxidant Supplementation Increases the Risk of Skin Cancers in Women but Not in Men. (Hercberg et al, 2007) (#French adults, 7,876 women and 5,141 men. Total # = 13,017) This research aimed to test whether supplementation with a combination of antioxidant vitamins and minerals could reduce the risk of skin cancers (SC). It was performed within the framework of the **Supplementation in Vitamins and Mineral Antioxidants study, a randomized, double-blinded, placebo-controlled, primary prevention trial** testing the efficacy of nutritional doses of antioxidants in reducing incidence of cancer and ischemic heart disease in the general population. French adults (7876 women and 5141 men) were randomized to take an oral daily capsule of antioxidants (120 mg vitamin C, 30 mg vitamin E, 6 mg β-carotene, 100 μg selenium,

and 20 mg zinc) or a matching placebo. The median time of follow-up was 7.5 y. A total of 157 cases of all types of SC were reported, from which 25 were melanomas. Because the effect of antioxidants on SC incidence varied according to gender, men and women were analyzed separately. *In women, the incidence of SC was higher in the antioxidant group* [adjusted hazard ratio (adjusted HR) = 1.68; P = 0.03]. **Conversely, in men, incidence did not differ between the 2 treatment groups.** Despite the small number of events, *the incidence of melanoma was also higher in the antioxidant group for women.* The incidence of nonmelanoma SC did not differ between the antioxidant and placebo groups. Our findings suggest that antioxidant supplementation affects the incidence of SC differentially in men and women, with *a 4-fold higher melanoma risk in women — but not in men — who used nutritionally appropriate doses of antioxidant supplements.* (Hercberg et al, 2007)

Antioxidant Vitamin Supplement Use and Risk of Dementia or Alzheimer's Disease in Older Adults (Gray et al, 2007) (#2,969)

Investigators examined whether use of vitamins C or E alone or in combination was associated with lower incidence of dementia or Alzheimer's disease (AD). Prospective cohort study. SETTING: Group Health Cooperative, Seattle, Washington. PARTICIPANTS: **Two thousand nine hundred sixty-nine participants** aged 65 and older without cognitive impairment at baseline in the Adult Changes in Thought study. MEASUREMENTS: Participants were followed biennially to identify incident dementia and AD diagnosed according to standard criteria. Participants were considered to be users of vitamins C or E if they self-reported use for at least 1 week during the month before baseline. RESULTS: Over a mean follow-up±standard deviation of 5.5±2.7 years, 405 subjects developed dementia (289 developed AD). The use of vitamin E was not associated with dementia or with AD. No association was found between vitamin C alone or concurrent use of vitamin C and E and either outcome. CONCLUSION: In this study, **the use of supplemental vitamin E and C, alone or in combination, did not reduce risk of AD or overall dementia over 5.5 years of follow-up.**

Beta carotene supplementation and age-related maculopathy in a randomized trial of US physicians (#22,071

apparently healthy US male physicians) (Christen et al. 2007). Investigators tested whether beta carotene supplementation affects the incidence of age-related maculopathy (ARM) in a large-scale randomized trial. **DESIGN:** Randomized, double-masked, placebo-controlled trial among **22,071 apparently healthy US male physicians** aged 40 to 84 years. Participants were randomly assigned to receive beta carotene (50 mg every other day) or placebo. Main Outcome Measure: Incident ARM responsible for a reduction in best-corrected visual acuity to 20/30 or worse. **RESULTS:** After 12 years of treatment and follow-up, there were 162 cases of ARM in the beta carotene group vs 170 cases in the placebo group. The results were similar for the secondary end points of ARM with or without vision loss and advanced ARM. **CONCLUSIONS:** These randomized data relative to 12 years of treatment among a large population of **apparently healthy men indicate that beta carotene supplementation has no beneficial or harmful effect on the incidence of ARM. Long-term supplemental use of beta carotene neither decreases nor increases the risk of ARM**.

Atherosclerosis and oxidant stress: the end of the road for antioxidant vitamin treatment? (Thomson et al. 2007) Experimental data, however, have not translated into clinical benefit: most antioxidant vitamin trials have failed to reduce cardiovascular morbidity and mortality. Moreover, recent clinical trials have suggested that **mono-therapy with certain antioxidant vitamins like vitamin E may, in fact, be detrimental.** As a result of the disappointing outcome of 'antioxidant' vitamin trials, some authors have questioned both the utility of 'antioxidant' treatment in CVD and the supposedly central role of oxidative stress in atherogenesis. **The clinical promise of antioxidant vitamins has failed to translate into clinical benefit.**

Effects of antioxidant supplementation on the aging process. (Fusco et al. 2007). Even if antioxidant supplementation is receiving growing attention and is increasingly adopted in Western countries, supporting evidence is still scarce and equivocal.

Effect of high-dose alpha-tocopherol supplementation on biomarkers of oxidative stress and inflammation and carotid atherosclerosis in patients with coronary artery disease.

(Devaraj et al. 2007) (#90 patients with CAD). Oxidative stress and inflammation are crucial in atherogenesis. alpha-Tocopherol is both an antioxidant and an antiinflammatory agent. OBJECTIVE: They evaluated the effect of RRR-alpha-tocopherol supplementation on carotid atherosclerosis in patients with stable coronary artery disease (CAD) on drug therapy. DESIGN: **Randomized, controlled, double-blind trial** compared RRR-alpha-tocopherol (1200 IU/d for 2 y) with placebo in **90 patients with CAD**. Intimal medial thickness (IMT) of both carotid arteries was measured by high-resolution B-mode ultrasonography at 0, 1, 1.5, and 2 y. At 6-mo intervals, plasma alpha-tocopherol concentrations, C-reactive protein (CRP), LDL oxidation, monocyte function (superoxide anion release, cytokine release, and adhesion to endothelium), and urinary F(2)-isoprostanes were measured. RESULTS: alpha-Tocopherol concentrations were significantly higher in the alpha-tocopherol group but not in the placebo group. High-sensitivity CRP concentrations were significantly lowered with alpha-tocopherol supplementation than with placebo. alpha-Tocopherol supplementation significantly reduced urinary F(2)-isoprostanes and monocyte superoxide anion and tumor necrosis factor release compared with baseline and placebo. **No significant difference was observed in the mean change in total carotid IMT in the placebo and alpha-tocopherol groups. In addition, no significant difference in cardiovascular events was observed.** CONCLUSIONS: **High-dose RRR-alpha-tocopherol supplementation in patients with CAD was safe and significantly reduced plasma biomarkers of oxidative stress and inflammation but had no significant effect on carotid IMT during 2 years.**

Multivitamins do not improve radiation therapy-related fatigue: results of a double-blind randomized crossover trial. (de Souza et al. 2007) (randomized 40 patients to either placebo or Centrum Silver) Fatigue is a common symptom in cancer patients receiving radiation therapy. PATIENTS AND METHODS: They conducted a **double-blind randomized crossover trial** of multivitamins versus placebo in patients with breast cancer undergoing radiation therapy to evaluate fatigue and quality of life. RESULTS: We **randomized 40 patients to either placebo or Centrum Silver**. At the middle of the radiation treatments, patients were switched from placebo to multivitamins and vice versa. Patients answered the EORTC QLQ C-30 quality of life (QOL) and Chalder fatigue questionnaires at the beginning, middle, and end of radiation therapy. Both groups experienced decreases in general and physical fatigue scores at the end of the course of placebo compared with the assessment prior to this treatment. They also observed significant improvements in functional and symptoms score scales of the QOL questionnaire in the patients on placebo. **No significant changes were elicited with the use of multivitamins.** They also observed significantly lower rates of fatigue in the patients who had just finished a course of placebo as compared with patients finishing a course of multivitamins. CONCLUSION: **Multivitamins do not improve radiation-related fatigue in patients with breast cancer.**

Combined vitamin C and E supplementation during pregnancy for preeclampsia prevention: a systematic review (Polyzos et al, 2007) (#4,680 pregnant women)

The effect of combined vitamin C and E supplementation during pregnancy on the prevention of preeclampsia and major adverse infant outcomes has been reviewed. We searched MEDLINE and the Central Library of Controlled Trials of the Cochrane Library through August 2006 for relevant clinical trials. Interstudy heterogeneity was evaluated using the chi(2) statistic (Q statistic) test. Pooled relative risks (RRs) and 95% confidence intervals (CIs) were calculated with a fixed or random-effects model as appropriate. Four trials that **collectively randomized 4680 pregnant women to either the combination of vitamin C and vitamin E or placebo were included in the analysis.** There were no significant differences between the vitamin and placebo groups in the risk of preeclampsia, 11% versus 11.4%, fetal or neonatal loss, 2.6% versus 2.3%, or small for gestational age (SGA) infant, 20.6% versus 20%. Although **there was a higher risk for preterm birth in the vitamin group, 19.5% versus 18%, this finding was not significant. Combined vitamin C and E supplementation during pregnancy does not reduce the risk of preeclampsia, fetal or neonatal loss, small for gestational age infant, or preterm birth. Such supplementation should be discouraged unless solid supporting data from randomized trials become available.** MEDLINE analysis of the literature questions the use of vitamin C and E supplements.

Antioxidant therapy to prevent preeclampsia: a randomized controlled trial (Spinnato et al. 2007) (#739)

Investigators studied whether antioxidant supplementation will reduce the incidence of preeclampsia among patients at increased risk. METHODS: A **randomized, placebo-**

controlled, double-blind clinical trial was conducted at four Brazilian sites. Women between 12 0/7 weeks and 19 6/7 weeks of gestation and diagnosed to have chronic hypertension or a prior history of preeclampsia were randomly assigned to daily treatment with both vitamin C (1,000 mg) and vitamin E (400 International Units) or placebo. Analyses were adjusted for clinical site and risk group (prior preeclampsia, chronic hypertension, or both). A sample size of 734 would provide 80% power to detect a 40% reduction in the risk of preeclampsia, assuming a placebo group rate of 21% and alpha=.05. The alpha level for the final analysis, adjusted for interim looks, was 0.0458. RESULTS: **Outcome data for 707 of 739 randomly assigned patients revealed no significant reduction in the rate of preeclampsia** compared with placebo. **There were no differences in mean gestational age at delivery or rates of perinatal mortality, abruptio placentae, preterm delivery, and small for gestational age or low birth weight infants.** *Among patients without chronic hypertension, there was a slightly higher rate of severe preeclampsia in the study group.* CONCLUSION: **This trial failed to demonstrate a benefit of antioxidant supplementation in reducing the rate of preeclampsia among patients with chronic hypertension and/or prior preeclampsia.**

The effect of vitamin E on blood pressure in individuals with type 2 diabetes: a randomized, double-blind, placebo-controlled trial (Ward et al, 2007) (#58 with type 2 diabetes)

Oxidative stress has been suggested to play a role in the development of diabetes, hypertension and vascular dysfunction. Vitamin E, a major lipid-soluble dietary antioxidant, has two major dietary forms, alpha-tocopherol and gamma-tocopherol. The potential importance of gamma-tocopherol has largely been overlooked. Our aim was to investigate the effect of alpha-tocopherol and gamma-tocopherol supplementation on 24-h ambulatory blood pressure (BP) and heart rate, vascular function and oxidative stress in individuals with type 2 diabetes. METHOD: **Fifty-eight individuals with type 2 diabetes were randomized in a double-blind, placebo-controlled trial. Participants were randomized to a daily dose of 500 mg/day RRR-alpha-tocopherol, 500 mg/day mixed tocopherols (60% gamma-tocopherol) or placebo for 6 weeks.** Primary endpoints were 24-h ambulatory BP and heart rate, endothelium-dependent and independent vasodilation and plasma and urinary F2-isoprostanes. RESULTS: Treatment with **alpha-tocopherol significantly increased systolic BP, diastolic BP,** pulse pressure and heart rate versus placebo. **Treatment with mixed tocopherols significantly increased systolic BP, diastolic BP, pulse pressure and heart rate versus placebo. Treatment with alpha-tocopherol or mixed tocopherols significantly reduced plasma F2-isoprostanes versus placebo,** but had no effect on urinary F2-isoprostanes. Endothelium-dependent and independent vasodilation was not affected by either treatment. CONCLUSION: **In contrast to our initial hypothesis,** *treatment with either alpha- or mixed tocopherols significantly increased BP, pulse pressure and heart rate in individuals with type 2 diabetes.*

Natonal Institutes of Health State-of-the-Science Conference Statement: Multivitamin/Mineral Supplements and Chronic Disease Prevention (NIH State-of-the Science Panel. 2007).

=== At least half of American adults take a dietary supplement, the majority of which are multivitamin/multimineral (MVM) supplements. As more and more Americans seek strategies for maintaining good health and preventing disease, and as the marketplace offers an increasing number of products to fulfill that desire, **it is important that consumers have the best possible information to make their choices.** Assessing the available scientific evidence on the benefits of MVM supplement use for chronic disease prevention, identifying the gaps in the evidence, and recommending an appropriate research agenda to meet the shortfalls are subjects considered in this report.

Most people assume that the ingredients in MVM supplements are safe. There is evidence, however, that certain ingredients in MVM supplements can produce adverse effects, including **reports from RCTs that noted excess lung cancer occurring in asbestos workers and smokers consuming β-carotene.** In addition, *esophageal cancer excess was found with long-term follow-up of older Chinese patients (the Linxian study by Blot et al.) treated with selenium, β-carotene, and vitamin E supplements* (Blot et al, 1993) (NIH State-of-the Science Panel. 2007).

Tumor-targeted induction of oxystress for cancer therapy

== (Fang, Nakamura, Iyer, 2007) (#) Reactive oxygen species (ROS), such as superoxide anion radicals (O_2^{*-}) and hydrogen peroxide (H_2O_2) are potentially harmful by-products of normal cellular metabolism that directly affect cellular functions. ROS is generated by all aerobic organisms and it seems to be **indispensable for signal transduction pathways that regulate cell growth and reduction-oxidation (redox) status.** However, overproduction of these highly reactive oxygen metabolites can initiate lethal chain reactions, which involve oxidation and damage to structures that are crucial for cellular integrity and survival. In fact, **many antitumor agents, such as vinblastine, cisplatin, mitomycin C, doxorubicin, camptothecin, inostamycin, neocarzinostatin and many others exhibit antitumor activity via (EMOD) ROS-dependent activation of apoptotic cell death, suggesting potential use of ROS as an antitumor principle. Thus, a unique anticancer strategy named "oxidation therapy" has been developed by inducing cytotoxic oxystress for cancer treatment. This goal could be achieved mainly by two methods, namely, (i) inducing the generation of ROS directly to solid tumors and (ii) inhibiting the antioxidative enzyme (defense) system of tumor cells.** Since 1950s, many strategies have been employed based on the first method, namely, administration of ROS per se (e.g. H_2O_2) or ROS generating enzyme to tumor bearing animals. However no successful and practical results were obtained probably because of the lack of tumor selective ROS delivery and hence resulting in subsequent induction of severe side effects. To overcome these obstacles, we developed polyethylene glycol (PEG) conjugated O_2^*- or H_2O_2-generating enzymes, xanthine oxidase (XO) and D-amino acid oxidase (DAO) (PEG-DAO) respectively. More recently, a pegylated (PEG) zinc protoporphyrin (PEG-ZnPP) and a highly water soluble micellar formulation of ZnPP based on amphiphilic styrene maleic acid (SMA) copolymer, SMA-ZnPP, are prepared, which are potent inhibitors of heme oxygenase-1 (HO-1). **HO-1 is a major antioxidative enzyme of tumors, that is different in mechanism of catalase or superoxide dismutase (SOD).** Consequently, both PEG-enzymes and PEG-ZnPP exhibited superior in vivo pharmacokinetics than their parental molecules, particularly in tumor delivery by taking advantage of the EPR effect of macromolecular nature, and thus showed remarkable antitumor effects suggesting the potentials of this anticancer therapeutic for clinical application. Furthermore, **it has been well known that many antioxidative enzymes such as catalase, SOD are down-regulated in most solid tumors in vivo.** On the contrary, **HO-1 is highly upregulated and it plays a very important role of antioxidation, because HO-1 generates biliverdin, which being converted to bilirubin exhibits a very potent antioxidative effect, and**

hence antiapoptosis in tumors. Thus this oxidation therapy, by inhibiting this HO-I dependent antioxidant (bilirubin) formation by ZnPP, and by enhancing ROS generation, is expected to offer a powerful therapeutic modality for future anti-cancer therapy. (Fang, Nakamura, Iyer, 2007).

I believe that a sufficiency or surfeit of EMODs should be referred to as "prooxidant sufficiency" or "prooxidant bliss" and avoidance of the old inaccurate term of "oxidative stress."

Diet and risk of ovarian cancer in the California Teachers Study cohort. 000 (Chang et al, 2007) (#97,275) Dietary phyto-chemical compounds, including isoflavones and isothiocyanates, may inhibit cancer development but have not yet been examined in prospective epide-miologic studies of ovarian cancer. The authors have investigated the association between consumption of these and other nutrients and ovarian cancer risk in a prospective cohort study. Among **97,275 eligible women in the Califor-nia Teachers Study cohort** who completed the baseline dietary assessment in 1995-1996, 280 women developed invasive or borderline ovarian cancer by December 31, 2003. Multivariable Cox proportional hazards regression, with age as the timescale, was used to estimate relative risks and 95% confidence inter-vals; all statistical tests were two sided. **Intake of isoflavones was associated with lower risk of ovarian cancer.** Compared with the risk for women who consumed less than 1 mg of total isoflavones per day, the relative risk of ovar-ian cancer associated with consumption of more than 3 mg/day was 0.56 (95% confidence interval: 0.33, 0.96). **Intake of isothiocyanates or foods high in isothiocyanates was not associated with ovarian cancer risk, nor was intake of macronutrients, antioxidant vitamins, or other micronutri-ents. Although dietary consumption of isoflavones may be associated with decreased ovarian cancer risk, most dietary factors are unlikely to play a major role in ovarian cancer development** (Chang et al, 2007). 000

Dietary carotenoids and risk of colorectal cancer in a pooled analysis of II cohort studies. (Mannisto et al, 2007) (#702,647 participants) Dietary carotenoids have been hypothesized to protect against epithelial cancers. The authors analyzed the associations between intakes of spe-cific carotenoids (alpha-carotene, beta-carotene, beta-cryptoxanthin, lutein +

zeaxanthin, and lycopene) and risk of colorectal cancer using the primary data from 11 cohort studies carried out in North America and Europe. Carotenoid intakes were estimated from food frequency questionnaires administered at baseline in each study. During 6-20 years of follow-up between 1980 and 2003, 7,885 incident cases of colorectal cancer were diagnosed among 702,647 participants. The authors calculated study-specific multivariate relative risks and then combined them using a random-effects model. In general, intakes of specific carotenoids were not associated with colorectal cancer risk. The pooled multivariate relative risks of colorectal cancer comparing the highest quintile of intake with the lowest ranged from 0.92 for lutein + zeaxanthin to 1.04 for lycopene; only for lutein + zeaxanthin intake was the result borderline statistically significant (95% confidence interval: 0.84, 1.00). The associations observed were generally similar across studies, for both sexes, and for colon cancer and rectal cancer. **These pooled data did not suggest that carotenoids play an important role in the etiology of colorectal cancer.** (Mannisto et al, 2007). 000

Phase II, randomized, controlled trial of high-dose N-acetylcysteine in high-risk cardiac surgery patients (Haase M, et al. 2007) (#60 cardiac surgery patients at higher risk of postoperative renal failure) To assess the effect of high-dose N-acetylcysteine on renal function in cardiac surgery patients at higher risk of postoperative renal failure. DESIGN: **Multiblind, placebo-controlled, randomized, phase II clinical trial**. SETTING: Operating rooms and intensive care units of two tertiary referral hospitals. PATIENTS: A total of **60 cardiac surgery patients at higher risk of postoperative renal failure**. INTERVENTIONS: Patients were allocated to either 24 hrs of high-dose **N-acetylcysteine infusion (300 mg/kg body weight in 5% glucose, 1.7 L)** or placebo (5% glucose, 1.7 L). MEASUREMENTS AND MAIN RESULTS: The primary outcome measure was the absolute change in serum creatinine from baseline to peak value within the first five postoperative days. Secondary outcomes included the relative change in serum creatinine, peak serum creatinine level, serum cystatin C, and in urinary output. Further outcomes were needed for renal replacement therapy, length of ventilation, and length of stay in the intensive care unit and hospital. Randomization was successful and patients were well balanced for preoperative and intraoperative characteristics. There was no significant attenuation in the increase in serum creatinine from baseline to peak when comparing

N-acetylcysteine with placebo. Also, there was no attenuation in the increase in serum cystatin C from baseline to peak for N-acetylcysteine compared with placebo. Likewise, there was no evidence for differences in any other clinical outcome. CONCLUSIONS: **In this phase II, randomized, controlled trial, high-dose N-acetylcysteine was no more effective than placebo in attenuating cardiopulmonary bypass-related acute renal failure in high-risk cardiac surgery patients**. Department of Intensive Care, Austin Hospital, University of Melbourne, Australia.

Circa 2008

Antioxidant supplements for prevention of mortality in healthy participants and patients with various diseases. (Bjelakovic, Nikolova, Gludd, Simonetti and Gludd, 2008 Apr) (#232,550 Cochrane Database Syst Rev.) OBJECTIVES: To assess the effect of antioxidant supplements on mortality in primary or secondary prevention randomized clinical trials. We included all primary and secondary prevention randomized clinical trials on antioxidant supplements (beta-carotene, vitamin A, vitamin C, vitamin E, and selenium) versus placebo or no intervention. Trials with adequate randomization, blinding, and follow-up were classified as having a low risk of bias. RESULTS: **Sixty-seven randomized trials** with **232,550 participants** were included. Forty-seven trials including 180,938 participants had low risk of bias. Twenty-one trials included 164,439 healthy participants. Forty-six trials included 68,111 participants with various diseases (gastrointestinal, cardiovascular, neurological, ocular, dermatological, rheumatoid, renal, endocrinological, or unspecified). **Overall, the antioxidant supplements had no significant effect on mortality in a random-effects meta-analysis**, but *significantly increased mortality in a fixed-effect model*. **In the trials with a low risk of bias, the antioxidant supplements significantly increased mortality**. When the different antioxidants were assessed separately, analyses including trials with a low risk of bias and excluding selenium trials found *significantly increased mortality by vitamin A, beta-carotene, and vitamin E,* but **no significant detrimental effect of vitamin C**. Low-bias risk trials on selenium found no significant effect on mortality. CONCLUSIONS: We found no evidence to support antioxidant supplements for primary or secondary prevention. *Vitamin A, beta-carotene, and vitamin E may increase mortality.* **Antioxidant supplements need to**

be considered medicinal products and should undergo sufficient evaluation before marketing.

Systematic review: primary and secondary prevention of gastrointestinal cancers with antioxidant supplements.

(Bjelakovic, Nikolova, Simonette and Gludd, 2008 Sept) (#211,818 participants) The evidence on whether antioxidant supplements prevent gastrointestinal cancers is contradictory. Using the Cochrane Collaboration methodology, we reviewed the randomized trials comparing antioxidant supplements with placebo or no intervention on the occurrence of gastrointestinal cancers. RESULTS: We identified 20 randomized trials (211,818 participants) assessing beta-carotene, vitamin A, vitamin C, vitamin E, and selenium. **The antioxidant supplements were without a significant effect on the occurrence of gastrointestinal cancers.** *Antioxidant supplements had no significant effect on mortality in a random-effects model meta-analysis but significantly increased mortality in a fixed-effect model meta-analysis. CONCLUSIONS: There was no evidence that the studied antioxidant supplements prevented gastrointestinal cancers. On the contrary, they seem to increase overall mortality.*

Vitamins E and C in the prevention of cardiovascular disease in men: the Physicians' Health Study II randomized controlled trial

(Sesso et al, 2008) (#14,641 US male physicians); a randomized, double-blind, placebo-controlled factorial trial of vitamin E and vitamin C that began in 1997 and continued until its scheduled completion on August 31, 2007. CONCLUSIONS: In this large, long-term trial of male physicians, **neither vitamin E nor vitamin C supplementation reduced the risk of major cardiovascular events. These data provide no support for the use of these supplements (vitamins E and C) for the prevention of cardiovascular disease in middle-aged and older men.**

Both {alpha}- and -Carotene, but Not Tocopherols and Vitamin C, Are Inversely Related to 15-Year Cardiovascular Mortality in Dutch Elderly Men

(Buijsse et al, 2008) (#559 men (mean age ~72 y) free of chronic diseases); intake of different carotenoids, *alpha-* and *gamma-*-tocopherol, and vitamin C with 15-y

CVD mortality in elderly men who participated in the Zutphen Elderly Study. **Carrots were the primary source of *alpha-* and *beta-*carotene and their consumption was related to a lower risk of death** from CVD. **Intakes of carotenoids other than α- and *beta*-carotene were not associated with CVD mortality, nor were vitamin C and *alpha-* and *gamma-*tocopherol.** CONCLUSION: **dietary intakes of α-carotene and *beta*-carotene are inversely associated with CVD mortality in elderly men. This study does not indicate an important role for other carotenoids, tocopherols, or vitamin C in lowering the risk of CVD death.**

Vitamin E and selenium supplementation and risk of prostate cancer in the Vitamins and lifestyle (VITAL) study cohort (Peters et al, 2008) (#35,242 men) RESULTS: **A 10-year average intake of supplemental vitamin E was not associated with a reduced prostate cancer risk overall.** CONCLUSIONS: In this prospective cohort, **long-term supplemental intake of vitamin E and selenium were not associated with prostate cancer risk overall**; however, risk of clinically relevant advanced disease was reduced with greater long-term vitamin E supplementation.

VITAL (VITamins And Lifestyle) study (2008) (Slatore et al, 2008) (#77,721 men and women); explore the association of supplemental multivitamins, vitamin C, vitamin E, and folate with incident lung cancer; Prospective cohort; Cases were identified through the Seattle–Puget Sound **SEER (Surveillance, Epidemiology, and End Results)** cancer registry. **There was no inverse association with any supplement. *Supplemental vitamin E was associated with a small increased risk of lung cancer.*** This risk of supplemental vitamin E was largely confined to current smokers and was greatest for non–small cell lung cancer. Conclusions: **Supplemental multivitamins, vitamin C, vitamin E, and folate were not associated with a decreased risk of lung cancer. *Supplemental vitamin E was associated with a small increased risk.*** Patients should be counseled against using these supplements to prevent lung cancer.

The Institute of Medicine states that most North American adults get enough vitamin E from their normal diets to meet current recommendations.

The recent announcement of early termination of the Selenium and Vitamin E Cancer Prevention Trial's (SELECT) interventions because of lack of efficacy and observation of possible adverse events (i.e., small and not statistically significant increases in type 2 diabetes in those receiving selenium alone and in prostate cancer incidence in the vitamin E (alone) group) should serve as a reminder that the unexpected can happen in these well-designed trials (Lippman et al, 2009).

Vitamin E for Alzheimers and mild cognitive impairment. Cochrane Database Syst Rev. (2008) (Isaac et al, 2008) (#769 participants); assess the efficacy of Vitamin E in the treatment of Alzheimer's disease and prevention of progression of Mild Cognitive Impairment to Alzheimer's disease. *More participants taking Vitamin E suffered a fall.* There was no significant difference in the probability of progression from MCI to AD between the Vitamin E group and the placebo group. CONCLUSIONS: There is no evidence of efficacy of Vitamin E in the prevention or treatment of people with AD or MCI.

Efficacy of Antioxidant Supplementation in Reducing Primary Cancer Incidence and Mortality: Systematic Review and Meta-analysis (Bardia et al, 2008) (#104,196 participants) Twelve eligible trials, 9 of high methodological quality, were identified (total subject population, 104,196). Antioxidant supplementation did not significantly reduce total cancer incidence or mortality or any site-specific cancer incidence. *Beta carotene supplementation was associated with an increase in the incidence of cancer among smokers and with a trend toward increased cancer mortality.* Selenium supplementation was associated with reduced cancer incidence in men but not in women and with reduced cancer mortality. Vitamin E supplementation had no apparent effect on overall cancer incidence or cancer mortality. CONCLUSION: Beta carotene supplementation appeared to increase cancer incidence and cancer mortality among smokers, whereas vitamin E supplementation had no effect. Selenium supplementation might have anticarcinogenic effects in men and thus requires further research.

Carotenoids and the risk of developing lung cancer: a systematic review. (Gallicchio et al, 2008) (Six randomized clinical trials & 25 prospectie observational studies) Carotenoids are thought to have anti-cancer properties, but findings from population-based research have been inconsistent. OBJECTIVE: They aimed to conduct a systematic review of the associations between carotenoids and lung cancer. DESIGN: They searched electronic databases for articles published through September 2007. **Six randomized clinical trials examining the efficacy of beta-carotene supplements and 25 prospective observational studies** assessing the associations between carotenoids and lung cancer were analyzed by using random-effects meta-analysis. RESULTS: The pooled relative risk (RR) for the studies comparing beta-carotene supplements with placebo was 1.10. Among the observational studies that adjusted for smoking, the pooled RRs comparing highest and lowest categories of total carotenoid intake and of total carotenoid serum concentrations were 0.79 and 0.70, respectively. For beta-carotene, highest compared with lowest pooled RRs were 0.92 for dietary intake and 0.84 for serum concentrations. For other carotenoids, the RRs comparing highest and lowest categories of intake ranged from 0.80 for beta-cryptoxanthin to 0.89 for alpha-carotene and lutein-zeaxanthin; for serum concentrations, the RRs ranged from 0.71 for lycopene to 0.95 for lutein-zeaxanthin. CONCLUSIONS: **beta-Carotene supplementation is not associated with a decrease in the risk of developing lung cancer.** Findings from prospective cohort studies suggest inverse associations between carotenoids and lung cancer; however, **the decreases in risk are generally small and not statistically** significant. These inverse associations may be the result of carotenoid measurements' function as a marker of a healthier lifestyle (higher fruit and vegetable consumption) or of residual confounding by smoking.

Antioxidant enriched enteral nutrition and oxidative stress after major gastrointestinal tract surgery. (van Stijn et al, 2008) (#21 undergoing major upper gastrointestinal tract surgery) Goal: To investigate the effects of an enteral supplement containing antioxidants on circulating levels of antioxidants and indicators of oxidative stress after major gastrointestinal surgery. METHODS: **Twenty-one patients** undergoing major upper gastrointestinal tract surgery were randomized in a single centre, open label study on the effect of postoperative enteral nutrition supplemented with antioxidants. The effect on circulating levels of antioxidants and indicators of oxidative stress, such as F2-isoprostane, was studied. RESULTS: The antioxidant enteral supplement showed no adverse effects and was well tolerated. After surgery a decrease in the circulating levels of antioxidant parameters was observed. Only selenium and glutamine levels were restored to pre-operative values one week after surgery. **F2-isoprostane increased in the first three postoperative days only in the antioxidant supplemented group.** Lipopolysaccharide binding protein (LBP) levels decreased faster in the antioxidant group after surgery. CONCLUSION: **Despite lower antioxidant levels there was no increase in the circulating markers of oxidative stress on the first day after major abdominal surgery.** The rise in F2-isoprostane in patients receiving the antioxidant supplement may be related to the conversion of antioxidants to oxidants which raises questions on antioxidant supplementation. Module AOX restored the postoperative decrease in selenium levels. The rapid decrease in LBP levels in the antioxidant group suggests a possible protective effect on gut wall integrity.

Vitamin E supplementation may transiently increase tuberculosis risk in males who smoke heavily and have high dietary vitamin C intake.

(Hemila and Kaprio, 2008 Oct) (#29,023 males aged 50-69 years, smoking at baseline, with no tuberculosis) Vitamin E and beta-carotene affect the immune function and might influence the predisposition of man to infections. To examine whether vitamin E or beta-carotene supplementation affects tuberculosis risk, we analysed data of the Alpha-Tocopherol Beta-Carotene Cancer Prevention (ATBC)Study, a randomised controlled trial which examined the effects of vitamin E (50 mg/d) and beta-carotene (20 mg/d) on lung cancer. The trial was conducted in the general community in Finland in 1985-93; the intervention lasted for 6.1 years (median). **The ATBC Study cohort consists of 29,023 males aged 50-69 years, smoking at baseline, with no tuberculosis diagnosis prior to randomization. Vitamin E supplementation had no overall effect on the incidence of tuberculosis nor had beta-carotene.** Nevertheless, **dietary vitamin C intake significantly modified the vitamin E effect.** *Among participants who obtained 90 mg/d or more of vitamin C in foods, vitamin E supplementation increased tuberculosis risk by 72%.* **This effect was restricted to participants who smoked heavily.** Finally, in participants not supplemented with vitamin E, dietary vitamin C had a negative association with tuberculosis risk so that the adjusted risk was 60 (95% CI 16, 81)% lower in the highest intake quartile compared with the lowest. *Our finding that vitamin E seemed to transiently increase the risk of tuberculosis in those who smoked heavily and had high dietary vitamin C intake should increase caution towards vitamin E supplementation for improving the immune system.*

Vitamin E supplementation and pneumonia risk in males who initiated smoking at an early age: effect modification by body weight and dietary vitamin C. (Hemila and Kaprio, 2008 Nov) (#21,657 ATBC Study participants who initiated smoking by the age of 20 years) *They had found a 14% higher incidence of pneumonia with vitamin E supplementation in a subgroup of the Alpha-Tocopherol Beta-Carotene Cancer Prevention (ATBC) Study cohort*: participants who had initiated smoking by the age of 20 years. In this study, they explored the modification of vitamin E effect by body weight, because the same dose could lead to a greater effect in participants with low body weight. METHODS: The ATBC Study recruited males aged 50-69 years who smoked at least 5 cigarettes per day at the baseline; it was conducted in southwestern Finland in 1985-1993. The current study was restricted to 21,657 ATBC Study participants who initiated smoking by the age of 20 years; the median follow-up time was 6.0 years. The hospital-diagnosed pneumonia cases were retrieved from the national hospital discharge register (701 cases). RESULTS: Vitamin E supplementation had no effect on the risk of pneumonia in participants with body weight in a range from 70 to 89 kg. *Vitamin E increased the risk of pneumonia in participants with body weight less than 60 kg, and in participants with body weight over 100 kg.* The harm of vitamin E supplementation was restricted to participants with dietary vitamin C intake above the median. CONCLUSION: **Vitamin E supplementation may cause harmful effects on health in certain groups of male smokers**. The dose of vitamin E used in the ATBC Study, 50 mg/day, is substantially smaller than conventional vitamin E doses that are considered safe. Our findings should increase caution towards taking vitamin E supplements.

Oral administration of vitamin C decreases muscle mitochondrial biogenesis and hampers training-induced adaptations in endurance performance. (Gomez-Cabrera et al, 2008)

(#14 men) Exercise practitioners often take vitamin C supplements because **intense muscular contractile activity can result in oxidative stress,** as indicated by altered muscle and blood glutathione concentrations and increases in protein, DNA, and lipid peroxidation. There is, however, considerable debate regarding the beneficial health effects of vitamin C supplementation. OBJECTIVE: This study was designed to study the effect of vitamin C on training efficiency in rats and in humans. DESIGN: The human study was **double-blind and randomized. Fourteen men** (27-36 y old) were trained for 8 wk. Five of the men were supplemented daily with an oral dose of 1 g vitamin C. In the animal study, 24 male Wistar rats were exercised under 2 different protocols for 3 and 6 wk. Twelve of the rats were treated with a daily dose of vitamin C (0.24 mg/cm2 body surface area). RESULTS: *The administration of vitamin C significantly (P=0.014) hampered endurance capacity.* The adverse effects of vitamin C may result from its capacity to reduce the exercise-induced expression of key transcription factors involved in mitochondrial biogenesis. These factors are peroxisome proliferator-activated receptor co-activator 1, nuclear respiratory factor 1, and mitochondrial transcription factor A. **Vitamin C also prevented the exercise-induced expression of cytochrome C (a marker of mitochondrial content) and of the antioxidant enzymes superoxide dismutase and glutathione peroxidase.** CONCLUSION: **Vitamin C supplementation decreases training efficiency because it prevents some cellular adaptations to exercise.**

Antioxidant vitamin and mineral supplements for preventing age-related macular degeneration. (Evans and Henshaw, 2008) (#23,099 participants) Some observational studies have suggested that people who eat a diet rich in antioxidant vitamins (carotenoids, vitamins C and E) or minerals (selenium and zinc) may be less likely to develop age-related macular degeneration (AMD). OBJECTIVES: The aim of this review was to examine the evidence as to whether or not taking vitamin or mineral supplements prevents the development of AMD. SEARCH STRATEGY: We searched the Cochrane Central Register of Controlled Trials (CENTRAL) (which contains the Cochrane Eyes and Vision Group Trials Register) in The Cochrane Library (2007, Issue 3), MEDLINE (1966 to August 2007), SIGLE (1980 to 2005/03), EMBASE (1980 to August 2007), National Research Register (2007, Issue 3), AMED (1985 to January 2006) and PubMed (on 24 January 2006 covering last 60 days), reference lists of identified reports and the Science Citation Index. We contacted investigators

321

and experts in the field for details of unpublished studies. SELECTION: They included all randomized trials comparing an antioxidant vitamin and/or mineral supplement (alone or in combination) to control. We included only studies where supplementation had been given for at least one year. DATA AND ANALYSIS: Both review authors independently extracted data and assessed trial quality. Data were pooled using a fixed-effect model. RESULTS: Three randomized controlled trials were included in this review (**23,099 people randomized**). These trials investigated **alpha-tocopherol and beta-carotene** supplements. **There was no evidence that antioxidant vitamin supplementation prevented or delayed the onset of AMD.** The pooled risk ratio for any age-related maculopathy (ARM) was 1.04, for AMD (late ARM) was 1.03. Similar results were seen when the analyses were restricted to beta-carotene and alpha-tocopherol. CONCLUSIONS: **There is no evidence to date that the general population should take antioxidant vitamin and mineral supplements to prevent or delay the onset of AMD.** There are several large ongoing trials. People with AMD should see the related Cochrane review "Antioxidant vitamin and mineral supplements for slowing the progression of age-related macular degeneration" written by the same author.

Multivitamin-multimineral supplement use and mammographic breast density. (Berube et al, 2008) (#Premenopausal (777) and postmenopausal (783) women; total 1,560) The effect of multivitamin-multimineral supplements on the occurrence of chronic diseases, such as breast cancer, is unclear. Breast density is increasingly used as a biomarker of breast cancer risk. **Objective:** The present study evaluated the association of multivitamin-multimineral supplement use with breast density. **Design: Premenopausal (*n* = 777) and postmenopausal (*n* = 783) women** were recruited at the time of screening mammography. Diet and multivitamin-multimineral and individual vitamin and mineral supplement use were assessed with a self-administered food-frequency questionnaire. Breast density from screening mammograms was measured using a computer-assisted method. Crude and adjusted means in breast density were evaluated according to multivitamin-multimineral supplement use using generalized linear models. **Results: Current multivitamin-multimineral supplement use was reported by 21.7% of women** (20.7% and 22.6% of premenopausal and postmenopausal women, respectively). **Premenopausal women who were currently using multivitamin-multimineral supple-**

ments had higher adjusted mean breast density (**45.5%**) than past (**42.9%**) or never (**40.2%**) users. Of the current users, breast density was not related to duration of multivitamin-multimineral supplement use. In postmenopausal women, multivitamin-multimineral supplement use was not associated with breast density. **Conclusion:** *Regular use of multivitamin-multimineral supplements may be associated with higher mean breast density among premenopausal women.*

Mammographic breast density is increasingly used as a biomarker of breast cancer risk because of its strong positive relation to the risk of the disease (Pike, 2005).

Breast density has been hypothesized to reflect the quantity of breast tissue and the population of breast cells at risk of carcinogenic transformation (Boyd et al, 2005) (Trichopoulos, Lagiou and Adami, 2005).

Antioxidant supplements to prevent or slow down the progression of AMD: a systematic review and meta-analysis.

(Evans, 2008) (#23,099 people were randomized in three trials) The aim of this review was to examine the evidence as to whether antioxidant vitamin or mineral supplements prevent the development of AMD or slow down its progression. METHODS: Randomized trials comparing antioxidant vitamin and/or mineral supplement to control were identified by systematic electronic searches (updated August 2007) and contact with investigators. Data were pooled after investigating clinical and statistical heterogeneity. RESULTS: **There was no evidence that antioxidant (vitamin E or beta-carotene) supplementation prevented AMD. A total of 23,099 people were randomized in three trials with treatment duration of 4-12 years.** There was evidence that antioxidant (beta-carotene, vitamin C, and vitamin E) and zinc supplementation slowed down the progression to advanced AMD and visual acuity loss in people with signs of the disease. The majority of people were randomized in one trial (AREDS, 3640 people randomized). There were seven other small trials (total randomized 525). CONCLUSIONS: **Current evidence does not support the use of antioxidant vitamin supplements to prevent AMD.** People with AMD, or early signs of the disease, may experience some benefit from taking supplements as used in the AREDS trial. *Potential harms of high-dose antioxidant*

323

supplementation must be considered. These may include an increased risk of lung cancer in smokers (beta-carotene), heart failure in people with vascular disease or diabetes (vitamin E) and hospitalization for genitourinary conditions (zinc).

High maternal plasma antioxidant (vit. C) concentrations associated with preterm delivery (Joshi et al, 2008) (#140 normotensive women)

Our earlier study has shown that **increased maternal oxidative stress and reduced antioxidants like vitamin E and C play an important role in fetal growth in preeclampsia.** However, the role of antioxidants and their effects on gestation and birth outcome in normotensive pregnancies are not conclusive. The present study examined plasma malondialdehyde as a marker of oxidative stress and antioxidant concentrations (vitamins E and C) in maternal as well as in cord blood samples in normotensive women who delivered both preterm and at term. METHODS: **140 normotensive pregnant women** were recruited at Bharati Medical Hospital, Pune, India, during the year 2007. Maternal and cord samples were examined for oxidative stress levels and vitamin C and E concentrations in women who delivered preterm (n=40) and at term (n=100). Mean values were compared with those of women delivering at term using the t test. RESULTS: Increased (p<0.05) oxidative stress was seen in preterm mothers as well as in cord samples. *Preterm mothers had higher vitamin C concentrations (p<0.05), and these were positively associated with oxidative stress* (p=0.02). **Vitamin E levels were comparable between groups.** CONCLUSIONS: Increased maternal circulating vitamin C concentrations and increased oxidative stress are associated with preterm delivery. **I believe that the increased vitamin C level in preterm (mothers) deliveries is significant and a cause of problems in the newborns.**

Vitamin E and age-related cataract in a randomized trial of women. (#39,876) (Christen et al. 2008).

Investigators studied whether vitamin E supplementation decreases the risk of age-related cataract in women. **DESIGN:** Randomized, double-masked, placebo-controlled trial. **PARTICIPANTS: Thirty-nine thousand eight hundred seventy-six (39,876)** apparently healthy female health professionals aged 45 years or older. **INTERVENTION:** Participants were assigned randomly to receive either 600

IU natural-source vitamin E on alternate days or placebo and were followed up for presence of cataract for an average of 9.7 years. **MAIN OUTCOME MEASURE:** Age-related cataract defined as an incident, age-related lens opacity, responsible for a reduction in best-corrected visual acuity to 20/30 or worse, based on self-report and confirmed by medical record review. **RESULTS:** There was no significant difference between the vitamin E and placebo groups in the incidence of cataract. In subgroup analyses of subtypes, there were no significant effects of vitamin E on the incidence of nuclear, cortical, or posterior subcapsular cataract. Results were similar for extraction of cataract and subtypes. There was no modification of the lack of effect of vitamin E on cataract by baseline categories of age, cigarette smoking, multivitamin use, or several other possible risk factors for cataract. **CONCLUSIONS: These data from a large trial of apparently healthy female health professionals with 9.7 years of treatment and follow-up indicate that 600 IU natural-source vitamin E taken every other day provides no benefit for age-related cataract or subtypes.**

Antioxidants in cardiovascular health and disease: key lessons from epidemiologic studies. (Icox et al. 2008). **Interventional trials have been controversial, with some positive findings, many null findings, and some suggestion of harm in certain high-risk populations.** *Because of the mismatch between the epidemiologic studies and the interventional trials, some researchers have advocated ending antioxidant work.* **Others have questioned the validity of the LDL oxidative hypothesis itself.**

Both alpha and beta-Carotene, but Not Tocopherols and Vitamin C, Are Inversely Related to 15-Year Cardiovascular Mortality in Dutch Elderly Men - the Zutphen Elderly Study (Buijsse et al, 2008) (#559 men) The role of beta-carotene, alpha-tocopherol, and vitamin C in the prevention of cardiovascular diseases (CVD) is controversial. Prospective studies on gamma-tocopherol and carotenoids other than beta-carotene are sparse. We assessed relations between the intake of different carotenoids, alpha- and gamma-tocopherol, and vitamin C with 15-y CVD mortality in elderly men who participated in the **Zutphen Elderly Study.** Information on diet and potential confounding factors was collected in 1985, 1990, and 1995. In 1985, **559 men** (mean age ~72 y) free of chronic diseases were included

in the current analysis. After 15 y of follow-up, comprising 5744 person-years, 197 men had died from CVD. After adjustment for age, smoking, and other potential lifestyle and dietary confounders, relative risks of CVD death for a 1-SD increase in intake were 0.81 for alpha-carotene and 0.80 for beta-carotene. Carrots were the primary source of alpha- and beta-carotene and their consumption was related to a lower risk of death from CVD. Intakes of carotenoids other than alpha- and beta-carotene were not associated with CVD mortality, nor were vitamin C and alpha- and gamma-tocopherol. In conclusion, dietary intakes of alpha-carotene and beta-carotene are inversely associated with CVD mortality in elderly men. **This study does not indicate an important role for other carotenoids, tocopherols, or vitamin C in lowering the risk of CVD death.**

Observational studies on the effect of dietary antioxidants on asthma: a meta-analysis. (Gao et al. 2008) (#13,653) It has

been suggested that the rapid increase in asthma prevalence may in part be due to a decrease in the intake of dietary antioxidants, including vitamin C, vitamin E and beta-carotene. Epidemiological studies investigating the association between dietary antioxidant intake and asthma have generated inconsistent results. A meta-analysis was undertaken to examine the association between dietary antioxidant intake and the risk of asthma.

METHODS: The MEDLINE database was searched for observational studies in English-language journals from 1966 to March 2007. Data were extracted using standardized forms. Pooled odds ratios (OR) with 95% confidence intervals (CI) were calculated using a random effects model. Ten studies were eligible for inclusion. Seven studies, comprising **13,653 subjects**, used asthma or wheeze as their outcome; three studies explored the effect of antioxidant intake on lung function. RESULTS: **A higher dietary intake of antioxidants was not associated with a lower risk of having asthma.** The pooled OR for having asthma were 1.06 for subjects with a higher dietary vitamin C intake compared with those with a lower intake; 0.88 for vitamin E; and 1.12 for beta-carotene. **There was no significant association between dietary antioxidant intake and lung function except for a positive association between vitamin C intake and an increase in FEV(1). CONCLUSIONS: This meta-analysis does not support the hypothesis that dietary intake of the antioxidants vitamins C and E and beta-carotene influences the risk of asthma.**

Dietary antioxidants and the long-term incidence of age-related macular degeneration: the Blue Mountains Eye Study (Tan et al. 2008) (#Of 3,654 baseline (1992-1994) participants initially 49 years or older, 2,454 were reexamined after 5 years, 10 years, or both) Investigators assessed the relationship between baseline dietary and supplement intakes of antioxidants and the long-term risk of incident age-related macular degeneration (AMD). DESIGN: Australian population-based cohort study. PARTICIPANTS: Of **3,654 baseline (1992-1994) participants initially 49 years or older, 2,454 were reexamined after 5 years, 10 years, or both**. METHODS: Stereoscopic retinal photographs were graded using the Wisconsin Grading System. Data on potential risk factors were collected. Energy-adjusted intakes of alpha-carotene; beta-carotene; beta-cryptoxanthin; lutein and zeaxanthin; lycopene; vitamins A, C, and E; and iron and zinc were the study factors. Discrete logistic models assessed AMD risk. Risk ratios (RRs) and 95% confidence intervals (CIs) were calculated after adjusting for age, gender, smoking, and other risk factors. MAIN OUTCOME MEASURES: Incident early, late, and any AMD. RESULTS: For dietary **lutein and zeaxanthin, participants in the top tertile of intake had a reduced risk of incident neovascular AMD**, and those with above median intakes had a reduced risk of indistinct soft or reticular drusen. For total zinc intake the RR comparing the top decile intake with the remaining population was 0.56 for any AMD and 0.54 for early AMD. The highest compared with the lowest tertile of total beta-carotene intake predicted incident neovascular AMD. Similarly, **beta-carotene intake from diet alone predicted neovascular AMD**. This association was evident in both ever and never smokers. **Higher intakes of total vitamin E predicted late AMD.** CONCLUSIONS: In this population-based cohort study, higher dietary lutein and zeaxanthin intake reduced the risk of long-term incident AMD. This study confirmed the Age-Related Eye Disease Study finding of protective influences from zinc against AMD. *Higher beta-carotene intake was associated with an increased risk of age-related macular degeneration (AMD).*

Vitamins E and C in the prevention of cardiovascular disease in men: the Physicians' Health Study II randomized controlled trial (Sesso et al, 2008) (#14,641 US male physicians) a randomized, double-blind, placebo-controlled factorial trial of vitamin

327

E and vitamin C that began in 1997 and continued until its scheduled completion on August 31, 2007. CONCLUSIONS: In this large, long-term trial of male physicians, **neither vitamin E nor vitamin C supplementation reduced the risk of major cardiovascular events. These data provide no support for the use of these supplements (vitamins E and C) for the prevention of cardiovascular disease in middle-aged and older men.** Neither vitamin E nor vitamin C had a significant effect on total mortality but *vitamin E was associated with an increased risk of hemorrhagic stroke.*

Mistletoe therapy in oncology (Horneber et al, 2008) (#21 included studies, overall comprising 3,484 randomized cancer patients). Mistletoe extracts are commonly used in cancer patients. It is claimed that they improve survival and quality of life (QOL) in cancer patients. **OBJECTIVES:** To determine the effectiveness, tolerability and safety of mistletoe extracts given either as monotherapy or adjunct therapy for patients with cancer. **SEARCH STRATEGY:** Search sources included the Cochrane Central Register of Controlled Trials (CENTRAL, Issue 3, 2007) Cochrane Complementary Medicine Field Registry of randomized clinical trials (RCTs) and controlled clinical trials, MEDLINE, EMBASE, HEALTHSTAR, INT. HEALTH TECHNOLOGY ASSESSMENT, SOMED, AMED, BIOETHICSLINE, BIOSIS, CancerLit, CATLINE, CISCOM (August 2007). For the search the Standard Operating Procedures of the Information System in Health Economics at the German Institute for Medical Documentation and Information (DIMDI) were utilized. Reference lists of relevant articles and authors extensive files were searched for additional studies. Manufacturers of mistletoe preparations were contacted. **SELECTION CRITERIA:** We included RCTs of adults with cancer of any type. **The interventions were mistletoe extracts as sole treatments or given concomitantly with chemo- or radiotherapy.** The outcome measures were survival times, tumor response, QOL, psychological distress, adverse effects from antineoplastic treatment and safety of mistletoe extracts. **DATA COLLECTION AND ANALYSIS:** Three review authors independently assessed trials for inclusion in the review. All review authors independently took part in the extraction of data and assessment of study quality and clinical relevance. Disagreements were resolved by consensus. Study authors were contacted where information was unclear. Methodological quality was narratively described and additionally assessed with the Delphi list and the Jadad score. High methodological quality was defined if six out of nine Delphi criteria, or four out of five Jadad criteria were fulfilled. Results were presented qualitatively. **MAIN RESULTS:** Eighty studies were identified. Fifty-eight were excluded for various reasons, usually **as there was no prospective trial design with randomized treatment allocation.** Of the **21 included studies** 13 provided data on survival, 7 on tumor response, 16 on measures of QOL or psychological outcomes, or prevalence of chemotherapy-related adverse effects and 12 on side effects of mistletoe treatment; **overall comprising 3,484 randomized cancer patients.** Interventions evaluated were 5 preparations of mistletoe extracts from 5 manufacturers and one commercially not available preparation. The general reporting of RCTs was poor. Of the 13 trials investigating survival, 6 showed some evidence of a benefit, but none of them was of high methodological quality. The results of two trials in patients with melanoma and head and neck cancer gave some evidence that the used mistletoe extracts are not effective for improving survival. Of the 16 trials investigating the efficacy of mistletoe extracts for either improving QOL, psychological measures, performance index, symptom scales or the reduction of adverse effects of chemotherapy, 14 showed some evidence of a benefit, but only 2 of them including breast cancer patients during chemotherapy were of higher methodological quality. Data on side effects indicated that, depending on the dose, mistletoe extracts were usually well tolerated and had few side effects. **CONCLUSIONS: The evidence from RCTs to support the view that the application of mistletoe extracts has impact on survival or leads to an improved ability to fight cancer or to withstand anticancer treatments is weak.** Nevertheless, there is some evidence that mistletoe extracts may offer benefits on measures of QOL during chemotherapy for breast cancer, but these results need replication. Overall, more high quality, independent clinical research is needed to truly assess the safety and effectiveness of mistletoe extracts. Patients receiving mistletoe therapy should be encouraged to take part in future trails. (Horneber et al, 2008). 000

Mistletoes (*Viscum album*) are highly specialized angiosperms of the family Loranthaceae. **The antioxidant activity of phenolics is mainly because of their redox properties**, which allow them to act as reducing agents, hydrogen donors, free radical scavenger, singlet oxygen quenchers and metal chelators. This present study has verified that *V. album* extracts can act as primary and/or secondary antioxidants, being free radical scavengers and potent

Fe chelators, and these properties of health benefit are host dependent. (Ade-dayo and Ganiyu, 2008).

Oxystress inducing antitumor therapeutics via tumor-targeted delivery of PEG-conjugated D-amino acid oxidase = (Fang et al, 2008) (#)

We had developed a H_2O_2 generating enzyme, polyethylene glycol conjugated D-amino acid oxidase (PEG-DAO), which exhibited potent antitumor activity by generating toxic reactive oxygen species, namely oxidation therapy, subsequently **showed remarkable antitumor effect on murine Sarcoma 180 solid tumor,** by taking advantage of the enhanced permeability and retention effect. Along this line, we report here the preparation of PEG-DAO by use of recombinant DAO and its antitumor activity by using various tumor cell lines and tumor models. Recombinant DAO (rDAO) was obtained from E. coli BL21 (DE3) carrying the porcine DAO expression vector with high yield (20 mg/l) and high enzyme activity (5.3 U/mg). Pegylated rDAO (PEG-rDAO) showed high stability against sonication, repeated freezing/thawing, lyophilization and exhibited superior in vivo pharmacokinetics. PEG-rDAO had a molecular size of 65 kDa and existed as nanoparticles in aqueous solution with mean particle diameter of 119 nm. In vitro experiments showed strong cytotoxicity of PEG-rDAO against various tumor cells, whereas less cytotoxicity was found against various normal cells. In vivo antitumor treatment was carried out **using 2 mice tumor models, namely colon 38 tumor and Meth A tumor model.** PEG-rDAO was administered i.v. and after an adequate lag time, D-proline (the substrate of DAO) was injected i.p. to the tumor-bearing mice. Consequently, preferential generation of H_2O_2 in the tumor was successfully achieved, which **resulted in remarkable suppression of tumor growth without any visible side effects.** These findings suggest a potential of PEG-rDAO as a novel anticancer strategy toward clinical development. (Fang et al, 2008).

Clinical outcomes of contrast-induced nephropathy in patients undergoing percutaneous coronary intervention: a prospective, multicenter, randomized study to analyze the effect of hydration and acetylcysteine (Chen SL, et al. 2008) (#936)

The potential role of hydration in prevention of contrast-induced nephropathy (CIN) still remains to be unclear. METHODS: **Nine-hundred and thirty-six patients scheduled for percutaneous coronary intervention**

(PCI) were enrolled into the present study, and divided into normal (serum creatinine<1.5 mg/dl) and abnormal (serum creatine> or =1.5 mg/dl) groups according to their baseline serum concentration of creatinine. Each group was further randomly divided into two subgroups: hydration and nonhydration. **All patients in abnormal group took twice orally loading dose of 1200 mg acetylcysteine** (ATLS) at 12 h before scheduled time for coronary angiogram and immediately after procedure. Creatinine concentration was remeasured at the time of admission (just before catheterization), every day for the following three days. The primary end point during 6-month follow-up included clinical driven revascularization (either PCI or CABG), death from all causes, and requiring emergency renal-replacement therapy. RESULTS: **The incidence of CIN was more commonly in abnormal group that in normal group** (6.52% vs. 37.68%, p<0.001). Hydration had potentials in prevention of CIN only in patients with elevated baseline concentration of creatinine. Multivariate analysis demonstrated that the following variables remained to be significant factors correlating with CIN: age> or =70 years, contrast volume> or =320 ml, diabetes mellitus, and peripheral arterial disease. Patients with CIN in abnormal group had worse clinical outcomes, compared to patients with CIN in normal group. CONCLUSION: Hydration with 0.45% sodium *Patients with CIN and preexisting renal insufficiency had worse clinical outcomes.* chloride alone had no potential effect on the occurrence of CIN in patients with normal renal function. Combination of hydration with ATLS could reduce the incidence of CIN in patients at high risk. Nanjing First Hospital of Nanjing Medical University, Department of Cardiology, 68# Changle Road, 210006, Nanjing, China.

NAC had no effect on disease progression in non-diabetic kidney failure patients (Renke M. et al. 2008) (#20 non-diabetic patients with proteinuria)

Inhibition of the renin-angiotensin-aldosterone system with angiotensin-converting enzyme inhibitors (ACEI) and/or angiotensin II subtype 1 receptor antagonists (ARB) constitutes a strategy in the management of patients with chronic kidney disease. There is still no optimal therapy which can stop the progression of chronic kidney disease. **Antioxidants such as N-acetylcysteine (NAC) have been reported as a promising strategy** in this field. METHODS: In a **placebo-controlled, randomized**, open, 2-period cross-over study, we evaluated the influence of NAC (1,200 mg/day) added to renin-angiotensin-aldosterone system blockade on proteinuria

331

and surrogate markers of tubular injury and renal fibrosis in **20 non-diabetic patients with proteinuria** (0.4-6.36 g/24 h) with normal or decreased kidney function (estimated glomerular filtration rate 61-163 ml/min). Subjects entered the **8-week** run-in period during which the therapy using ACEI and/or ARB was established with blood pressure below 130/80 mm Hg. Next, patients were randomly assigned to 1 of 2 treatment sequences: NAC/washout/placebo or placebo/washout/NAC. Clinical evaluation and laboratory tests were performed at the randomization point and after each period of the study. RESULTS: **No significant changes in laboratory tests were observed.** CONCLUSION: **NAC had no effect on proteinuria, surrogate markers of tubular injury or renal fibrosis in non-diabetic patients with chronic kidney disease.**

N-acetylcysteine to reduce renal failure after cardiac surgery: a systematic review and meta-analysis (Naughton et al. 2008) (#Seven randomized controlled trials (RCTs, n = 1000)

To assess the effect of N-acetylcysteine (NAC) on acute renal failure and important clinical outcomes after cardiac surgery. METHODS: Two reviewers performed literature searches, using EMBASE and PubMed, of randomized controlled trials investigating the renoprotective effect of N-acetylcysteine in cardiac surgery. Treatment effects were calculated as relative risks (RR) with 95% confidence intervals (CI). Heterogeneity and publication bias were assessed using the I(2) test and funnel plots, respectively. Meta regression was performed to assess the effect of baseline renal function and the use of aprotinin on renal function. RESULTS: **Seven randomized controlled trials (RCTs) (n = 1000)** were identified. No study could demonstrate, either independently or meta-analytically, an improvement in the postoperative increase in creatinine, mortality, renal failure requiring renal replacement therapy, myocardial infarction, atrial fibrillation, or stroke. *There was a small, though significant increase in postoperative blood loss among patients treated with NAC* (weighted mean difference 119 mL 95% CI 51, 187). After meta regression neither increase in postoperative creatinine nor renal replacement therapy was associated with the baseline creatinine or with NAC dose. CONCLUSION: **This analysis did not find that treatment with NAC was associated with clinical renal protection during cardiac surgery, or improvement in other clinical outcomes.** Department of Anesthesia and Pain Management, University Health Network, Toronto General Hospital, University of Toronto, Ontario, Canada.

Circa 2009

Is there a role for supplemented antioxidants in the prevention of atherosclerosis? (Katsiki and Manes, 2009) (#22 trials (N=134,590 subjects) BACKGROUND: Oxidative stress is thought to play a substantial role in the pathogenesis of atherosclerosis. Supplementation of antioxidants has been studied as a strategy in the prevention of occurrence and progression of atherosclerosis. METHOD: We searched the MEDLINE and PubMed databases (up to February 2008) for **randomized, double-blind, placebo-controlled trials** of antioxidant (and in particular vitamins E, C and/or beta-carotene) supplementation, published in English. RESULTS: We identified **22 trials (N=134,590 subjects)** of antioxidant supplementation for the prevention of atherosclerosis (7 primary, 13 secondary and 2 both primary and secondary). Of these studies, 10 examined the effect of a single antioxidant supplementation on primary or secondary prevention of cardiovascular disease, while 12 the effect of a combination of antioxidants. CONCLUSION: **As the majority of studies included in this review does not support a possible role of antioxidant supplementation in reducing the risk of cardiovascular disease, no definite conclusion can be drawn to justify the use of antioxidant vitamin supplements for the prevention of atherosclerotic events.**

Plasma carotenoids, retinol, and tocopherols and postmenopausal breast cancer risk in the Multiethnic Cohort Study: a nested case-control study. (Epplein et al, 2009) (#286 incident postmenopausal breast cancer cases were matched to 535 controls)

Assessments by the handful of prospective studies of the association of serum antioxidants and breast cancer risk have yielded inconsistent results. This multiethnic nested case-control study sought to examine the association of plasma carotenoids, retinol, and tocopherols with postmenopausal breast cancer risk. Methods: From the biospecimen subcohort of the Multiethnic Cohort Study, **286 incident postmenopausal breast cancer cases were matched to 535 controls** on age, sex, ethnicity, study location (Hawaii or California), smoking status, date/time of collection and hours of fasting. We measured prediagnostic circulating levels of individual carotenoids, retinol, and tocopherols. Conditional

333

logistic regression was used to compute odds ratios and 95% confidence intervals. **Results: Women with breast cancer tended to have lower levels of plasma carotenoids and tocopherols than matched controls, but the differences were not large or statistically significant and the trends were not monotonic. No association was seen with retinol.** A sensitivity analysis excluding cases diagnosed within 1 year after blood draw did not alter the findings. Conclusions: **The lack of significant associations in this multiethnic population is consistent with previously observed results from less racially-diverse cohorts and serves as further evidence against a causal link between plasma micronutrient concentrations and postmenopausal breast cancer risk.** Women with breast cancer tended to have lower levels of plasma carotenoids and tocopherols than matched controls, but the differences were not large or statistically significant and the trends were not monotonic.

Vitamins E and C in the prevention of prostate and total cancer in men: the Physicians' Health Study II randomized controlled trial (2009) (Gazziano et al, 2009) (#14,641 male physicians); evaluate whether long-term vitamin E or C supplementation decreases risk of prostate and total cancer events among men; a randomized, double-blind, placebo-controlled factorial trial of vitamins E and C that began in 1997 and continued until its scheduled completion on August 31, 2007. **Neither vitamin E nor vitamin C had a significant effect on prostate, colorectal, lung, or other site-specific cancers.** CONCLUSIONS: In this large, long-term trial of male physicians, neither vitamin E nor C supplementation reduced the risk of prostate or total cancer. These **data provide no support for the use of these supplements for the prevention of cancer in middle-aged and older men.**

Advertisers suggest that taking certain vitamin or mineral supplements can lower prostate cancer risk. While some studies have found that there might be a protective benefit from some supplements, recent **results from 2 large studies didn't find any.** (Lippman et al, 2009) (Gaziano et al, 2009)

In 2001, researchers from the National Cancer Institute (NCI) and the Southwest Oncology Group (SWOG) launched the massive **SELECT study (short for**

Selenium and Vitamin E Cancer Prevention Trial) to find out whether taking selenium and vitamin E supplements could protect men from prostate cancer. **In October 2008, researchers halted the trial after early analysis showed the supplements weren't working, and in fact, in some cases, may have been doing more harm than good.**

In another large, long-term trial, called **the Physicians' Health Study II**, researchers from Brigham and Women's Hospital and Harvard Medical School studied whether taking vitamin E or vitamin C could reduce the risk of prostate cancer. Nearly **15,000 male doctors** participated in the trial. **After an average of 8 years, neither vitamin E nor vitamin C seemed to lower the risk of prostate cancer.**

Vitamin E, vitamin C, beta carotene, and cognitive function among women with or at risk of cardiovascular disease: The Women's Antioxidant and Cardiovascular Study (2009) (Kang et al, 2009) (#2,824 participants); Vitamin E supplementation and beta carotene supplementation were not associated with slower rates of cognitive change. Although vitamin C supplementation was associated with better performance at the last assessment, it was not associated with cognitive change over time. CONCLUSIONS: **Antioxidant supplementation (vitamin E, C and beta carotene) did not slow cognitive change among women with pre-existing cardiovascular disease or cardiovascular disease risk factors.**

Effects of vitamins C and E and beta-carotene on the risk of type 2 diabetes in women at high risk of cardiovascular disease: Women's Antioxidant Cardiovascular Study (2009) (Song et al, 2009) (#8,171 female health professionals); CONCLUSION: Our randomized trial data showed **no significant overall effects of vitamin C, vitamin E, and beta-carotene on risk of developing type 2 diabetes in women at high risk of CVD.**

Vitamins C and E and Beta Carotene Supplementation and Cancer Risk: Women's Antioxidant Cardiovascular Study (2009) (Lin et al, 2009) (#7,627 female health professionals); vitamin C (500 mg of ascorbic acid daily), natural-source vitamin E (600 IU of α-tocopherol every other day), and beta carotene (50 mg every other day). **There were no statistically significant effects of use of any antioxidant on total cancer incidence.** Conclusions: **Supplementation with vitamin C, vitamin E, or beta carotene offers no overall benefits in the primary prevention of total cancer incidence or cancer mortality.**

Effect of selenium and vitamin E on risk of prostate cancer and other cancers: the Selenium and Vitamin E Cancer Prevention Trial (SELECT) (2009) (Lippman et al, 2009) (#35,533 men) **There were statistically nonsignificant increased risks of prostate cancer in the vitamin E group** but not in the selenium + vitamin E group. CONCLUSION: **Selenium or vitamin E, alone or in combination at the doses and formulations used, did not prevent prostate cancer in this population of relatively healthy men.** *The trial was stopped ahead of its original 12 year deadline because of a lack of any noticeable benefit.*

"The largest prostate cancer prevention trial has found that selenium is no more effective than a placebo," said David Schardt, a senior nutritionist. "Bayer is ripping people off when it suggests otherwise in these dishonest ads."

Multivitamin Use and Risk of Cancer and Cardiovascular Disease in the Women's Health Initiative Cohorts (Neuhouser et al, 2009) (#161,808 postmenopausal women taking part in the Women's Health Initiative clinical trials); **Conclusion:** After a median follow-up of 8.0 and 7.9 years in the clinical trial and observational study cohorts, respectively, the Women's Health Initiative study provided convincing evidence that **multivitamin use has little or no influence on the risk of common cancers, CVD, or total mortality in postmenopausal women.**

There was no evidence that multivitamins confer meaningful benefit or harm in relation to cancer or cardiovascular disease. The risk for invasive cancers of the breast, colon/rectum, endometrium, lung, bladder, and ovary was no different among women who used multivitamin compared with those who did not use multivitamins. Similarly, risk of myocardial infarction, stroke, venous thrombosis, and death from any cause was no different for multivitamin users than for nonusers.

Multivitamins do not appear to be effective for the prevention of cancer or cardiovascular disease.

Neuhouser said: "To our surprise, we found that multivitamins did not lower the risk of the most common cancers and also had no impact on heart disease."

Thus, the largest multivitamin study shows that they do nothing to protect against cancer. Marian L. Neuhouser, the lead author and a nutritional epidemiologist with the Fred Hutchinson Cancer Research Center in Seattle, said, "Consumers spend money on dietary supplements with the thought that they are going to improve their health, but there's no evidence for this. Buying more fruits and vegetables might be a better choice."

In a separate statement, Neuhouser said: "Dietary supplements are used by more than half of all Americans, who spend more than 20 billion dollars on these products each year. However, scientific data are lacking on the long-term health benefits of supplements." Neuhouser suggested that women concentrate on getting their nutrients from food rather than supplements.

Researchers found the 41.5% of women who regularly took multivitamins were no more likely to avoid a range of cancers, heart disease, stroke or blood clots than those who didn't. In short, this large US study of over 160,000 postmenopausal women that found no convincing evidence that long term use of multivitamins changed their risk of developing common cancers, cardiovascular disease or dying prematurely.

Effects of antioxidant supplements on cancer prevention: meta-analysis of randomized controlled trials (Myung et al, 2009) (#161,045 total subjects); searched Medline (PubMed), Excerpta Medica database, and the Cochrane Review in October 2007; Among 3327 articles searched, 31 articles on 22 randomized controlled trials, which included 161,045 total subjects, 88,610 in antioxidant supplement groups and 72,435 in placebo or no-intervention groups, were included in the final analyses. In a fixed-effects meta-analysis of all 22 trials, antioxidant supplements were found to have no preventive effect on cancer. Conclusions: The

meta-analysis of randomized controlled trials indicated that **there is no clinical evidence to support an overall primary and secondary preventive effect of antioxidant supplements on cancer.**

MULTIVITAMINS: There was no evidence that multivitamins confer meaningful benefit or harm in relation to cancer or cardiovascular disease. The risk for invasive cancers of the breast, colon/rectum, endometrium, lung, bladder, and ovary was no different among women who used multivitamins compared with those who did not use multivitamins. Similarly, risk of myocardial infarction, stroke, venous thrombosis, and death from any cause was no different for multivitamin users than for nonusers. Multivitamins do not appear to be effective for the prevention of cancer or cardiovascular disease.

Decision Analysis Supports the Paradigm That Indiscriminate Supplementation of Vitamin E Does More Harm than Good (Dotan et al, 2009) the **major randomized clinical trials have yielded disappointing results on the effects of vitamin E on both mortality and morbidity.** Recent meta-analyses have concluded that **vitamin E supplementation increases mortality.** This conclusion has raised much criticism, most of it relating to three issues: (1) the choice of clinical trials to be included in the meta-analyses; (2) the end point of these meta-analyses (only mortality); and (3) the heterogeneity of the analyzed clinical trials with respect to both population and treatment. Our goal was to bring this controversy to an end by using a Markov-model approach, which is free of most of the limitations involved in using meta-analyses. The researchers examined data from *more than 300,000 subjects in the US, Europe and Israel. This "disappointing" study warns that indiscriminate use of high-dose Vitamin E supplementation does more harm than good and that indiscriminate supplementation of high doses of vitamin E is not beneficial in preventing CVD.*

CONCLUSIONS: Their study demonstrated that in terms of QALY, **indiscriminate supplementation of high doses of vitamin E is not beneficial in preventing CVD.**

Their objective was to reassess the outcome of nondiscriminatory supplementation of vitamin E with respect to its effects on cardiovascular-related events and mortality. Their analysis, applying a Markov model, revealed that *supplementing the general public with vitamin E results in loss of quality-adjusted life years.*

Further, Dotan states, " **Unfortunately, major randomized clinical trials yielded disappointing results and recent meta-analyses concluded that indiscriminate, high dose vitamin E supplementation results in increased mortality.** Our major finding was that the average quality-adjusted life years (QALY) of vitamin E- supplemented individuals was 0.30 QALY less than that of untreated people. In our view, this supports the view that indiscriminate supplementation of high dose vitamin E can not be recommended to the general public. In short, we adopt the view that **vitamin E is a "double-edge sword" that should not be consumed until criteria are defined to predict who is likely to benefit from high dose supplementation of vitamin E."** (Dotan, Lichtenberg and Pinchuk, 2009)

Effects of long-term antioxidant supplementation and association of serum antioxidant concentrations with risk of metabolic syndrome in adults (SU.VI.MAX) (Czernichow et al, 2009) (#5,220 adults) Adults (n = 5,220) participating in the SUpplementation en VItamines et Minéraux AntioXydants (SU.VI.MAX) primary prevention trial were randomly assigned to receive a supplement containing a combination of antioxidants (vitamins C and E, β-carotene, zinc, and selenium) at nutritional doses or a placebo. **Antioxidant supplementation for 7.5 y did not affect the risk of metabolic syndrome (MetS);** *Baseline serum antioxidant concentrations of β-carotene and vitamin C, however, were negatively associated with the risk of MetS*; the adjusted odds ratios, respectively. **Baseline serum zinc concentrations were positively associated with the risk of developing MetS. Conclusions: The experimental finding of no beneficial effects over seven-plus years of antioxidant supplementation** in a generally well-nourished population is **consistent with recent reports of a lack of efficacy of antioxidant supplements.**

Metabolic syndrome refers to a collection of risk factors for type 2 diabetes, heart disease and stroke -- including high blood pressure,

abdominal obesity, low levels of "good" HDL cholesterol, elevated tri-glycerides and high blood sugar. The condition is diagnosed when a person has at least three of those risk factors.

In a study of nutrients in older American women (Women's Health Study cohort), for most nutrients, no decline in intake was observed, as might have been expected in an aging cohort. Instead, intake of many nutrients increased, primarily because of the rising use of dietary supplements. Use of dietary supplements by older individuals is of particular importance because of the potential risk to benefit ratio of synthetic supplement intake levels despite the possibility of declining food intake. However, possible risks from obtaining a large proportion of required nutrients from dietary supplements rather than deriving them from foods should be approached with caution (Park et al, 2009).

Total and Cancer Mortality After Supplementation With Vitamins and Minerals: 10 year Follow-up of the Linxian General Population Nutrition Intervention Trial. (Qiao et al, 2009)

(#29,584 adult participants) The General Population Nutrition Intervention Trial was a randomized primary esophageal and gastric cancer prevention trial conducted from 1985 to 1991, in which **29, 584 adult participants** in Linxian, China, were given daily vitamin and mineral supplements. Treatment with "factor D," a combination of 50 μg selenium, 30 mg vitamin E, and 15 mg beta-carotene, led to decreased mortality from all causes, cancer overall, and gastric cancer. **Here, they present a 10-year follow-up after the end of active intervention**. Hazard ratios (HRs) and 95% confidence intervals (CIs) for the cumulative effects of four vitamin and mineral supplementation regimens were calculated using adjusted proportional hazards models. Results: **Participants who received factor D had lower overall mortality and gastric cancer mortality; reduction in cumulative gastric cancer mortality from 4.28% to 3.84%, than subjects who did not receive factor D**. Reductions were mostly attributable to benefits to subjects younger than 55 years. **Esophageal cancer deaths between those who did and did not receive factor D were not different overall**; however, decreased 17% among participants younger than 55 but *esophageal cancer deaths increased 14% among those aged 55 years or older. Vitamin A and zinc supplementation was associated with increased total and stroke mortality*; vitamin C and molybdenum supplementation, with decreased stroke mortality.

Conclusion: The beneficial effects of selenium, vitamin E, and beta-carotene on mortality were still evident up to 10 years after the cessation of supplementation and were consistently greater in younger participants. Late effects of other supplementation regimens were also observed. **This study illustrates the confusion in the data, especially with the Linxian studies. They represent an exception to most of the other studies and should be viewed with caution.**

Vitamin A and retinol intakes and the risk of fractures among participants of the Women's Health Initiative Observational Study. (Caire-Juvera et al. 2009) (#75,747 women from the Women's Health Initiative Observational Study) Excessive intakes of vitamin A have been shown to have adverse skeletal effects in animals. High vitamin A intake may lead to an increased risk of fracture in humans. OBJECTIVE: The objective was to evaluate the relation between total vitamin A and retinol intakes and the risk of incident total and hip fracture in postmenopausal women. DESIGN: A total of **75,747 women from the Women's Health Initiative Observational Study** participated. The risk of hip and total fractures was determined using Cox proportional hazards models according to different intakes of vitamin A and retinol. RESULTS: In the analysis adjusted for some covariates, **the association between vitamin A intake and the risk of fracture was not statistically significant.** Analyses for retinol showed similar trends. When the interaction term was analyzed as categorical, the highest intake of retinol with vitamin D was significant (P = 0.033). **Women with lower vitamin D intake** (< or =11 microg/d) in the highest quintile of intake of both vitamin A **and retinol had a modest increased risk of total fracture.** CONCLUSIONS: No association between vitamin A or retinol intake and the risk of hip or total fractures was observed in postmenopausal women. *Only a modest increase in total fracture risk with high vitamin A and retinol intakes was observed in the low vitamin D-intake group.*

Modification of the effect of vitamin E supplementation on the mortality of male smokers by age and dietary vitamin C. (Hemila and Kaprio, 2009 Apr) (#29,133) The Alpha-Tocopherol, Beta-Carotene Cancer Prevention (ATBC) Study (1985-1993) recruited **29,133 Finnish male cigarette smokers,** finding that vitamin E supplementation had

no overall effect on mortality. The authors of this paper found that the effect of vitamin E on respiratory infections in ATBC Study participants was modified by age, smoking, and dietary vitamin C intake; therefore, they examined whether the effect of vitamin E supplementation on mortality is modified by the same variables. During a median follow-up time of 6.1 years, 3,571 deaths occurred. Age and dietary vitamin C intake had a second-order interaction with vitamin E supplementation of 50 mg/day. *Among participants with a dietary vitamin C intake above the median of 90 mg/day, vitamin E increased mortality among those aged 50-62 years by 19%, whereas vitamin E decreased mortality among those aged 66-69 years by 41%. Vitamin E had no effect on participants who had a dietary vitamin C intake below the median.* Smoking quantity did not modify the effect of vitamin E. This study provides strong evidence that the effect of vitamin E supplementation on mortality varies between different population groups. Further study is needed to confirm this heterogeneity.

Vitamin E supplement use and the incidence of cardiovascular disease and all-cause mortality in the Framingham Heart Study: Does the underlying health status play a role?

(Dietrich et al, 2009) (#4,270 Framingham Study participants) Observational studies generally showed beneficial associations between supplemental vitamin E intake and cardiovascular disease (CVD) risk whereas intervention trials reported adverse effects of vitamin E supplements. We hypothesize that these discordant findings result from differing underlying health status of study participants in observational and intervention studies. Objective: Determine if the relation between supplemental vitamin E intake and CVD and all-cause mortality (ACM) depends on pre-existing CVD. Design: Proportional hazards regression to relate supplemental vitamin E intake to the 10-year incidence of CVD and ACM in **4,270 Framingham Study participants** stratified by baseline CVD status. Results: Eleven percent of participants used vitamin E supplements at baseline. In participants with pre-existing CVD, there were 28 (44%) and 20 (32%) incident cases of CVD and ACM in the vitamin E supplement users versus 249 (47%) and 202 (38%) in the non-users, respectively. In participants without pre-existing CVD, there were 51 (13%) and 47 (12%) cases of CVD and ACM in the vitamin E supplement group versus 428 (13%) and 342 (10%) in the non-vitamin E supplement group, respectively. Conclusion: **CVD status has no apparent influence on the association of supplemental vitamin E intake and risk for**

CVD and ACM in this large, community-based study. Further research is needed to clarify the basis for the discrepant results between intervention and observational studies of supplemental vitamin E intake.

Antioxidants prevent health-promoting effects of physical exercise in humans. (Ristow et al, 2009) (#39 healthy young men)

Exercise promotes longevity and ameliorates type 2 diabetes mellitus and insulin resistance. However, **exercise also increases mitochondrial formation of presumably harmful reactive oxygen species (ROS).** They evaluated the effects of a combination **of vitamin C (1000 mg/day) and vitamin E (400 IU/day)** on insulin sensitivity as measured by glucose infusion rates (GIR) during a hyperinsulinemic, euglycemic clamp in previously untrained (n = 19) and pretrained (n = 20) healthy young men. Before and after a 4 week intervention of physical exercise, GIR was determined, and muscle biopsies for gene expression analyses as well as plasma samples were obtained to compare changes over baseline and potential influences of vitamins on exercise effects. *Exercise increased parameters of insulin sensitivity (GIR and plasma adiponectin) only in the absence of antioxidants in both previously untrained (P < 0.001) and pretrained (P < 0.001) individuals.* This was **paralleled by increased expression of ROS-sensitive** transcriptional regulators of insulin sensitivity and ROS defense capacity, peroxisome-proliferator-activated receptor gamma (PPARgamma), and PPARgamma coactivators PGC1alpha and PGC1beta only in the absence of antioxidants (P < 0.001 for all). **Molecular mediators of endogenous ROS defense (superoxide dismutases 1 and 2; glutathione peroxidase) were also induced by exercise, and this effect too was blocked by antioxidant supplementation.** Consistent with the concept of mitohormesis, exercise-induced oxidative stress ameliorates insulin resistance and causes an adaptive response promoting endogenous antioxidant defense capacity. **Supplementation with antioxidants may preclude these health-promoting effects of exercise in humans.**

Long-term use of beta-carotene, retinol, lycopene, and lutein supplements and lung cancer risk: results from the VITamins And Lifestyle (VITAL) study. (Satia et al, 2009)

(#77,126 (VITAL) cohort Study in Washington State) **High-dose beta-carotene supplementation in high-risk persons has been linked to increased**

lung cancer risk in clinical trials; whether effects are similar in the general population is unclear. The authors examined associations of supplemental beta-carotene, retinol, vitamin A, lutein, and lycopene with lung cancer risk among participants, aged 50-76 years, in the VITamins And Lifestyle (VITAL) cohort Study in Washington State. In 2000-2002, eligible persons (n = **77,126**) completed a 24-page baseline questionnaire, including detailed questions about supplement use (duration, frequency, dose) during the previous 10 years from multivitamins and individual supplements/mixtures. Incident lung cancers (n = 521) through December 2005 were identified by linkage to the Surveillance, Epidemiology, and End Results cancer registry. *Longer duration of use of individual beta-caro-tene, retinol, and lutein supplements (but not total 10-year average dose) was associated with statistically significantly elevated risk of total lung cancer and histologic cell types.* There was little evidence for effect modification by gender or smoking status. **Long-term use of individual beta-carotene, retinol, and lutein supplements should not be recommended for lung cancer prevention, particularly among smokers.**

Although this book is primarily about multivitamins and antioxidant vitamins, current discussions center around the antioxidant potential of lycopene. The following abstract makes a key observation concerning any contribution of lycopene as an effective antioxidant: **Are the health attributes of lycopene related to its antioxidant function?** (Erdman, Ford and Lindshield, 2009) In Erdman's review of human and animal trials with lycopene, or lycopene-containing extracts, **there is limited support for the in vivo antioxidant function for lyco-pene. Moreover, tissue levels of lycopene appear to be too low to play a meaningful antioxidant role. We conclude that there is an overall shortage of supportive evidence for the "antioxidant hypothesis" as lycopene's major in vivo mechanism of action.** Erdman's laboratory has postulated that metabolic products of lycopene, the lycopenoids, may be responsible for some of lycopene's reported bioactivity.

Concentrations of antioxidant vitamins in maternal and cord serum and their effect on birth outcomes. (Wang et al. 2009) (#143 mother-neonate pairs) Emerging evidence indicates that maternal oxidative stress during pregnancy could impair fetal growth and that antioxidant vitamins (e.g. vitamins A, E and C) have a significant role in maintaining physiological processes of pregnancy and growth. AIMS: To determine the concentrations of vitamins A, E, and C in pair-matched maternal and cord serum samples of neonate, and thus to investigate the relationship between maternal serum levels of these vitamins at delivery and birth outcomes. METHODS: A total of **143 mother-neonate pairs** were recruited into the cross-sectional descriptive study. Demographic information was investigated by questionnaire. After delivery, both cord and maternal blood were collected for quantification of serum levels of vitamins A, E and C by HPLC. RESULTS: **Maternal serum levels of vitamins A and E were significantly higher than those in cord serum.** In contrast, vitamin C level in cord serum was significantly higher than that in maternal serum. Further, **we found that maternal vitamin A status was significantly correlated to both birth weight and birth height, and these were manifested by these findings: (i) per 250.2 g reduction in birth weight concomitant with 1 micromol/L increase in maternal serum vitamin A level;** and (ii) per 1% increase in the ratio of serum vitamin A level of neonate to mother concomitant with 0.8 cm increase in birth height. CONCLUSION: **Maternal vitamin A, but not vitamins E and C, during pregnancy had a significant effect on birth outcomes.** Further studies are necessary to investigate the role of these antioxidant vitamins in fetal growth at various gestation stages.

WHO Vitamin C and Vitamin E trial group. World Health Organisation multicentre randomised trial of supplementation with vitamins C and E among pregnant women at high risk for pre-eclampsia in populations of low nutritional status from developing countries. (Villar et al, 2009) (#687 women)

Supplementation was not associated with a reduction of pre-eclampsia, eclampsia, gestational hypertension, nor any other maternal outcome. Low birthweight (RR:, small for gestational age and perinatal deaths were also unaffected. **Vitamins C and E at the doses used did not prevent pre-eclampsia in these high-risk women.**

Vitamin and mineral use and risk of prostate cancer: the case-control surveillance study. (Zhang et al. 2009) (#1,706

prostate cancer cases and 2,404 matched controls). *Men who used zinc for ten years or more, either in a multivitamin or as a supplement, had an approximately two-fold increased risk of prostate cancer. The finding that long-term zinc intake from multivitamins or single supplements was associated with a doubling in risk of prostate cancer adds to the growing evidence for an unfavorable effect of zinc on prostate cancer carcinogenesis.*

"Is the oxidative stress theory of aging dead?" (STUDY) (Pérez et al. 2009). Because only one (the deletion of the Sod1 gene) of the 18 genetic manipulations we studied had an effect on lifespan, our data calls into serious question the hypothesis that alterations in oxidative damage/stress play a role in the longevity of mice. This 2009 review of experiments in mice concluded that almost all manipulations of antioxidant systems had no effect on aging.

The oxidative stress menace to coronary vasculature: any place for antioxidants? (Briasoulis et al. 2009). Interventional trials have been controversial, with some positive findings, many null findings, and some suggestion of harm in certain high-risk populations. Therefore, treatment with antioxidant vitamins C and E should not be recommended for the prevention or treatment of coronary atherosclerosis.

Oral antioxidant supplementation does not prevent acute mountain sickness: Double blind, randomized placebo-controlled trial. (Baillie et al. 2009) (#83). daily dose of I g l-ascorbic acid, 400 IU of alpha-tocopherol acetate and 600 mg of alpha-lipoic acid. There was no difference in **AMS** incidence or severity between the antioxidant and placebo groups using the **LLS** at any time at high altitude.

Total dietary antioxidant index and survival in patients with glioblastoma multiforme. (Il'yasova et al. 2009) (#814 glioblastoma multiforme cases). Overall, our results indicated no consistent, significant association of survival with dietary antioxidant intake or its combination with vitamin supplements.

Glioblastoma multiforme (**GBM'**) is the most common and most aggressive type of primary brain tumor in humans, involving glial cells and accounting for 52% of all parenchymal brain tumor cases and 20% of all intracranial tumors.

Associations between alpha-tocopherol, beta-carotene, and retinol and prostate cancer survival (Watters et al. 2009)

(#29,133) Previous studies suggest that carotenoids and tocopherols (vitamin E compounds) may be inversely associated with prostate cancer risk, yet little is known about how they affect prostate cancer progression and survival. We investigated whether serum alpha-tocopherol, beta-carotene, and retinol concentrations, or the alpha-tocopherol and beta-carotene trial supplementation, affected survival of men diagnosed with prostate cancer during the **alpha-Tocopherol, beta-Carotene Cancer Prevention Study, a randomized, double-blind, placebo-controlled primary prevention trial** testing the effects of beta-carotene and alpha-tocopherol supplements on cancer incidence in adult male smokers in southwestern Finland (n = **29,133).** Prostate cancer survival was examined using the Kaplan-Meier method with deaths from other causes treated as censoring, and using Cox proportional hazards regression models with hazard ratios (HR) and 95% confidence intervals (CI) adjusted for family history of prostate cancer, age at randomization, benign prostatic hyperplasia, age and stage at diagnosis, height, body mass index, and serum cholesterol. As of April 2005, 1,891 men were diagnosed with prostate cancer and 395 died of their disease. **Higher serum alpha-tocopherol at baseline was associated with improved prostate cancer survival,** especially among cases who had received the alpha-tocopherol intervention of the trial and who were in the highest quintile of alpha-tocopherol at baseline or at the 3-year follow-up measurement. **Serum beta-carotene, serum retinol, and supplemental beta-carotene had no apparent effects on survival**. These findings suggest that higher alpha-tocopherol (and not beta-carotene or retinol) status increases overall prostate cancer survival. Further investigations, possibly including randomized studies, are needed to confirm this observation.

Antioxidant Supplementation and Risk of Incident Melanomas. Results of a Large Prospective Cohort Study. (Asgari et al, 2009) (#69,671 men and women) Objective To examine

whether *antioxidant supplement use is associated with melanoma risk in*

347

light of recently published data from the Supplementation in Vitamins and Mineral Antioxidants (SUVIMAX) study, which reported a 4-fold higher melanoma risk in women randomized to receive a supplement with **nutritionally appropriate doses of antioxidants. Design:** Population-based prospective study (Vitamins and Lifestyle [VITAL] cohort). **Setting:** Western Washington State. **Participants:** A total of **69,671 men and women** who self-reported (1) intake of multivitamins and supplemental antioxidants, including selenium and beta carotene, during the past 10 years and (2) melanoma risk factors on a baseline questionnaire. **Main Outcome Measure:** Incident melanoma identified through linkage to the Surveillance, Epidemiology, and End Results (SEER) cancer registry. **Results:** Cox proportional hazards regression models were used to estimate multivariable relative risks (RRs) and 95% confidence intervals (CIs) for multivitamin, supplemental selenium, and supplemental beta carotene use. **After adjusting for melanoma risk factors, we did not detect a significant association between multivitamin use and melanoma risk in women or in men.** Moreover, **we did not observe increased melanoma risk with the use of supplemental beta carotene** or selenium at doses comparable with those of the SUVIMAX study. **Conclusion: Antioxidants taken in nutritional doses do not seem to increase melanoma risk.**

Folic acid and risk of prostate cancer: results from a randomized clinical trial. (Figueiredo et al, 2009) (643 randomly assigned men).

Data regarding the association between folate status and risk of prostate cancer are sparse and conflicting. We studied prostate cancer occurrence in the Aspirin/Folate Polyp Prevention Study, a placebo-controlled randomized trial of aspirin and folic acid supplementation for the chemoprevention of colorectal adenomas conducted between July 6, 1994, and December 31, 2006. Participants were followed for up to 10.8 years and asked periodically to report all illnesses and hospitalizations. Aspirin alone had no statistically significant effect on prostate cancer incidence, but there were marked differences according to folic acid treatment. Among the **643 men who were randomly assigned** to placebo or supplementation with folic acid, *the estimated probability of being diagnosed with prostate cancer over a 10-year period was 9.7% in the folic acid group and 3.3% in the placebo group*. In contrast, baseline dietary folate intake and plasma folate in nonmultivitamin users were inversely associated with risk of prostate cancer, although these associations did not attain statistical significance in adjusted analyses. These findings highlight the potential complex role of folate in prostate cancer and the possibly different effects of folic acid-containing supplements vs natural sources of folate. (Figueiredo et al, 2009).

Green tea, black tea consumption and risk of lung cancer: a meta-analysis (Tang et al, 2009) (#meta-analysis included 22 studies) Studies investigating the association of green tea and black tea consumption with lung cancer risk have reported inconsistent findings. To provide a quantitative assessment of this association, we conducted a meta-analysis on the topic. Studies were identified by a literature search in PubMed from 1966 to November 2008 and by searching the reference lists of relevant studies. Summary relative risk (RR) estimates and their corresponding 95% confidence intervals (CIs) were calculated based on random-effects model. Our meta-analysis included 22 studies provided data on consumption of green tea or black tea, or both related to lung cancer risk. **For green tea, the summary RR indicated a borderline significant association between highest green tea consumption and reduced risk of lung cancer.** Furthermore, **an increase in green tea consumption of two cups/day was associated with an 18% decreased risk of developing lung cancer. For black tea, no statistically significant association was observe through the meta-analysis.** In conclusion, our data suggest that high or an increase in consumption of green tea but not black tea may be related to the reduction of lung cancer risk. (Tang et al, 2009).

Green tea consumption and risk of stomach cancer: a meta-analysis of epidemiologic studies (Myung, Int J Cancer. et al, 2009) (#13 epidemiologic studies) This meta-analysis investigated the quantitative association between the consumption of green tea and the risk of stomach cancer in epidemiologic studies using crude data and adjusted data. We searched MEDLINE, EMBASE and the Cochrane Review in August 2007. All the articles searched were independently reviewed and selected by 3 evaluators according to predetermined criteria. A total of **13 epidemiologic studies** were included. When all the case-control and cohort studies were pooled, the odds ratios (OR) [corrected] of stomach cancer for the highest level of green tea consumption when compared with the lowest level of consumption were shown to be 1.10 using the crude data and 0.82 using the adjusted data. In the meta-analyses of case-control studies, **no significant association was seen between green tea**

consumption and stomach cancer using the crude data, but green tea was shown to have a preventive effect on stomach cancer using the adjusted data. *In the meta-analyses of the recent cohort studies, the highest green tea consumption was shown to significantly increase stomach cancer risk using the crude data,* but no significant association between them was seen when using the adjusted data. Unlike the case-control studies, no preventive effect on stomach cancer was seen for the highest green tea consumption in the meta-analysis of the recent cohort studies. Further clinical trials are needed. (Myung, Int J Cancer. et al, 2009).

Green tea consumption and gastric cancer in Japanese: a pooled analysis of six cohort studies. (Inoue et al, 2009) (# 219,080 subjects, 3,577 cases of gastric cancer) Previous experimental studies have suggested many possible anti-cancer mechanisms for green tea, but epidemiological evidence for the effect of green tea consumption on gastric cancer risk is conflicting.

OBJECTIVE: To examine the association between green tea consumption and gastric cancer. **METHODS:** We analysed original data from six cohort studies that measured green tea consumption using validated questionnaires at baseline. Hazard ratios (HRs) in the individual studies were calculated, with adjustment for a common set of variables, and combined using a random-effects model. **RESULTS:** During 2 285 968 person-years of follow-up for **a total of 219,080 subjects, 3,577 cases of gastric cancer were identified.** Compared with those drinking <1 cup/day, **no significant risk reduction for gastric cancer was observed with increased green tea consumption in men, even in stratified analyses by smoking status and subsite. In women, however, a significantly decreased risk was observed for those with consumption of > or =5 cups/day.** This decrease was also significant for the distal subsite. **In contrast, a lack of association for proximal gastric cancer was consistently seen in both men and women.**

CONCLUSIONS: Green tea may decrease the risk of distal gastric cancer in women. (Inoue et al, 2009).

Green tea (Camellia sinensis) for the prevention of cancer. Cochrane Database Syst Rev. 2009 Jul 8;(3):CD005004). (Boehm et al, 2009) (#Fifty-one studies with more than 1.6 million participants were included) Tea is one of the most commonly consumed beverages worldwide. Teas from the plant Camellia sinensis can be grouped into green, black and oolong tea. Cross-culturally tea drinking habits vary. **Camellia sinensis contains the active ingredient polyphenol, which has a subgroup known as catechins. Catechins are powerful antioxidants.** It has been suggested that green tea polyphenol may inhibit cell proliferation and observational studies have suggested that green tea may have cancer-preventative effects.

OBJECTIVES: To critically assess any associations between green tea consumption and the risk of cancer incidence and mortality.

SEARCH STRATEGY: We searched eligible studies up to January 2009 in the Cochrane Central Register of Controlled Trials (CENTRAL), MEDLINE, EMBASE, Amed, CancerLit, Psych INFO and Phytobase and reference lists of previous reviews and included studies. **SELECTION CRITERIA:** We included all prospective, controlled interventional studies and observational studies, which either assessed the associations between green tea consumption and risk of cancer incidence or that reported on cancer mortality. **DATA COLLECTION AND ANALYSIS:** At least two review authors independently applied the study criteria, extracted data and assessed methodological quality of studies. Due to the nature of included studies, which were mainly epidemiological, results were summarized descriptively according to cancer diagnosis.

MAIN RESULTS: Fifty-one studies with more than 1.6 million participants were included. Twenty-seven of them were case-control studies, 23 cohort studies and one randomized controlled trial (RCT). **Twenty-seven studies tried to establish an association between green tea consumption and cancer of the digestive tract, mainly of the upper gastrointestinal tract, five with breast cancer, five with prostate cancer, three with lung cancer, two with ovarian cancer, two with urinary bladder cancer one with oral cancer, three further studies included patients with various cancer diagnoses.** The methodological quality was measured with the

Newcastle-Ottawa scale (NOS). The 9 nested case-control studies within prospective cohorts were of high methodological quality, 13 of medium, and 1 of low. One retrospective case-control study was of high methodological quality and 21 of medium and 5 of low. **Results from studies assessing associations between green tea and risk of digestive tract cancer incidence were highly contradictory. There was limited evidence that green tea could reduce the incidence of liver cancer. The evidence for esophageal, gastric, colon, rectum, and pancreatic cancer was conflicting.** In prostate cancer, observational studies with higher methodological quality and the only included RCT suggested a decreased risk in men consuming higher quantities green tea or green tea extracts. **However, there was limited to moderate evidence that the consumption of green tea reduced the risk of lung cancer, especially in men, and *urinary bladder cancer or that it could even increase the risk of the latter.* There was moderate to strong evidence that green tea consumption does not decrease the risk of dying from gastric cancer. There was limited moderate to strong evidence for lung, pancreatic and colorectal cancer.**

AUTHORS' CONCLUSIONS: There is insufficient and conflicting evidence to give any firm recommendations regarding green tea consumption for cancer prevention. The results of this review, including its trends of associations, need to be interpreted with caution and their generalisability is questionable, as the majority of included studies were carried out in Asia (n = 47) where the tea drinking culture is pronounced. **Desirable green tea intake is 3 to 5 cups per day (up to 1200 ml/day), providing a minimum of 250 mg/day catechins. If not exceeding the daily recommended allowance, those who enjoy a cup of green tea should continue its consumption.** Drinking green tea appears to be safe at moderate, regular and habitual use. (Boehm et al, 2009)

Lack of genoprotective effect of phytosterols and conjugated linoleic acids on Caco-2 cells (and did not exhibit potential anti-carcinogenic activity) (Daly et al, 2009) (#) Much

interest has focused on the cholesterol-lowering effects of phytosterols (plant sterols) but limited data suggests they may also possess anti-carcinogenic activity. **Conjugated linoleic acids (CLA),** sourced from meat and dairy products of ruminant animals, has also received considerable attention as a potential anti-cancer agent. Therefore, the aims of this project were to (i) examine the effects of phytosterols and CLA on the viability and growth of **human intestinal Caco-2 cells** and (ii) determine their potential genoprotective (comet assay), COX-2 modulatory (ELISA) and apoptotic (Hoechst staining) activities. **The Caco-2 cell line is a continuous line of heterogeneous human epithelial colorectal adenocarcinoma cells**, developed by the Sloan-Kettering Institute for Cancer Research through research conducted by Dr. Jorgen Fogh. Caco-2 cells were supplemented with the phytosterols campesterol, β-sitosterol, or β-sitostanol, or a CLA mixture, or individual CLA isomers (c10t12-CLA, t9t11-CLA) for 48 h. The three phytosterols, at the highest levels tested, were found to reduce both the viability and growth of Caco-2 cells while CLA exhibited isomer-specific effects. **None of the phytosterols protected against DNA damage.** At a concentration of 25 µM, both c10t12-CLA and t9t11-CLA enhanced ($P < 0.05$) oxidant-induced, but not mutagen-induced, DNA damage. Neither the phytosterols nor CLA induced apoptosis or modulated COX-2 production. In conclusion, **campesterol, β-sitosterol, β-sitostanol, c10t12-CLA, and t9t11-CLA were not toxic to Caco-2 cells, at the lower levels tested, and did not exhibit potential anti-carcinogenic activity.** (Daly et al, 2009). 000

Pre-radiotherapy plasma carotenoids and markers of oxidative stress are associated with survival in head and neck squamous cell carcinoma patients: a prospective study

(Sakhi et al, 2009) (#178 total; 78 HNSCC patients and 100 healthy controls). The purpose of this study was to compare plasma levels of antioxidants and oxidative stress biomarkers in **head and neck squamous cell carcinoma (HNSCC)** patients with healthy controls. Furthermore, the effect of radiotherapy on these biomarkers and their association with survival in HNSCC patients were investigated. METHODS: **Seventy-eight HNSCC patients and 100 healthy controls** were included in this study. Follow-up samples at the end of radiotherapy were obtained in 60 patients. **Fifteen antioxidant biomarkers (6 carotenoids, 4 tocopherols, ascorbic acid, total antioxidant capacity, glutathione redox potential, total glutathione and total cysteine) and four oxidative stress biomarkers (total hydroperoxides, gamma-glutamyl transpeptidase, 8-isoprostagladin F2alpha and ratio of oxidized/total ascorbic acid) were measured in plasma samples.** Analysis

353

of Covariance was used to compare biomarkers between patients and healthy controls. Kaplan-Meier plots and Cox' proportional hazards models were used to study survival among patients. RESULTS: Dietary antioxidants (carotenoids, tocopherols and ascorbic acid), **ferric reducing antioxidant power (FRAP)** and modified FRAP were lower in HNSCC patients compared to controls and **dietary antioxidants decreased during radiotherapy. Total hydroperoxides (d-ROMs), a marker for oxidative stress**, were higher in HNSCC patients compared to controls and increased during radiotherapy. Among the biomarkers analyzed, **high levels of plasma carotenoids before radiotherapy are associated with a prolonged progression-free survival**. Additionally, **high relative increase in plasma levels of d-ROMs and high relative decrease in FRAP during radiotherapy are also positively associated with survival.** CONCLUSIONS: Biomarkers of antioxidants and oxidative stress are unfavorable in HNSCC patients compared to healthy controls, and radiotherapy affects many of these biomarkers. **Increasing levels of antioxidant biomarkers before radiotherapy and increasing oxidative stress during radiotherapy may improve survival indicating that different factors/mechanisms may be important for survival before and during radiotherapy in HNSCC patients.** Thus, the therapeutic potential of optimizing antioxidant status and oxidative stress should be explored further in these patients. *I believe that this important study shows that increased plasma EMODs and lowered plasma antioxidants are associated with increased survival in HNSCC.* (Sakhi et al, 2009).

The protective effect of silibinin against mitomycin C-induced intrinsic apoptosis in human melanoma A375-S2 cells. (Jiang et al, 2009) Silibinin, a natural flavonoid, is known for its hepatoprotective, anti-inflammatory, and anti-carcinogenic effects. We found that silibinin exhibited a protective effect against chemotherapeutic reagent mitomycin C-induced cell death in A375-S2 cells in a p53-dependent manner, which contradicted the findings of previous studies investigating the anti-neoplastic activity of silibinin and developing silibinin as a potential anti-neoplastic drug in clinical therapy. Mitomycin C administration triggered a time- and dose-dependent cell death in A375-S2 cells. Apoptotic morphology, DNA fragmentation, and caspase-3 activation demonstrated that the major cause of A375-S2 cell death by mitomycin C was apoptosis. This was associated with a marked increase of p53 level and changes in mito-

chondria associated proteins. However, **preincubation with silibinin prior to mitomycin C treatment substantially suppressed cell apoptosis,** attenuated the change of p53 and Bcl-2 expressions, blocked the translocation of Bax to mitochondrial outer membrane, and ameliorated the loss of mitochondrial membrane potential, but mitomycin C stimuli led to few changes in the protein levels of caspase 8, Fas ligand, and Fas-associated death domain protein, indicating that silibinin protected cells from mitomycin C-induced apoptosis mainly via suppressing the mitochondria-mediated intrinsic apoptosis pathway, but not in an extrinsic manner. (Jiang et al, 2009).

Are the health attributes of lycopene related to its antioxidant function? (Erdman, Ford and Lindshield, 2009) A variety of epidemiological trials have suggested that higher intake of lycopene-containing foods (primarily tomato products) or blood lycopene concentrations are associated with decreased cardiovascular disease and prostate cancer risk. Of the carotenoids tested, lycopene has been demonstrated to be the most potent in vitro antioxidant leading many researchers to conclude that the antioxidant properties of lycopene are responsible for disease prevention. In our review of human and animal trials with lycopene, or lycopene-containing extracts, **there is limited support for the in vivo antioxidant function for lycopene. Moreover, tissue levels of lycopene appear to be too low to play a meaningful antioxidant role. We conclude that there is an overall shortage of supportive evidence for the "antioxidant hypothesis" as lycopene's major in vivo mechanism of action.** Our laboratory has postulated that metabolic products of lycopene, the lycopenoids, may be responsible for some of lycopene's reported bioactivity.

Does antioxidant vitamin supplementation protect against muscle damage? (McGinley et al. 2009). The antioxidant vitamins C (ascorbic acid) and E (tocopherol) are among the most commonly used sport supplements, and are often taken in large doses by athletes and other sports persons because of their potential protective effect against muscle damage. Although there is some evidence to show that both antioxidants can reduce indices of oxidative stress, **there is little evidence to support a role for vitamin C and/or vitamin E in protecting against muscle damage.** Indeed, *antioxidant supplementation may actually interfere with the cellular signalling*

functions of ROS, thereby adversely affecting muscle performance. Since the potential for long-term harm does exist, the casual use of high doses of antioxidants by athletes and others should perhaps be curtailed.

The oxidative stress menace to coronary vasculature: any place for antioxidants? (Briasoulis et al. 2009). Interventional trials have been controversial, with some positive findings, many null findings, and some suggestion of harm in certain high-risk populations. Therefore, **treatment with antioxidant vitamins C and E should not be recommended for the prevention or treatment of coronary atherosclerosis.**

Use of Supplements of Multivitamins, Vitamin C, and Vitamin E in Relation to Mortality (Pocobelli et al, 2009) (#77,719 subjects aged 50–76 years) Washington State residents aged 50–76 years who completed a mailed self-administered questionnaire in 2000–2002. **Multivitamin use was not related to total mortality.** However, **vitamin C and vitamin E use were associated with small decreases in risk.** In cause-specific analyses, use of multivitamins and use of vitamin E were associated with decreased risks of CVD mortality. In contrast, **vitamin C use was not associated with CVD mortality. Multivitamin and vitamin E use were not associated with cancer mortality.** Some of the associations we observed were small and may have been due to unmeasured healthy behaviors that were more common in supplement users.

N-acetylcysteine does not prevent contrast-induced nephropathy after cardiac catheterization in patients with diabetes mellitus and chronic kidney disease: a randomized clinical trial (Amini et al. 2009) (#90 patients) Patients with **diabetes mellitus (DM) and chronic kidney disease (CKD) constitute to be a high-risk population for the development of contrast-induced nephropathy (CIN), in which the incidence of CIN is estimated to be as high as 50%.** We performed this trial to assess the efficacy of N-acetylcysteine (NAC) in the prevention of this complication. METHODS: In a prospective, double-blind, placebo controlled, randomized clinical trial, we studied **90 patients undergoing elective diagnostic coronary angiography with DM**

and **CKD** (serum creatinine > or = 1.5 mg/dL for men and > or = 1.4 mg/dL for women). The patients were **randomly assigned** to receive either oral NAC (600 mg BID, starting 24 h before the procedure) or placebo, in adjunct to hydration. Serum creatinine was measured prior to and 48 h after coronary angiography. The primary end-point was the occurrence of CIN, defined as an increase in serum creatinine > or = 0.5 mg/dL (44.2 micromol/L) or > or = 25% above baseline at 48 h after exposure to contrast medium. RESULTS: Complete data on the outcomes were available on 87 patients, 45 of whom had received NAC. **There were no significant differences between the NAC and placebo groups in baseline characteristics, amount of hydration, or type and volume of contrast used**, except in gender and the use of statins. CIN occurred in 5 out of 45 (11.1%) patients in the NAC group and 6 out of 42 (14.3%) patients in the placebo group. CONCLUSION: **There was no detectable benefit for the prophylactic administration of oral NAC over an aggressive hydration protocol in patients with DM and CKD**. Department of cardiology, Tehran Heart Center, Tehran University of Medical Sciences, Tehran, Iran.

N-acetylcysteine in cardiovascular-surgery-associated renal failure: a meta-analysis (Nigwekar SU, Kandula P. 2009) (#Twelve studies comprising 1,324 patients)

Clinical trials with N-acetylcysteine (NAC) in perioperative cardiovascular settings have shown inconsistent effects for renal endpoints. We aimed to systematically review these trials to ascertain its role in prevention of post-cardiovascular surgery acute renal failure. METHODS: We searched MEDLINE, EMBASE, Cochrane Renal Health Library, and Google Scholar for randomized controlled studies that evaluated NAC in adult patients undergoing cardiovascular surgery. Acute renal failure, acute renal failure requiring dialysis, and mortality were the primary outcomes. Additional outcomes studied were length of intensive care unit stay, postoperative serum creatinine, creatinine clearance, renal biomarkers, and adverse effects of NAC. RESULTS: **Twelve studies comprising 1,324 patients** were found to be eligible. **Meta-analytic estimates showed that NAC was not associated with reduction in acute renal failure, acute renal failure requiring dialysis or mortality. N-acetylcysteine was well tolerated but was not associated with any reduction in the length of intensive care unit stay**. It had inconsistent effects on postoperative serum creatinine, creatinine clearance, and renal biomarkers. Subgroup analysis restricted to studies using

357

intravenous NAC preparation showed a nonsignificant trend toward reduction in acute renal failure without any significant change in other outcomes. CONCLU-SIONS: **Overall analysis of the existent literature shows that NAC is not beneficial in the prevention of post-cardiovascular surgery renal dysfunction. Routine use of NAC for this indication should be avoided.** Department of Internal Medicine, Rochester General Hospital and University of Rochester School of Medicine, Rochester, New York 14621, USA.

Meta-analysis of N-acetylcysteine to prevent acute renal failure after major surgery (Ho M, Morgan DJ. 2009) (#10 studies involving a total of 1,193 adult patients)

Acute renal failure after major surgery is associated with significant mortality and morbidity that theoretically may be attenuated by N-acetylcysteine. DESIGN: Meta-analysis of relevant studies sourced from the Cochrane Controlled Trial Register (2007 issue 4), EMBASE, and MEDLINE databases (1966 to February 1, 2008) with-out language restriction. SETTING & POPULATION: Adult patients undergoing major surgery without the use of radiocontrast. SELECTION CRITERIA FOR STUDIES: **Randomized controlled studies** comparing N-acetylcysteine with a placebo perioperatively. DATA ANALYSIS: Categorical variables are reported as odds ratio (OR) with 95% confidence interval (CI), and continuous variables are reported as weighted-mean-difference (WMD) with 95% CI. OUTCOME MEA-SURES: Effects of N-acetylcysteine on mortality and acute renal failure requir-ing dialysis were the main outcomes of interest. Additional outcome measures included an incremental increase in serum creatinine concentration greater than 25% above baseline, surgical reexploration for bleeding, amount of allogeneic blood transfusion, and length of intensive care unit stay. RESULTS: **10 studies involving a total of 1,193 adult patients undergoing major surgery were considered. N-Acetylcysteine use was not associated with a decrease in mortality,** acute renal failure requiring dialysis, incremental increase in serum creatinine concentration greater than 25% above baseline, **or length of inten-sive care unit stay.** N-acetylcysteine did not appear to increase the risk of surgi-cal reexploration for bleeding or amount of allogeneic blood transfusion required. LIMITATIONS: Most studied patients had cardiac surgery and normal renal func-tion preoperatively. CONCLUSIONS: **There is no current evidence that N-acetylcysteine used perioperatively can alter mortality or renal out-**

comes when radio-contrast is not used. Intensive Care Unit, Royal Perth Hospital, Perth, WA 6000, Australia.

Efficacy of N-acetylcysteine in preventing renal injury after heart surgery: a systematic review of randomized trials (Adabag et al. 2009) (#1163)
The aim of this study was to assess whether perioperative N-acetylcysteine (NAC), an antioxidant, prevents **acute renal injury (ARI)** after cardiac surgery. METHODS AND RESULTS: We performed **a systematic review** of **randomized controlled trials (RCTs) of NAC in adult cardiac surgery patients**. The RCTs were identified by searching MEDLINE (1960-2008), clinicaltrials.gov website, and hand-searching references of relevant publications. Primary outcome was ARI (absolute increase >0.5 mg/dL or relative increase >25%, in serum creatinine from baseline within 5 days after surgery). Random effects model was used to perform **a meta-analysis**. Forest plots and I(2) test were used to assess heterogeneity among studies. **Ten RCTs (n = 1163 patients)** were included. Mean age was 70 +/- 7.4 years, 71% were male, and 66% underwent coronary artery bypass surgery. **N-Acetylcysteine did not reduce ARI incidence.** Overall, 3.3% of patients required and 3% died. There was a trend towards reduced ARI incidence among patients with baseline chronic kidney disease assigned to intravenous NAC. CONCLUSION: This **meta-analysis of RCTs showed that prophylactic perioperative NAC in cardiac surgery does not reduce ARI, hemodialysis, or death.** Division of Cardiology (111 C), Minneapolis Veterans Affairs Medical Center, The Minneapolis VA Center for Chronic Disease Outcomes Research, and the University of Minnesota.

Circa 2010

Vitamin C supplements and the risk of age-related cataract: a population-based prospective cohort study in women.
(Rautiainen et al, 2010) (#24,593 women) Experimental animal studies have shown adverse effects of high-dose vitamin C supplements on age-related cataract. **Objective:** We examined whether **vitamin C supplements (≈1000 mg) and multivitamins containing vitamin C (≈60 mg)** are associated with the incidence of age-related cataract extraction in a population-based, prospective cohort of women. **Design:** Our study included **24,593 women** aged 49–83

y from the Swedish Mammography Cohort (follow-up from September 1997 to October 2005). We collected information on dietary supplement use and lifestyle factors with the use of a self-administrated questionnaire. Cataract extraction cases were identified by linkage to the cataract extraction registers in the geographical study area. **Results:** During the **8.2 y of follow-up** (184,698 person-years), we identified 2497 cataract extraction cases. The multivariable hazard ratio (HR) for vitamin C supplement users compared with that for nonusers was 1.25 (95% CI: 1.05, 1.50). The HR for the duration of >10 y of use before baseline was 1.46 (95% CI: 0.93, 2.31). The HR for the use of multivitamins containing vitamin C was 1.09 (95% CI: 0.94, 1.25). *Among women aged ≥65 y, vitamin C supplement use increased the risk of cataract by 38% (95% CI: 12%, 69%). Vitamin C use among hormone replacement therapy users compared with that among nonusers of supplements or of hormone replacement therapy was associated with a 56% increased risk of cataract* (95% CI: 20%, 102%). Vitamin C use among corticosteroid users compared with that among nonusers of supplements and corticosteroids was associated with an HR of 1.97 (95% CI: 1.35, 2.88). **Conclusion**: Our results indicate that *the use of vitamin C supplements may be associated with higher risk of age-related cataract among women.*

The harmful effects with the antioxidant vitamins raises the possibility the antioxidant enzymes may also be harmful. The following paper by Delcourt et al **demonstrates an association between high levels of superoxide dismutase and glutathione peroxidase and an increased risk of cataract formation.**

Associations of cataract with antioxidant enzymes and other risk factors: the French Age-Related Eye Diseases (POLA) Prospective Study. (Delcourt et al, 2003) (#1,947 survivors)

PURPOSE: To determine the association of potential risk factors, including antioxidant enzymes, with the incidence of cataract. DESIGN: Cohort study. PARTICIPANTS: At baseline, the Age-Related Eye Diseases (Pathologies Oculaires Liées à l'Age, POLA) Study included **2,584 residents** of Sète (southern France) aged 60 years or older. From September 1998 to May 2000, **a 3-year follow-up examination was performed on 1947 of the 2436 surviving participants** (79.9%). METHODS: Cataract classification was based on a standardized lens examination at the slit lamp, according to Lens Opacities Classification System III. Biologic mea-

surements were performed at baseline from fasting blood samples. MAIN OUT-
COME MEASURES: At baseline and follow-up, the presence of cataract was defined
as: NC or nuclear opalescence (NO) > or = 4 for nuclear cataract, C > or = 4 for
cortical cataract, and P > or = 2 for posterior cataract (PSC) opacities, using opacity grades corrected for interobserver variability. Incidence rates were assessed
separately for right and left eyes and for each type of cataract. RESULTS: In the multivariate model, **the incidence of cortical cataract was increased in subjects
with high red blood cell superoxide dismutase activity. The incidence of
PSC cataract was increased in subjects with a high level of plasma glutathione peroxidase. In addition to age, gender, and opacities at baseline,
significant risk factors for incident cataract were: long-duration diabetes
and lifetime heavy smoking.** CONCLUSIONS: *Consistent with the baseline analysis, the results of this prospective study suggest that **antioxidant enzymes might be
implicated in the etiology of cataract.*** **This supports my contention that
cataracts are due to an EMOD insufficiency, as is the large group of coex-
istent diseases, such as cancer, heart disease, stroke, arthritis, diabetes,
and cataract formation.**

Micronutrient concentrations and subclinical atherosclero-
sis in adults with HIV. (Falcone et al, 2010) (#298 Nutrition for
Healthy Living participants) Extremes in micronutrient intakes are common
in HIV-infected patients in developed countries and may affect the progression
of atherosclerosis in this population. **Objective:** They completed a cross-sectional study examining the association between serum micronutrient concentrations and surrogate markers of atherosclerosis in a cohort of HIV-infected adults.
**Design: They measured serum selenium, zinc, vitamin A, and vitamin
E concentrations as well as carotid intima-media thickness (c-IMT)
and coronary artery calcium (CAC) in 298 Nutrition for Healthy Living participants.** They performed multivariate regression of c-IMT and CAC
with each micronutrient with adjustment for HIV-related and cardiovascular disease risk factors. **Results:** In the multivariate analysis, **the highest tertile of
serum vitamin E concentration was associated with higher common
and internal c-IMT and CAC scores. Participants with higher vitamin
E concentrations were more likely to have detectable CAC and common c-IMT >0.8 mm.** Other than vitamin E, micronutrients had no association
with markers of atherosclerosis. **Conclusions:** Our study showed that *elevated*

serum vitamin E concentrations are associated with abnormal markers of atherosclerosis and may increase the risk of cardiovascular complications in HIV-infected adults.

Multivitamin use and breast cancer incidence in a prospective cohort of Swedish women. (Larsson et al, 2010) (#35,329 cancer-free women)

Many women use multivitamins in the belief that these supplements will prevent chronic diseases such as cancer and cardiovascular disease. However, whether the use of multivitamins affects the risk of breast cancer is unclear. **Objective:** They prospectively examined the association between multivitamin use and the incidence of invasive breast cancer in the **Swedish Mammography Cohort. Design:** In 1997, **35,329 cancer-free women** completed a self-administered questionnaire that solicited information on multivitamin use as well as other breast cancer risk factors. Relative risks (RRs) and 95% CIs were calculated by using Cox proportional hazard models and adjusted for breast cancer risk factors. **Results:** During a mean follow-up of 9.5 y, 974 women were diagnosed with incident breast cancer. *Multivitamin use was associated with a statistically significant increased risk of breast cancer.* The association did not differ significantly by hormone receptor status of the breast tumor. **Conclusions:** These results suggest that *multivitamin use is associated with an increased risk of breast cancer. Use of multivitamins was linked to a statistically significant 19 per cent increased risk of breast cancer* (after adjusting for lifestyle and risk factors like weight, diet, smoking, exercise, and family history of breast cancer.

Women and men use multivitamins in the belief that they will protect them from chronic diseases like cancer and heart disease but that is not proven. Please remember that in **February 2009, the *Archives of Internal Medicine* published details of a large US study of over 160,000 postmenopausal women that found no convincing evidence that long term use of multivitamins changed their risk of developing common cancers, cardiovascular disease or dying prematurely.**

Other research has found that women who take multivitamins have increased breast density, which is linked to a relatively higher risk of

breast cancer (Berube et al, 2008). **The widespread use of multiviamins points out an important public health concern.**

Vitamins C and E to prevent complications of pregnancy-associated hypertension (Roberts et al, 2010) (#10,154) Oxidative

stress has been proposed as a mechanism linking the poor placental perfusion characteristic of preeclampsia with the clinical manifestations of the disorder. We assessed the effects of **antioxidant supplementation with vitamins C and E**, initiated early in pregnancy, on the risk of serious adverse maternal, fetal, and neonatal outcomes related to pregnancy-associated hypertension. METHODS: They conducted a multicenter, **randomized, double-blind trial** involving nulliparous women who were at low risk for preeclampsia. Women were randomly assigned to begin daily supplementation with **1000 mg of vitamin C and 400 IU of vitamin E or matching placebo between the 9th and 16th weeks of pregnancy.** The primary outcome was severe pregnancy-associated hypertension alone or severe or mild hypertension with elevated liver-enzyme levels, thrombocytopenia, elevated serum creatinine levels, eclamptic seizure, medically indicated preterm birth, fetal-growth restriction, or perinatal death. RESULTS: A total of **10,154 women** underwent randomization. The two groups were similar with respect to baseline characteristics and adherence to the study drug. Outcome data were available for 9,969 women. **There was no significant difference between the vitamin and placebo groups in the rates of the primary outcome or in the rates of preeclampsia. Rates of adverse perinatal outcomes did not differ significantly between the groups.** CONCLUSIONS: **Vitamin C and E supplementation initiated in the 9th to 16th week of pregnancy in an unselected cohort of low-risk, nulliparous women did not reduce the rate of adverse maternal or perinatal outcomes related to pregnancy-associated hypertension.**

Vitamin E and age-related macular degeneration in a randomized trial of women. (Christen et al. 2010) (#39,876) a

large-scale randomized trial of female health professionals, long-term alternate-day use of 600 IU of natural-source vitamin E had no large beneficial or harmful effect on risk of AMD. Investigators tested whether alternate day vitamin E affects the incidence of age-related macular degeneration (AMD) in a large-scale randomized trial of women.

DESIGN: Randomized, double-masked, placebo-controlled trial. **PARTICI-PANTS: Thirty-nine thousand eight hundred seventy-six (39,876)** apparently healthy female health professionals aged 45 years or older. **INTER-VENTION:** Participants were assigned randomly to receive either 600 IU of natural-source vitamin E on alternate days or placebo. **MAIN OUTCOME MEASURES:** Incident AMD responsible for a reduction in best-corrected visual acuity to 20/30 or worse based on self-report confirmed by medical record review. **RESULTS:** After 10 years of treatment and follow-up, there were 117 cases of AMD in the vitamin E group and 128 cases in the placebo group.

CONCLUSIONS: In a **large-scale randomized trial of female health professionals, long-term alternate-day use of 600 IU of natural-source vitamin E had no large beneficial or harmful effect on risk of AMD**.

Vitamins C and E for prevention of pre-eclampsia in women with type 1 diabetes (DAPIT) (McCance et al. 2010) (#762 women)

Rates of pre-eclampsia did not differ between vitamin (15%, n=57) and placebo (19%, 70) groups. **Supplementation with vitamins C and E did not reduce risk of pre-eclampsia in women with type 1 diabetes.**

Daily intake of antioxidants in relation to survival among adult patients diagnosed with malignant glioma. (DeLorenze et al. 2010).

Geometric mean values for 11 fat-soluble and 6 water-soluble individual antioxidants, antioxidant index and 3 macronutrients were virtually the same when comparing all cases (n=748) to self-reported cases only (n=450). For patients diagnosed with Grade II and Grade III histology, moderate (915.8-2118.3 mcg) intake of fat-soluble lycopene was associated with poorer survival when compared to low intake (0.0-914.8 mcg), for self-reported cases only. High intake of vitamin E and moderate/high intake of secoisolariciresinol among Grade III patients indicated greater survival for all cases. In Grade IV patients, moderate/high intake of cryptoxanthin and high intake of secoisolariciresinol were associated with poorer survival among all cases. Among Grade II patients, moderate intake of water-soluble folate was associated with greater survival for all cases; high intake of vitamin C and genistein and the highest level of the antioxidant index were associated with poorer survival for all cases.

CONCLUSIONS: The associations observed in our study suggest that the influence of some antioxidants on survival following a diagnosis of malignant glioma are inconsistent and vary by histology group.

Effects of vitamin on stroke subtypes: meta-analysis of randomized controlled trials. (Schurks et al. 2010) (#118,765)

Systematic review and meta-analysis of randomized, placebo controlled trials published until January 2010. **Data sources** Electronic databases (Medline, Embase, Cochrane Central Register of Controlled Trials) and reference lists of trial reports. **Selection criteria** Randomized, placebo controlled trials with ≥1 year of follow-up investigating the effect of vitamin E on stroke. **Review methods and data extraction** Two investigators independently assessed eligibility of identified trials. Disagreements were resolved by consensus. Two different investigators independently extracted data. Risk ratios (and 95% confidence intervals) were calculated for each trial based on the number of cases and non-cases randomized to vitamin E or placebo. Pooled effect estimates were then calculated. **Results** Nine trials investigating the effect of vitamin E on incident stroke were included, **totalling 118,765 participants (59,357 randomized to vitamin E and 59 408 to placebo).** Among those, seven trials reported data for total stroke and five trials each for hemorrhagic and ischemic stroke. Vitamin E had no effect on the risk for total stroke. In contrast, the risk for hemorrhagic stroke was increased, while the risk of ischemic stroke was reduced. There was little evidence for heterogeneity among studies. Meta-regression did not identify blinding strategy, vitamin E dose, or morbidity status of participants as sources of heterogeneity. In terms of absolute risk, this translates into one additional hemorrhagic stroke for every 1,250 individuals taking vitamin E, in contrast to one ischemic stroke prevented per 476 individuals taking vitamin E. **Conclusion** In this meta-analysis, *vitamin E increased the risk for hemorrhagic stroke by 22% and reduced the risk of ischemic stroke by 10%.* This differential risk pattern is obscured when looking at total stroke. Given the relatively small risk reduction of ischemic stroke and the generally more severe outcome of hemorrhagic stroke, *indiscriminate widespread use of vitamin E should be cautioned against.*

Multivitamin/Mineral supplementation does not affect standardized assessment of academic performance in elementary

school children (Perlman et al. 2010) (#students in grades three through six, approximate age range=8 to 12 years old) Limited research suggests that micronutrient supplementation may have a positive effect on the academic performance and behavior of school-aged children. To determine the effect of multivitamin/mineral supplementation on academic performance, **students in grades three through six (approximate age range=8 to 12 years old)** were recruited from 37 parochial schools in northern New Jersey to participate in a **double-blind, placebo-controlled clinical trial** conducted during the 2004-2005 academic school year. Participants were randomized to receive either a standard children's multivitamin/mineral supplement (MVM) or a placebo. MVM or placebo was administered in school only during lunch or snack period by a teacher or study personnel who were blinded to group assignment. The main outcome measured was change in scores on **Terra Nova, a standardized achievement test administered by the State of New Jersey**, at the beginning of March 2005 compared to March 2004. **Compared with placebo, participants receiving MVM supplements showed no statistically significant improvement for Terra Nova National Percentile total scores by treatment assignment or for any of the subject area scores using repeated measures analysis of variance. No significant improvements were observed in secondary end points: number of days absent from school, tardiness, or grade point average.** In conclusion, the in-school **daily consumption of an MVM supplement by third- through sixth-grade inner-city children did not lead to improved school performance** based upon standardized testing, grade point average, and absenteeism.

Differential effects of concomitant use of vitamins C and E on trophoblast apoptosis and autophagy between normoxia and hypoxia-reoxygenation (Hung et al, 2010) *Concomitant supplementation of vitamins C and E during pregnancy has been reportedly associated with low birth weight, the premature rupture of membranes and fetal loss or perinatal death in women at risk for preeclampsia*; however, the cause is unknown. They surmise that hypoxia-reoxygenation (HR) within the intervillous space due to abnormal placentation is the mechanism and hypothesize that concomitant administration of aforementioned vitamin antioxidants detrimentally affects trophoblast cells during HR. METHODOLOGY/PRINCIPAL FINDINGS: Using villous explants, concomitant

administration of 50 microM of vitamins C and E was observed to reduce apoptotic and autophagic changes in the trophoblast layer at normoxia (8% oxygen) but to cause more prominent apoptosis and autophagy during HR. Furthermore, increased levels of Bcl-2 and Bcl-xL in association with a decrease in the autophagy-related protein LC3-II were noted in cytotrophoblastic cells treated with vitamins C and E under standard culture conditions. In contrast, vitamin treatment decreased Bcl-2 and Bcl-xL as well as increased mitochondrial Bak and cytosolic LC3-II in cytotrophoblasts subjected to HR.

CONCLUSIONS/SIGNIFICANCE: **Our results indicate that concomitant administration of vitamins C and E has differential effects on the changes of apoptosis, autophagy and the expression of Bcl-2 family of proteins in the trophoblasts between normoxia and HR.** *These changes may probably lead to the impairment of placental function and suboptimal growth of the fetus.*

Green Tea Drinking and Subsequent Risk of Breast Cancer in a Population to Based Cohort of Japanese Women/No effect. (Iwasaki et al, 2010) (#53,793).

Although many *in vitro* and animal studies have demonstrated a protective effect of green tea against breast cancer, findings from epidemiological studies have been inconsistent, and whether high green tea intake reduces the risk of breast cancer remains unclear. **Methods:** In this Japan Public Health Center-based Prospective Study, **581 cases of breast cancer were newly diagnosed in 53,793 women during 13.6 years' follow-up** from the baseline survey in 1990 to 1994. After the five-year follow-up survey in 1995 to 1998, 350 cases were newly diagnosed in 43,639 women during 9.5 years' follow-up. The baseline questionnaire assessed the frequency of total green tea drinking while the five-year follow-up questionnaire assessed that of two types of green tea, *Sencha* and *Bancha/Genmaicha*, separately. **Results:** Compared with women who drank less than one cup of green tea per week, the adjusted hazard ratio (HR) for women who drank five or more cups per day was 1.12 in the baseline data. Similarly, compared with women who drank less than one cup of *Sencha* or *Bancha/Genmaicha* per week, adjusted HRs for women who drank 10 or more cups per day were 1.02 for *Sencha* and 0.86 for *Bancha/Genmaicha*. No inverse association was found regardless of hormone receptor-defined subtype or menopausal status. **Conclusions:** In this **population-based**

367

prospective cohort study in Japan we found no association between green tea drinking and risk of breast cancer (Iwasaki et al, 2010).

Specifically, (-)-epigallocatechin-3-gallate (EGCG), the most abundant and biologically active catechin in green tea, might play an important role in cancer prevention, but this is not supported by this study.

Green tea consumption and breast cancer risk or recurrence: a meta-analysis. (Ogunleye et al, 2010) (#5,617 breast cancer cases).

Green tea is a commonly consumed beverage in Asia and has been suggested to have anti-inflammatory and possible anti-carcinogenic properties in laboratory studies. We sought to examine the association between green tea consumption and risk of breast cancer incidence or recurrence, using all available epidemiologic evidence to date. We conducted a systematic search of five databases and performed a meta-analysis of studies of breast cancer risk and recurrence published between 1998 and 2009, encompassing 5,617 cases of breast cancer. Summary relative risks (RR) were calculated using a fixed effects model, and tests of heterogeneity across combined studies were conducted. We identified two studies of breast cancer recurrence and seven studies of breast cancer incidence. Increased green tea consumption (more than three cups a day) was inversely associated with breast cancer recurrence. An analysis of case-control studies of breast cancer incidence suggested an inverse association with a pooled RR of 0.81 while **no association was found among cohort studies of breast cancer incidence. Combining all studies of breast cancer incidence resulted in significant heterogeneity.** Available epidemiologic evidence supports the hypothesis that increased green tea consumption may be inversely associated with risk of breast cancer recurrence. **The association between green tea consumption and breast cancer incidence remains unclear based on the current evidence** (Ogunleye et al, 2010). 000

Long-Term Use of supplemental vitamin C, vitamin D and vitamin E does not reduce the risk of urothelial cell carcinoma of the bladder in the VITamins and Lifestyle Study. (Hotaling 2010) (#77,050 eligible VITAL participants).

Urothelial carcinoma has the highest lifetime treatment cost of any cancer, making it an ideal target for preventative therapies. Previous work has suggested that certain vitamin and mineral supplements may reduce the risk of urothelial carcinoma. We used the prospective VITamins And Lifestyle cohort to examine the association of all commonly taken vitamin and mineral supplements as well as 6 common anti-inflammatory supplements with incident urothelial carcinoma in a United States population. **Long-Term Use of supplemental vitamin C, vitamin D and vitamin E does not reduce the risk of urothelial cell carcinoma of the bladder in the VITamins and Lifestyle Study. 000**

The vitamin C: vitamin K3 system - enhancers and inhibitors of the anticancer effect. (Lamson DW et al, 2010) The

oxidizing anticancer system of vitamin C and vitamin K3 (VC:VK3, producing hydrogen peroxide via superoxide) was combined individually with melatonin, curcumin, quercetin, or cholecalciferol (VD3) to determine interactions. Substrates were LNCaP and **PC-3 prostate cancer cell lines**. Three of the tested antioxidants displayed differences in cell line cytotoxicity. **Melatonin combined with VC:VK3 quenched the oxidizing effect,** while VC:VK3 applied 24 hours after melatonin showed no quenching. With increasing curcumin concentrations, an apparent combined effect of VC:VK3 and curcumin occurred in LNCaP cells, but not PC-3 cells. Quercetin alone was cytotoxic on both cell lines, but demonstrated an additional 50-percent cytotoxicity on PC-3 cells when combined with VC:VK3. **VD3 was effective against both cell lines, with more effect on PC-3.** This effect was negated on LNCaP cells with the addition of VC:VK3. In conclusion, **a natural antioxidant can enhance or decrease the cytotoxicity of an oxidizing anticancer system in vitro, but generalizations about antioxidants cannot be made.** (Lamson et al, 2010).

Age-related cataract in a randomized trial of vitamins E and C in men. (Christen et al. 2010. men) (#11,545). OBJECTIVE:

To test whether supplementation with alternate-day vitamin E or daily vitamin C affects the incidence of age-related cataract in a large cohort of men. METHODS: In a randomized, double-masked, placebo-controlled trial, **11,545 apparently healthy US male physicians** 50 years or older without a diagnosis of cataract at baseline were randomly assigned to receive **400 IU of vitamin E or placebo on alternate days and 500 mg of vitamin C** or placebo daily.

MAIN OUTCOME MEASURE: Incident cataract responsible for a reduction in best-corrected visual acuity to 20/30 or worse based on self-report confirmed by medical record review. APPLICATION TO CLINICAL PRACTICE: Long-term use of vitamin E and C supplements has no appreciable effect on cataract. RESULTS: **After 8 years of treatment and follow-up,** 1174 incident cataracts were confirmed. There were 579 cataracts in the vitamin E-treated group and 595 in the vitamin E placebo group. For vitamin C, there were 593 cataracts in the treated group and 581 in the placebo group.

CONCLUSION: **Long-term alternate-day use of 400 IU of vitamin E and daily use of 500 mg of vitamin C had no notable beneficial or harmful effect on the risk of cataract.** (Christen et al. 2010).

Effects of vitamin E on plasma lipid status and oxidative stress in Chinese women with metabolic syndrome. (Wang et al, 2010) (#not available)

Following the change of dietary structure and living style, metabolic syndrome (MetS) has become increasingly common in China, especially in women, who have abnormal plasma lipid profiles with increased levels of oxidative stress. **Vitamin E (VitE) is a powerful chain-breaking antioxidant,** which may be a protective factor against oxidative stress-related diseases. This study investigated the effects of three different dosages of tocopherol supplementation (100 IU /day, 200 IU /day, 300 IU /day) for 4 months in Chinese women with MetS. The **plasma VitE concentrations increased significantly after the 4 months of supplementation** ($p < 0.01$). **The protective decreases in plasma total cholesterol were significant in 200 IU/day and 300 IU/day VitE groups** ($p < 0.05$), **but decreases in high-density lipoprotein cholesterol were also significant in all the supplementation groups** ($p < 0.05$). **Plasma triglycerides were unaltered** ($p > 0.05$). The indicators of oxidative stress decreased substantially in all of the VitE supplementation groups: malondialdehyde (MDA) was reduced by nearly 50 percent (all groups, $p < 0.001$), erythrocyte hemolysis was decreased by nearly 40 percent (all groups, $p < 0.05$); among which the 300IU/day VitE group showed the most significant effect. **However, the activity of superoxide dismutase (SOD) decreased after the trial** ($p < 0.001$). **VitE provided marked benefits in reducing oxidative stress levels and improving lipid status in women with MetS.** Although **no dose-effect relationship was observed,** 300 IU VitE per day showed the optimal effect. Research is needed to identify potential protective mechanisms or utilization of vitamin E during MetS. (Wang et al, 2010).

Serum selenium and prognosis in cardiovascular disease: results from the AtheroGene study. (Lubos et al, 2010) (#1,731)

Experimental data suggest a protective role of the essential trace element selenium against cardiovascular disease (CVD), whereas epidemiological data remains controversial. We aimed to investigate the impact of serum selenium concentration in patients presenting with **stable angina pectoris (SAP)** or acute coronary syndrome (ACS) on long term prognosis. METHODS: Baseline selenium concentration was measured in **1731 individuals** (852 with SAP, and 879 with ACS). During a median follow-up of 6.1 years, 190 individuals died from cardiovascular causes. RESULTS: In those ACS patients who subsequently died of cardiac causes, selenium levels were lower compared to survivors (61.0microg/L versus 71.5microg/L; P<0.0001). In a fully adjusted model, patients in the highest tertile of selenium concentration had a hazard ratio of 0.38 (95% CI: 0.16-0.91; P=0.03) as compared with those in the lowest. **No association between selenium levels and cardiovascular outcome was observed in SAP.** CONCLUSIONS: **Low selenium concentration was associated with future cardiovascular death in patients with ACS.** (Lubos, 2010). It must be kept in mind that elemental selenium has prooxidant character.

Low Total and Nonheme Iron Intakes Are Associated with a Greater Risk of Hypertension. (Galan et al, 2010) (#2,895) The

relationship between iron intake and blood pressure (BP) status has not been well established. Only 1 cross-sectional study has suggested an inverse association of dietary total iron intake and nonheme iron intake with BP. We investigated the relationship between total, heme, and nonheme iron intakes, markers of iron status, 5.4-y changes in BP, and the incidence of hypertension. A total of 2895 participants included in the Supplémentation en Vitamines et Minéraux Antioxydants cohort were followed up for 5.4 y. At least 3 repeated 24-h dietary records were performed at baseline and 5.4 y later. Hemoglobin and serum ferritin concentrations were assessed at baseline. **Low nonheme iron intake at baseline was associated with a greater increase in systolic BP (SBP) and pulse pressure over time after adjustment for multiple possible confounding factors** (P-trend = 0.002 and 0.0005, respectively). Conversely, participants in the 3rd tertile of nonheme iron intake at baseline had a 37% lower risk of hypertension after 5.4 y of follow-up compared with those in the first tertile (P-trend = 0.04). Heme iron intake was not associated with BP changes or risk of hypertension. **Meat intake was positively associated with an increase in SBP** (P-trend = 0.04). **However, that relation became nonsignificant after adjusting for dietary pattern scores.** Baseline hemoglobin and ferritin concentrations were not associated with changes in BP or incidental hypertension. **Their data support a possible role of low nonheme iron intake, independent of heme iron intake, in the development of hypertension.** (Galan et al, 2010)

Multivitamin use and the risk of myocardial infarction: a population-based cohort of Swedish women. (Rautiainen et al, 2010) (#31,671 women with no history of cardiovascular disease (CVD) and 2,262 women with a history of CVD; total number - **33,933**). Dietary supplements are widely used in industrialized countries. **Objective:** The objective was to examine the association between multivitamin use and myocardial infarction (MI) in a prospective, population-based cohort of women. **Design:** The study included 31,671 women with no history of cardiovascular disease (CVD) and 2262 women with a history of CVD aged 49–83 y from Sweden. Women completed a self-administered questionnaire in 1997 regarding dietary supplement use, diet, and lifestyle factors. Multivitamins were estimated to contain nutrients close to recommended daily allowances: vitamin A (0.9 mg), vitamin C (60 mg), vitamin D (5 µg), vitamin E (9 mg), thiamine (1.2 mg), riboflavin (1.4 mg), vitamin B-6 (1.8 mg), vitamin B-12 (3 µg), and folic acid (400 µg). **Results:** During an average of 10.2 y of follow-up, 932 MI cases were identified in the CVD-free group and 269 cases in the CVD group. In the CVD-free group, use of multivitamins only, compared with no use of supplements, was associated with a multivariable-adjusted hazard ratio (HR) of 0.73 (95% CI: 0.57, 0.93). The HR for multivitamin use together with other supplements was 0.70 (95% CI: 0.57, 0.87). The HR for use of supplements other than multivitamins was 0.93 (95% CI: 0.81, 1.08). The use of multivitamins for ≥5 y was associated with an HR of 0.59 (95% CI: 0.44, 0.80). **In the CVD group, use of multivitamins alone or together with other supplements was not associated with MI. Conclusions: The use of multivitamins was inversely associated with MI, especially long-term use among women with no CVD.** (Rautiainen et al, 2010)

Impact of high-dose N-acetylcysteine versus placebo on contrast-induced nephropathy and myocardial reperfusion injury in unselected patients with ST-segment elevation myocardial infarction undergoing primary percutaneous coronary intervention. The LIPSIA-N-ACC (Prospective, Single-Blind, Placebo-Controlled, Randomized Leipzig Immediate PercutaneouS Coronary Intervention Acute Myocardial Infarction N-ACC) Trial (Thiele H, et al. 2010) (#251) The aim of this randomized, single-blind, controlled trial was to assess N-acetylcysteine effects on contrast-induced nephropathy and reperfusion injury in ST-segment elevation myocardial infarction patients undergoing primary angioplasty with moderate contrast volumes. BACKGROUND: High-dose N-acetylcysteine reduced the incidence of contrast-induced nephropathy (CIN) in patients with high contrast volumes and reduced reperfusion injury in animal trials. METHODS: Patients undergoing primary angioplasty were randomized to either high-dose N-acetylcysteine (2 x 1,200 mg/day for 48 h; n = 126) or placebo plus optimal hydration (n = 125). The 2 primary end points were: 1) the occurrence of >25% increase in serum creatinine level <72 h after randomization; and 2) a reduction in reperfusion injury measured as myocardial salvage index by magnetic resonance imaging. RESULTS: The median volume of an iso-osmolar contrast agent during angiography was 180 ml in the N-acetylcysteine and 160 ml in the placebo group. The primary end point contrast-induced nephropathy occurred in 14% of the N-acetylcysteine group and in 20% of the placebo group. The myocardial salvage index was also not different between both treatment groups. Activated oxygen protein products and oxidized low-density lipoprotein as markers for oxidative stress were reduced by as much as 20% in the N-acetylcysteine group, whereas no change was evident in the placebo group. CONCLUSIONS: High-dose intravenous N-acetylcysteine reduces oxidative stress. However, it does not provide an additional clinical benefit to placebo with respect to CIN and myocardial reperfusion injury in nonselected patients undergoing angioplasty with moderate doses of contrast medium and optimal hydration. The LIPSIA-N-ACC. Department of Internal Medicine-Cardiology, University of Leipzig Heart Center, Leipzig, Germany.

NAC had no effect on blood pressure and surrogate markers of cardiovascular injury in non-diabetic patients with

CKD (Renke et al. 2010) (#20 non-diabetic patients with albuminuria) Cardiovascular complications in patients with chronic kidney disease (CKD) are frequent. They show increased cardiovascular mortality and morbidity attributable to accumulation of several risk factors; e.g., hypertension, oxidative stress and elevated plasma homocysteine concentration. Despite recent progress in their management, there is still no optimal therapy that can stop progression of CKD and decrease cardiovascular outcome in these patients. **Antioxidants, e.g., N-acetylcysteine (NAC), have been suggested as a promising medicament** in this field. MATERIAL/METHODS: In a **placebo-controlled, randomized**, two-period cross-over study we evaluated the influence of **eight weeks** of **NAC therapy (1200 mg/day)** added to pharmacological renin-angiotensin system blockade on ambulatory blood pressure and surrogate markers of cardiovascular risk and injury in **20 non-diabetic patients with albuminuria** [30-915 mg per creatinine mg] and normal or slightly decreased kidney function [eGFR 61-163 ml/min]. After eight weeks run-in period during which the therapy using angiotensin converting enzyme inhibitors and/or angiotensin receptor blockers was settled, patients were randomly assigned to one of two treatment sequences: NAC/washout/placebo or placebo/washout/ NAC. RESULTS: **No significant changes in blood pressure, albuminuria and homocysteine plasma level were observed.** CONCLUSIONS: **NAC had no effect on blood pressure and surrogate markers of cardiovascular injury in non-diabetic patients with CKD.**

Quercetin and Ferulic Acid Aggravate Renal Carcinoma in Long-Term Diabetic Victims. (Chiu-Lan Hsieh et al, 2010) Many phytoantioxidants have therapeutic drawbacks due to their potent prooxidant bioactivity. It is hypothesized that phytoantioxidants (PAO) are beneficial only to the early-stage diabetes mellitus (DM) and will become ineffective once renopathy occurs. Gallic acid, rutin, EGCG, ferulic acid (FA), and quercetin were tried on the streptozotocin (STZ)-induced **DM rat model** for a 28 week experimental period. All of these PAO were shown to be ineffective for hypoglycemic action. The incidence of cataract (50%), injured glomerules, and renal cell carcinoma (RCC) was very common, among which the most severely affected involved the quercetin- and the FA-treated groups. The tumorigenicity of ferulic acid is still unclear. However, for quercetin, this can be attributted to (i) the prooxidant effect, (ii) the insulin—secretagogue bioactivity, and (iii) the competitive and noncompetitive inhibition on the O-methyltransferase to enhance the estradiol-induced tumorigenesis. Conclusively, the phytoantioxidants *quercetin and ferulic acid are able to aggravate, if not induce, nephrocarcinoma in mice.* It is time to reevaluate the tumorigenic detrimental effect of PAO, especially those exhibiting prooxidant bioactivity.

No beneficial effects of pine bark extract on cardiovascular disease risk factors (Drieling et al. 2010) (#130 individuals with increased cardiovascular disease risk)

Although modifiable cardiovascular disease risk factors are common, some patients eschew conventional drug treatments in favor of natural alternatives. **Pine bark extract, a dietary supplement source of antioxidant oligomeric proanthocyanidin complexes,** has multiple putative cardiovascular benefits. Studies published to date about the supplement have notable methodological limitations. METHODS: They **randomized 130 individuals with increased cardiovascular disease risk** to take **200 mg of a water-based extract of pine bark** (n = 64; Toyo-FVG, Toyo Bio-Pharma, Torrance, California; Shinyaku Co, Ltd, Saga, Japan; also marketed as Flavagenol in Japan) or placebo (n = 66) **once per day**. Blood pressure, our primary outcome, and other cardiovascular disease risk factors were measured at baseline and at **6 and 12 weeks**. Statistical analyses were conducted using regression models. RESULTS: Baseline characteristics did not differ between the study groups. Over the 12-week intervention, the sum of systolic and diastolic blood pressures decreased by 1.0 mm Hg in the pine bark extract-treated group and by 1.9 mm Hg. Other outcomes were likewise not significantly different, including body mass index, lipid panel measures, liver transaminase test results, lipoprotein cholesterol particle size, and levels of insulin, lipoprotein(a), fasting glucose, and high-sensitivity C-reactive protein. **There were no subgroups for whom intake of pine bark extract affected cardiovascular disease risk factors.** CONCLUSIONS: **This pine bark extract (at a dosage of 200 mg/d) was safe but was not associated with improvement in cardiovascular disease risk factors.** Although variations among participants, dosages, and chemical preparations could contribute to different findings compared with past studies, **our results are consistent with a general failure of antioxidants to demonstrate cardiovascular benefits.** (Drieling et al. 2010)

Dietary phytocompounds and risk of lymphoid malignancies in the California Teachers Study cohort. (Chang et al, 2010) (#110,215) Investigators examined whether dietary intake of isoflavones, lignans, isothiocyanates, antioxidants, or specific foods rich in these compounds is associated with reduced risk of B-cell non-Hodgkin lymphoma (NHL), multiple myeloma (MM), or Hodgkin lymphoma (HL) in a large, prospective cohort of women. **METHODS:** Between 1995-1996 and 31 December 2007, among **110,215 eligible members of the California Teachers Study cohort,** 536 women developed incident B-cell NHL, 104 developed MM, and 34 developed HL. Cox proportional hazards regression, with age as the time scale, was used to estimate adjusted rate ratios (RRs) with 95% confidence intervals (CIs) for risk of lymphoid malignancies. **RESULTS: Weak inverse associations with risk of diffuse large B-cell lymphoma were observed for isothiocyanates** and an antioxidant index measuring hydroxyl radical absorbance capacity. **Risk of other NHL subtypes, overall B-cell NHL, MM, or HL was not generally associated with dietary intake of isoflavones, lignans, isothiocyanates, antioxidants, or major food sources of these compounds. CONCLUSIONS: Isoflavones, lignans, isothiocyanates, and antioxidant compounds are not associated with risk of most B-cell malignancies,** but some phytocompounds may decrease the risk of selected subtypes (Chang et al, 2010).

Circa 2011

Vitamin E and All-cause Mortality: A Meta-Analysis (Abner EL, et al, 2011) (#246,371 subjects and 29,295 all-cause deaths). The current analysis reexamines the relationship between supplemental vitamin E and all-cause mortality. All **randomized, controlled trials** testing the treatment effect of vitamin E supplementation in adults for at least one year were sought. **MEDLINE, the Cochrane Library, and Biological Abstracts databases were searched using the terms "vitamin E," "alpha-tocopherol," "antioxidants," "clinical trial," and "controlled trial" for studies published through April 2010**; results were limited to English, German, or Spanish language articles. Studies were also obtained through reference mining. All randomized controlled trials using vitamin E, with a supplementation period of at least one year, to prevent or treat disease in adults were identified and abstracted independently by two raters. Mortality data from trials with a supplementation period of at least one year were pooled. The selected trials (n = 57) were published between 1988 and 2009. Sample sizes range from 28 to 39,876 (median = 423), **yielding 246,371 subjects and 29,295 all-cause deaths. Duration of supplementation for the 57 trials range from one to 10.1 years** (median = 2.6 years). A random effects meta-analysis produce an overall risk ratio of 1.00 (95% confidence interval: 0.98, 1.02); **additional analyses suggest no relationship between dose and risk of mortality. Based on the present meta-analysis, supplementation with vitamin E appears to have no effect on all-cause mortality at doses up to 5,500 IU/d.** (Abner et al, 2011).

Effect of vitamins C and E on antioxidant status of breast-cancer patients undergoing chemotherapy (Suhail et al, 2011) (#forty untreated breast-cancer patients (stage II) and compared with those of healthy controls).

Reactive oxygen/nitrogen species generated by chemotherapy. What is known and Objective: antineoplastic agents are prime suspects for the toxic side-effects of acute or chronic chemotherapy. The present study was undertaken to test whether **vitamins C and E (VCE) supplementation** protect against some of the harmful effects of treatment. In a commonly used anticancer drugs in breast-cancer patients. Methods: **randomized** 5-month study, the activity of various antioxidant enzymes (superoxide dismutase, catalase, glutathione-S-transferase and glutathione reductase) and the levels of malondialdehyde and reduced gluta-thione were measured in **forty untreated breast-cancer patients (stage II) and compared with those of healthy controls.** The degree of DNA damage was also assessed in the peripheral lymphocytes of the patients by alkaline single cell gel electrophoresis. The untreated patients were then randomly assigned to either mg/m(2) i.v. day 1, treatment with chemotherapy alone (5-fluorouracil 500 mg/m(2) i.v. day 1, mg/m(2) i.v. day 1 and cyclophosphamide 500 doxorubicin 50 weeks for six cycles) or to the same chemotherapy regimen supplemented every 3 mg gelatin capsule). On mg tablet and vitamin E 400 with VCE (vitamin C 500 completion of the treatments, both the groups were studied again for the levels of the markers measured prior to treatment. **Results and Discussion: untreated group showed significantly lower levels of antioxidant enzymes (P<0·001) and reduced glutathione (P<0·001), and more extensive lipid peroxidation (P<0·001) and DNA damage than healthy** controls. **Similar but less pronounced patterns were observed in the patients receiving chemotherapy alone**. The group of patients receiving VCE supplementation had all the marker levels moving towards normal values. Activities of superoxide dismutase, catalase, glutathione-S-transferase and glutathione reductase, and the levels of reduced glutathione were significantly increased ($P<0.01$) while, **the levels of malondialdehyde and DNA damage were significantly ($P<0.01$) reduced in the VCE supplemented group relative to those of patients receiving chemotherapy alone as well as relative to the pretreatment levels. Conclusion:** the presence of breast-cancer and chemotherapy. **DNA damage was also reduced by VCE.** The results suggest that VCE should be useful in protecting against chemotherapy-related side-effects and a randomized control trial to evaluate the effectiveness of VCE in breast-cancer patients using clinical outcomes would be appropriate. (Suhail et al, 2011). **I believe this shows that antioxidant supplementation restores harmful levels of antioxidants and this is verified by the reduced levels of DNA damage. In short, they are counteracting EMOD effects and apoptosis and protecting cancer cells.**

Long-Term Use of supplemental vitamins and minerals does not reduce the risk of urothelial cell carcinoma of the bladder in the VITamins and Lifestyle Study. (Hotaling et al, 2011) (#77,050 eligible VITAL participants)

Urothelial carcinoma has the highest lifetime treatment cost of any cancer, making it an ideal target for preventative therapies. Previous work has suggested that certain vitamin and mineral supplements may reduce the risk of urothelial carcinoma. We used the prospective VITamins And Lifestyle cohort to examine the association of all commonly taken vitamin and mineral supplements as well as 6 common anti-inflammatory supplements with incident urothelial carcinoma in a United States population. Materials and Methods. A total of **77,050 eligible VITAL participants** completed a detailed questionnaire at baseline on supplement use and cancer risk factors. After 6 years of follow-up 330 incident urothelial carcinoma cases in the cohort were identified via linkage to the Seattle-Puget Sound SEER cancer registry. **We analyzed use of supplemental vitamins (multivitamins, beta-carotene, retinol, folic acid, and vitamins B1, B3, B6, B12, C, D and E), minerals (calcium, iron, magnesium, zinc and selenium) and anti-inflammatory supplements (glucosamine, chondroitin, saw pal-**

metto, ginkgo biloba, fish oil and garlic). For each supplement the hazard ratios (risk ratios) for urothelial carcinoma comparing each category of users to nonusers, and 95% CIs, were determined using Cox proportional hazards regression, adjusted for potential confounders. Results: None of the vitamin, mineral or anti-inflammatory supplements was significantly associated with urothelial carcinoma risk in age adjusted or multivariate models. Conclusions: **The results of this study do not support the use of commonly taken vitamin or mineral supplements or 6 common anti-inflammatory supplements for the chemoprevention of urothelial carcinoma.** (Hotaling et al, 2011).

The effect of supplemental vitamins and minerals on the development of prostate cancer: a systematic review and meta-analysis. (Stratton, Godwin, 2011) (#Fourteen articles were included) Vitamin supplementation is used for many purposes with mainly alleged benefits. One of these is the use of various vitamins for the prevention of prostate cancer. METHODS: We conducted a systematic review and meta-analysis on this topic. Pubmed, Embase and the Cochrane Database were searched; as well, we hand searched the references in key articles. Randomized controlled trials (RCTs), cohort studies and case-control studies were included. The review assessed the effect of supplemental vitamins on the risk of prostate cancer and on disease severity and death in men with prostate cancer. RESULTS: **Fourteen articles were included in the final assessment**. Individually, a few of these studies showed a relationship between the ingestion of supplemental vitamins or minerals and the incidence or severity of prostate cancer, especially in smokers. However, **neither the use of multivitamin supplementation nor the use of individual vitamin/mineral supplementation affected the overall occurrence of prostate cancer or the occurrence of advanced/metastatic prostate cancer or death from prostate cancer when the results of the studies were combined in a meta-analysis**. We also conducted several sensitivity analyses by running meta-analysis using just the higher quality studies and just the RCTs. There were still no associations found. CONCLUSIONS: **There is no convincing evidence that the use of supplemental multivitamins or any specific vitamin affects the occurrence or severity of prostate cancer.** There was high heterogeneity among the studies so it is possible that unidentified subgroups may benefit or be harmed by the use of vitamins (Stratton, Godwin, 2011).

The outcome of 5-ALA-mediated photodynamic treatment in melanoma cells is influenced by vitamin C and heme oxygenase-1. (Grimm et al, 2011) (#) Photodynamic therapy (PDT) is an important clinical approach for cancer treatment. It involves the administration of a photosensitizer, followed by its activation with light and induction of cell death. **The underlying mechanism is an increased production of reactive oxygen species (ROS) leading to oxidative stress, which is followed by cell death. However, effectiveness of PDT is limited due to an initiation of endogenous rescue response systems like heme oxygenase-1 (HO-1) in tumor cells.** In recent years, **consuming of antioxidant supplements has become widespread,** but the effect of exogenously applied antioxidants on cancer therapy outcome remains unclear. Thus, this study was aimed to investigate if exogenous antioxidants might decrease ROS-induced cytotoxicity in photodynamic treatment. **Lycopene, β-carotene, vitamin C, N-acetylcysteine, trolox, and N-tert-butyl-α-phenylnitrone in different doses were administered to human melanoma cells prior exposure to photodynamic treatment. Supplementation with vitamin C resulted in a significant decrease of the cell death rate,** whereas the other tested antioxidants had no effect on cell viability and oxidative stress markers. **The simultaneous application of vitamin C with the HO-1 activity inhibitor zinc protoporphyrine IX (ZnPPIX) caused a considerable decrease of photodynamic treatment-induced cytotoxicity compared to ZnPPIX alone.** It can be summarized that exogenously applied antioxidants do not have a leading role in the protective response against photodynamic treatment. However, further studies are necessary to investigate more antioxidants and other substances, which might affect the outcome of photodynamic treatment in cancer therapy (Grimm et al, 2011).

The protective effects of nutritional antioxidant therapy on Ehrlich solid tumor-bearing mice depend on the type of antioxidant therapy chosen: histology, genotoxicity and hematology evaluations. (Miranda-Vilela AL, et al, 2011)

Strong evidence indicates that reactive oxygen species (ROS) play an important role in the initiation as well as the promotion phase of carcinogenesis. **Studies support the role of ROS in cancer, in part, by showing that dietary anti-**

oxidants act as cancer-preventive agents. Although results are promising, the research on this topic is still controversial. Thus, the aim of this study was to investigate whether vitamins C, E and pequi oil can, individually, provide prevention and/or be used afterward as an adjuvant in cancer therapy. Ehrlich solid tumor-bearing mice received antioxidant as follows: before tumor inoculation, before and after tumor inoculation (continuous administration), and after tumor inoculation; morphometric analyses of tumor, genotoxicity and hematology were then carried out. Antioxidant administrations before tumor inoculation effectively inhibited its growth in the three experimental protocols, but administrations after the tumor's appearance accelerated tumor growth and favored metastases. Continuous administration of pequi oil inhibited the tumor's growth, while the same protocol with vitamins E and C accelerated it (tumor growth), favoring metastasis and increasing oxidative stress on erythrocytes. Except for continuous administration with vitamin E, the development of ascites tumor metastases was linked with increased inflammation. Results suggest that the efficiency and applicability of antioxidants in the medical clinic can depend not only on the nature of the antioxidant, the type and stage of cancer being treated and the prevailing oxygen partial pressure in the tissues, but also on the type of antioxidant therapy chosen (Miranda-Vilela et al, 2011).

Prenatal exposure to flavonoids: Implication for cancer risk (Vanhees et al, 2011) Flavonoids are potent antioxidants, freely available as high-dose dietary supplements. However, they can induce DNA double-strand breaks (DSB) and rearrangements in the mixed-lineage leukemia (MLL) gene, which are frequently observed in childhood leukemia. We hypothesize that a deficient DSB repair, as a result of an Atm mutation, may reinforce the clastogenic effect of dietary flavonoids and increase the frequency of Mll rearrangements. Therefore, we examined the effects of in vitro and transplacental exposure to high, but biological amounts of flavonoids in mice with different genetic capacities for DSB repair (homozygous/heterozygous knock-in for human Atm mutation [Atm-ΔSRI] vs. wild type [wt]). In vitro exposure to genistein/quercetin induced higher numbers of Mll rearrangements in bone marrow cells of Atm-ΔSRI mutant mice compared with wt mice. Subsequently, heterozygous Atm-ΔSRI mice were placed on either a flavonoid-poor or a genistein-enriched (270 mg/kg) or quercetin-enriched (302 mg/kg) feed throughout pregnancy. Prenatal exposure to flavonoids associated

with higher frequencies of MII rearrangements and a slight increase in the incidence of malignancies in DNA repair-deficient mice. These data suggest that *prenatal exposure to both genistein and quercetin supplements could increase the risk on MII rearrangements especially in the presence of compromised DNA repair* (Vanhees et al, 2011). **I had posed the question some time ago, "Can antioxidants given during pregnancy (prenatal) cause the diseases associated with an EMOD insufficiency?" The answer may be "yes."**

Peroxiredoxin 6 overexpression attenuates cisplatin-induced apoptosis in human ovarian cancer cells. (Pak et al, 2011) Investigators examined the involvement of peroxiredoxin 6 (Prdx 6) in providing chemoprotection against cisplatin cytotoxicity in **SKOV-3 ovarian cancer cells. Treatment of SKOV-3 cells with cisplatin-induced cytotoxicity that was associated with increased accumulation of intracellular reactive oxygen species (ROS) and apoptosis mediated by proteolytically activated caspase 3 and 9. Overexpression of Prdx 6 protein or exposure to N-acetylcysteine (NAC) reversed the apoptotic effect of cisplatin by reducing ROS levels and suppressing the caspase signaling pathway.** These results indicate that targeting Prdx 6 may sensitize cancer cells to ROS-producing therapeutic treatments, such as anticancer drugs and radiation (Pak et al, 2011).

Black and green tea consumption and the risk of coronary artery disease: a meta-analysis (Wang et al, 2011) (#18 studies)

Epidemiologic studies are inconsistent regarding the association between tea consumption and the risk of coronary artery disease (CAD). **OBJECTIVE:** The objective was to perform a meta-analysis to determine whether an association exists between tea consumption and total CAD endpoints in observational studies. **DESIGN:** We searched PUBMED and EMBASE databases for studies conducted from 1966 through November 2009. Study-specific risk estimates were combined by using a random-effects model. **RESULTS:** A total of **18 studies were included in the meta-analysis: 13 studies on black tea and 5 studies on green tea. For black tea, no significant association was found through the** meta-analysis [highest compared with lowest, summary relative risk (RR): 0.92; 95% CI: 0.82, 1.04; an increment of 1 cup/d, summary RR: 0.98; 95% CI: 0.94, 1.02]. **For green tea, the summary RR indicated a significant association between the highest green tea consumption and reduced risk of CAD** (summary RR: 0.72; 95% CI: 0.58, 0.89). Furthermore, **an increase in green tea consumption of 1 cup/d was associated with a 10% decrease in the risk of developing CAD** (summary RR: 0.90; 95% CI: 0.82, 0.99). **CONCLUSIONS: Our data do not support a protective role of black tea against CAD.** The limited data available on green tea support a tentative association of green tea consumption with a reduced risk of CAD. However, additional studies are needed to make a convincing case for this association. (Wang et al, 2011).

No Convincing Evidence Shows That Selenium Supplements Protect Against Cancer

The 2011 **Cochrane meta-analysis reviewed 49 prospective observational studies, which included more than 1,078,000 participants, and 6 randomized controlled trials with a total of 43,408 participants (94% men).**

Overall, **some of the data showed that individuals with higher levels of selenium levels or a higher intake experienced a lower frequency of certain cancers, including bladder or prostate cancers, but there was no difference in the incidence of other malignancies such as breast cancer.** However, the authors point out that it was not possible to determine from these data whether selenium levels or intake was the actual reason for the lower cancer risk seen in some persons.

Other factors could influence cancer risk independent of selenium, they note. These include a healthier nutritional intake or lifestyle and more favorable living conditions.

However, **the randomized controlled trials with the most reliable results showed that organic selenium did not prevent prostate cancer in men and elevated the risk for**

nonmelanoma skin cancer in persons of both sexes. Although other trials showed a decrease in liver cancer among persons using selenium salt or organic supplements, "this evidence was less convincing" because of the trials' methodological shortcomings.

"We advise further investigation of selenium for liver cancer prevention before translating results into public health recommendations," the authors conclude. "We also recommend that there should be further evaluation of the effects of selenium supplements in populations according to their nutritional status as they may differ between undernourished and adequately nourished groups of people."

The study was partially funded by the Dr. Ernst and Anita Bauer Foundation, the EU-CAM-Cancer Project, the German Cancer Aid, and National Center for Complementary and Alternative Medicine. The authors have disclosed no relevant financial relationships.

Cochrane Database Syst Rev. 2011;5:CD005195

The findings from VITAL contrast those from the Supplementation in Vitamins and Mineral Antioxidants study, which *found a 4-fold higher melanoma risk in women — but not in men — who used nutritionally appropriate doses of antioxidant supplements (J Nutr. 2007;137:2098-2105).*

The results of another study that appeared in 2009 in the *Journal of Clinical Oncology* suggested that there was potential harm from selenium supplementation for men who already have prostate cancer (2009;27:3577-3583). *A high level of selenium in the blood was associated with a slightly elevated risk for aggressive prostate cancer.*

Other factors could influence cancer risk independent of selenium, they note. These include a healthier nutritional intake or lifestyle and more favorable living conditions.

However, *the randomized controlled trials with the most reliable results showed that organic selenium did not prevent prostate cancer in men and elevated the risk for nonmelanoma skin cancer in persons of both sexes*. Although other trials showed a decrease in liver cancer among persons using selenium salt or organic supplements, "this evidence was less convincing" because of the trials' methodological shortcomings.

Analysis of the RCTs showed that taking organic selenium supplements did not reduce the risk for prostate cancer in men, while *results of the Nutritional Prevention of Cancer Trial showed that selenium supplementation (200 μg per day) increased the risk for non-melanoma skin cancer by 17% and the risk for squamous cell carcinoma by 25% in both genders after a mean follow-up of 7.4 years.*

Study results for the prevention of liver cancer suggested a decreased risk with selenium supplements, but the findings were inconsistent. "Therefore, we cannot conclude that there is strong support for selenium supplements as agents for the prevention of liver cancer," write Dennert and co-authors in the *Cochrane Database Systematic Reviews*.

They conclude: "Currently, regular intake of selenium supplements for cancer prevention cannot be recommended to either the selenium-replete or deficient populations."

CHAPTER TWELVE

HARMFUL EFFECTS

90 STUDY REPORTS SHOWING
HARMFUL EFFECTS

SUMMARY OF 90 STUDY REPORTS
SHOWING ADVERSE, HARMFUL EFFECTS
R.M. Howes M.D., Ph.D.

316 Antioxidant studies: 90 studies showing harmful effects

1. **Isotretinoin-Basal Cell Carcinoma Study Group** (#981 patients with two or more previously treated basal cell carcinomas) low-dose regimen of isotretinoin not only is **ineffective in reducing the occurrence of basal cell carcinoma at new sites** in patients with two or more previously treated basal cell carcinomas but *also is associated with significant adverse systemic effects. (adverse mucocutaneous effects and serum triglyceride elevations)*

2. α-**Tocopherol,** β-**Carotene Cancer Prevention Study (ATBC study)** (Heinonen et al, 1994) (#29,133 men); *50% increase in hemorrhagic stroke deaths among vitamin E group; 11% increase in ischemic heart disease deaths among* β-*carotene group; 18% increase in lung cancer among* β-*carotene group.*

This study was stopped 21 months earlier than planned. The incidence of lung cancer was 18% higher among men who took the beta-carotene supplement and *eight percent more men in this group died, as compared to those receiving other treatments or placebo.* (Albanes et al, 1996)

3. The β-Carotene and Retinol Efficacy Trial (CARET) (Omenn et al, 1996) (#14,254 heavy smokers and 4,060 asbestos workers) (total #18,314 men and women); *28% increase in lung cancer; 26% increase in CVD (nonsignificant); 17% increase in total mortality among treatment group. This study was stopped 21 months earlier than planned.*

4. **Energy, nutrient intake and prostate cancer risk: a population-based case-control study in Sweden. (Andersson et al. 1996) (#1,062)** *In age-adjusted analyses, there were positive associations of prostate cancer (all stages combined) risk with total energy intake as well as intake of total fat (saturated and monounsaturated), protein, retinol and zinc.* The positive association with energy intake was stronger for advanced cancer, with an excess risk of 70% for the highest quartile vs. the lowest. *After adjustment for energy intake, there was no apparent association of prostate cancers (all stages combined) with any of the investigated nutrients. However, a weak positive association between intake of retinol and advanced cancer was observed.*

5. **Alpha-Tocopherol and beta-carotene supplements and lung cancer incidence in the alpha-tocopherol, beta-carotene cancer prevention study: effects of base-line characteristics and study compliance.** (#29,133 men, smokers) *beta-Carotene supplementation was associated with increased lung cancer risk. beta-Carotene supplementation at pharmacologic levels may modestly increase lung cancer incidence in cigarette smokers,* and this effect may be associated with heavier smoking and higher alcohol intake.

6. **ATBC Sub-Study Shows Increased CVD Deaths** (Rapola et al, 1997) (#1,862 men, with prior myocardial infarction); there **were no significant differences** in major coronary events but *significantly*

more deaths from fatal coronary heart disease. There were no significant differences in the number of major coronary events between any supplementation group and the placebo group. There were *significantly more deaths from fatal coronary heart disease in the beta-carotene and combined alpha-tocopherol and beta-carotene groups* than in the placebo group. *The risk of fatal coronary heart disease increased in the groups that received either beta-carotene or the combination of alpha-tocopherol and beta-carotene.* They do not recommend the use of alpha-tocopherol or beta-carotene supplements in this group of patients.

A report from the **Atherosclerosis Risk In Communities (ARIC)** group found individuals with *the highest carotid IMT to have lower levels of plasma carotenoids but higher alpha-tocopherol and retinol levels compared to controls* (Iribarren et al, 1997).

7. **The influence of antioxidant nutrients on platelet function in healthy volunteers** (Calzada et al, 1997) (#40 healthy volunteers) **Supplementation of healthy volunteers with** *vitamin E decreased platelet function* whereas supplementation with vitamin c or beta-carotene had no significant effects.

8. **The Multivitamins and Probucol Study** (Tardif et al, 1997) (#317 participants); *Probucol has been pulled off the market due harmful effects and the likelihood of cardiac arrhythmias.*

9. **Vitamin E Worsens Metabolic Parameters in Type 2 Diabetics. (Skrha et al. 1997) (#12)** *decreases of glucose disposal rate, metabolic clearance rate of glucose, and insulin receptor number were found after vitamin E administration as compared with pretreated values. A worsening of diabetes control as observed by an increase of HbAIC was present.*

10. **The Nurses' Health Study and Folic Acid and Colon Cancer** (Giovannucci et al, 1998) (#88,756 women taking vitamin

C and B-carotene, for 8 years); Dr. Andy Ness, of Bristol University, reported in the British Medical Journal in Dec. 2004, that there is *the possibility of increased risk of breast cancer in women taking folic acid supplements throughout pregnancy*. The researchers followed up **2,928 pregnant women** who had taken part in a supplemental trial in the 1960s. *The risk of death from breast cancer was much higher in women who had received high doses of the supplement* than in those who had been given a placebo.

11. **The effects of antioxidant supplementation during Percoll preparation on human sperm DNA integrity** (Hughes et al, 1998) (#150 patients) *acetyl cysteine or ascorbate and alpha tocopherol together induced further DNA damage to human sperm*.

12. **Dietary intake of antioxidant (pro)-vitamins, respiratory symptoms and pulmonary function: the MORGEN study**

(Grievink et al. 1998) (#6,555 adults) *Vitamin E intake showed no association with most symptoms and lung function, but had a positive association with productive cough. The intake of beta-carotene was not associated with most symptoms but had a positive association with wheeze.*

13. **Antioxidant nutrient intake and diabetic retinopathy: the San Luis Valley Diabetes Study. (Mayer-Davis et al. 1998) (#387 participants with type 2 diabetes)** *An increase over time in vitamin C intake from the first to ninth deciles was associated with a risk for increased severity of diabetic retinopathy (DR)*, although excess risk was not observed for the tenth decile or the second through fourth quintiles compared to the first quintile. *Increased intake of vitamin E was associated with increased severity of DR among those not taking insulin. Among those taking insulin, increased intake of beta-carotene was associated with a risk for severity of diabetic retinopathy (DR). Depending*

on insulin use, there appeared to be a potential for deleterious effects of nutrient antioxidants.

14. **Familial hypercholesterolemia, intima-to-media thickness (FH IMT study)** (Raal et al, 1999) (#15 with homozygous familial hypercholesterolemia); *homozygous familial hypercholesterolemia, intima-to-media thickness (FH IMT study) increased with vitamin E supplements* (400 mg/day) for 2 years.

15. **Antioxidant supplementation in vitro does not improve human sperm motility** (Donnelly et al, Fertil Steril. 1999) (#60 patients) *progressive motility, average path velocity, curvilinear velocity, straight-line velocity, and linearity were decreased significantly, with the greatest inhibition observed with the highest concentrations of antioxidants. In short, sperm motility was significantly decreased by antioxidants vitamin C and alpha-tocopherol.*

16. **The effect of ascorbate and alpha-tocopherol supplementation in vitro on DNA integrity and hydrogen peroxide-induced DNA damage in human spermatozoa** (Donnelly et al, Mutagenesis. 1999) (#Semen samples with normozoospermic and asthenozoospermic profiles (n = 15 for each control and antioxidant group) **Addition of** *both ascorbate and alpha-tocopherol in combination to sperm preparation medium actually induced DNA damage and intensified the damage induced by H_2O_2.*

17. **Controlled trial of alpha-tocopherol and beta-carotene supplements on stroke incidence and mortality in male smokers. (Leppala et al. 2000) (#28,519 male cigarette smokers).** *alpha-Tocopherol supplementation increased the risk of subarachnoid hemorrhage 50%* but decreased that of cerebral infarction 14%, whereas *beta-carotene supplementation increased the risk of intracerebral hemorrhage 62%. alpha-Tocopherol supplementation also increased the risk of fatal subarachnoid hemorrhage 181%.* The overall net effects of either supplementation on the

incidence and mortality from total stroke were nonsignificant. *alpha-Tocopherol supplementation increases the risk of fatal hemor-rhagic strokes* but prevents cerebral infarction. The effects may be due to the antiplatelet actions of alpha-tocopherol. *beta-Carotene supplementation increases the risk of intracerebral hemorrhage*.

18. **Multivitamin use and mortality in a large prospective study.** (Watkins et al, 2000) (#1,063,023 adults) **Multivitamin users had heart disease and cerebrovascular disease mortality risks similar to those of nonusers, whereas combination users had mortality risks that were 15% lower than those of nonusers**. Multivitamin and combination use had minimal effect on cancer mortality overall, *although mortality from all cancers combined was increased among male current smokers who used multivitamins alone or in combination with vitamin A, C, or E.*

19. **Randomized Trial of Supplemental ß-Carotene to Prevent Second Head and Neck Cancer** (Mayne et al, 2001) (#264 patients who had been curatively treated for a recent early-stage squamous cell carcinoma of the oral cavity, pharynx, or larynx.); **Supplemental ß-carotene had no significant effect on second head and neck cancer or lung cancer**. Whereas none of the effects were statistically significant, the *point estimates suggested a possible decrease in second head and neck cancer risk but a possible increase in lung cancer risk.*

20. **HDL Atherosclerosis Treatment study (HATS)** (Brown et al, 2001) (#160 participants); an antioxidant cocktail (vitamin E, ß-carotene, vitamin C, and selenium) had a 0.7% progression in steno-sis after 3 years, compared with 0.4% regression in the group on only simvastatin/niacin. Thus, *antioxidant supplements may have inter-fered with the efficacy of statin-plus-niacin therapy.* **No clinical or angiographically measurable benefit from antioxidants was found.** *When used in combination with simvastatin/niacin, antioxidants negated the benefit of the latter on plasma lipid profile and stenosis progression.*

21. **Women's Angiographic Vitamin and Estrogen (WAVE) Trial** (Waters et al, 2002) (#423 postmenopausal women, with at least one 15% to 75% coronary stenosis); **neither HRT nor antioxidant vitamin supplements (vitamins C & E) provided any cardiovascular benefit.** Instead, *a potential for harm was suggested. There is some **evidence of potentially adverse effects of antioxidant supplements on CVD as assessed by angiographic end points.** In the Women's Angiographic Vitamin and Estrogen Study, postmenopausal women with coronary disease on hormone replacement therapy given vitamin E plus vitamin C had an unexpected significantly higher all-cause mortality rate and a trend for an increased cardiovascular mortality rate* compared with the vitamin placebo women.

22. **Mega-dose vitamins and minerals in the treatment of non-metastatic breast cancer: an historical cohort study** (Lesperance et al, 2002) (#90 patients with non-metastatic breast cancer who received conventional treatment) ***Breast cancer–specific survival (i.e., patients censored only at death from breast cancer) and disease-free survival were shorter in the nutrient-supplemented group* than in the non-supplemented group, but the differences were not statistically significant.**

Investigators stated that, "**It is troubling that both** (Lesperance et al, 2002 and Ferreira et al, 2004**) reported results suggesting poorer survival with concurrent administration of antioxidants and cytotoxic therapy.**"

23. **Retinol intake and bone mineral density in the elderly: the Rancho Bernardo Study.** (Promislow et al, 2002) (#570 women and 388 men) In both sexes, *increasing retinol became negatively associated with skeletal health at intakes not far beyond the recommended daily allowance (RDA), intakes reached predominately by supplement users.*

393

24. **Vitamin A intake and hip fractures among postmeno-pausal women.** (Feskanich et al, 2002) (#72,337 postmenopausal women) After controlling for confounding factors, *women in the highest quintile of total vitamin A intake (≥3000 µg/d of retinol equivalents [RE]) had a significantly elevated relative risk (RR) of hip fracture* compared with women in the lowest quintile of intake (<1250 µg/d of RE). *This increased risk was attributable primarily to retinol. Long-term intake of a diet high in retinol may pro-mote the development of osteoporotic hip fractures in women.*

25. **Selenium and vitamin E supplements for prostate cancer: evidence or embellishment? (Moyad et al. 2002) (# not available) Selenium supplements provided a benefit only for those individuals who had lower levels of baseline plasma selenium.** *Other subjects, with normal or higher selenium levels, did not benefit and may have an increased risk for prostate cancer. Vitamin E supplements in higher doses (> or =100 IU) were also associated with a higher risk of aggressive or fatal prostate cancer in nonsmokers from a past prospective study.*

26. **Vitamins E & A fail to reduce incidence or mortality of lung cancer: <u>Cochrane Database Syst Rev.</u> 2003.** (Caraballoso et al., 2003) (#109,394 participants); *When beta-carotene was combined with retinol, data from a single study showed that there was a statistically significant, **increased risk of lung cancer incidence and mortality** in people with risk factors for lung cancer who took both vitamins.*

27. **Neoplastic and Antineoplastic Effects of Beta Carotene on Colorectal Adenoma Recurrence: Results of a Randomized Trial** (Baron et al, 2003) (#864 subjects who had had an adenoma removed and were polyp-free); *For participants who smoked ciga-rettes and also drank more than one alcoholic drink per day, beta carotene doubled the risk of adenoma recurrence.*

28. **Serum retinol levels and the risk of fracture.** (Michealsson et al, 2003) (#2,322 men) *The risk of fracture was highest among*

men with the highest levels of serum retinol. The risk of fracture was further increased within the highest quintile for serum retinol. Men with retinol levels in the 99th percentile (>103.12 µg per deciliter [3.60 µmol per liter]) had an overall risk of fracture that exceeded the risk among men with lower levels by a factor of seven.

29. **Plasma carotenoids and tocopherols and risk of myocardial infarction in a low-risk population of US male physicians. (Hak et al. 2003) (#531 physicians diagnosed with MI)** *that men with high plasma gamma-tocopherol levels tended to have an increased risk of nonfatal and fatal MI.*

30. **Selenium supplementation and secondary prevention of nonmelanoma skin cancer in a randomized trial.** (Duffield-Lillico, 2003) (#1,312). *selenium supplementation was associated with statistically significantly elevated risk of squamous cell carcinoma* and of total nonmelanoma skin cancer. **Results from the Nutritional Prevention of Cancer Trial conducted among individuals at high risk of nonmelanoma skin cancer continue to demonstrate that selenium supplementation is ineffective at preventing basal cell carcinoma and that** *it increases the risk of squamous cell carcinoma and total nonmelanoma skin cancer.*

31. **Supplemental vitamin C increase cardiovascular disease risk in women with diabetes** (Iowa Women's Health study) (Lee et al, 2004) (#1,923 postmenopausal women who reported being diabetic) *A high vitamin C intake from supplements is associated with an increased risk of cardiovascular disease mortality in postmenopausal women with diabetes.*

32. **Cochrane Database Syst Rev. 2004: Vitamins E & A fail to reduce incidence or mortality of gastrointestinal cancer.** (Cochrane Database Syst Rev.) (Bjelakovic et al, 2004) (#170,525 participants); *antioxidant supplements significantly increased*

mortality. *Beta-carotene and vitamin A and beta-carotene and vitamin E significantly increased mortality,* while *beta-carotene alone only tended to do so.* **Selenium showed significant beneficial effect on gastrointestinal cancer incidences.** *When the selenium trials were excluded, both analyses showed a statistically significant increase in mortality, which was particularly strong in patients taking beta carotene and vitamin A.* **CONCLUSIONS: They could not find evidence that antioxidant supplements prevent gastrointestinal cancers. On the contrary, they seem to increase overall mortality.**

33. **ATBC 6-year followup study (2004)** (Thornwall et al., 2004) (#29,133 male smokers); *ß-Carotene seemed to increase the post-trial risk of first-ever non-fatal MI.*

34. **Meta-analysis: high-dosage vitamin E supplementation may increase all-cause mortality** (Miller et al., 2004) (#135,967 subjects); *high doses of vitamin E increased mortality.*

Vitamin E (alpha-tocopherol) alone in doses of 400 units is of questionable value, and larger doses may cause *intracranial hemorrhage* or interact negatively with lipid-lowering drugs. Vitamin E should not be used in patients who have *bleeding disorders* or patients on anti-coagulants or acetylsalicylic acid (ASA).

35. **Vitamin C worsens coronary atherosclerosis in those with two copies of the haptoglobin 2 gene.** (Levy et al, 2004) (#299 postmenopausal women) *antioxidant therapy (1,000 mg/day of vitamin C + 800 IU/day of vitamin E) was associated with improvement of coronary atherosclerosis in diabetic women with two copies of the haptoglobin 1 gene but worsening of coronary atherosclerosis in those with two copies of the haptoglobin 2 gene.*

36. **Early Infant Multivitamin Supplementation Is Associated With Increased Risk for Food Allergy and Asthma (Milner**

et al. 2004) (#over 8,000) *In multivariate logistic analyses, a history of vitamin use within the first 6 months of life was associated with a higher risk for asthma in black infants. Early vitamin use was also associated with a higher risk for food allergies in the exclusively formula-fed population. Vitamin use at 3 years of age was associated with increased risk for food allergies.*

37. **Use of multivitamins and prostate cancer mortality in a large cohort of US men.** (Stevens et al, 2005) (#475,726 men who were cancer-free) *Regular multivitamin use was associated with a small increase in prostate cancer death rates.*

38. **Vitamin A Supplementation for Reducing the Risk of Mother-to-child Transmission of HIV Infection: Cochrane systematic review 2005**. (Wiysonge et al, 2005) (#3,033 females) *the trial in Tanzania found evidence that vitamin A supplementation increased the risk of MTCT of HIV* compared with placebo and multivitamins (excluding vitamin A).

39. **HOPE-TOO Extension** (Lonn et al, 2005) (#3,994 original study enrollees) *Another subgroup finding in HOPE-TOO was a vitamin E–associated increased risk of heart failure incidence that appeared in a secondary end point analysis in the 4.5-year report and persisted in the 7-year extended follow-up, as did the risk of hospitalization for heart failure. Patients in the vitamin E group had a higher risk of heart failure and hospitalization for heart failure.*

40. **A randomized trial of antioxidant vitamins to prevent second primary cancers in head and neck cancer patients** (Bairati et al, 2005 Apr 6) (#540 patients with stage I or II head and neck cancer treated by radiation therapy) Compared with patients receiving placebo, *patients receiving alpha-tocopherol supplements had a higher rate of second primary cancers during the supplementation period* but a lower rate after supplementation was discontinued. Similarly, **the rate of having a recurrence or**

second primary cancer was higher during but lower after supplementation with alpha-tocopherol. CONCLUSIONS: *alpha-Tocopherol supplementation produced unexpected adverse effects on the occurrence of second primary cancers and on cancer-free survival.* **Note:** *Patients taking an antioxidant were 1.65 times more likely to suffer a return of their original cancer during the three years they were on the supplement. The risk was highest among those taking only vitamin E (1.86 times higher).*

41. **Randomized trial of antioxidant vitamins to prevent acute adverse effects of radiation therapy in head and neck cancer patients** (Bairati et al, 2005 Aug 20) (#540 patients with stage I or II head and neck cancer treated by radiation therapy) During the course of the trial, *supplementation with beta-carotene was discontinued because of ethical concerns.* **Quality of life was not improved by the supplementation.** *The rate of local recurrence of the head and neck tumor tended to be higher in the supplement arm of the trial.* **Note:** Researchers were concerned to find that **the rate of local recurrence (that is, a return of the original cancer) was 54 percent higher among patients on the combination pill than those on placebo.** There was a smaller but still worrisome increase among those on vitamin E only. *This trial suggests that use of high doses of antioxidants as adjuvant therapy might compromise radiation treatment efficacy.*

42. **Antioxidants for preventing pre-eclampsia. (Rumbold et al, Apr 18, 2005. CD004072) (#35,812 women and 37,353 pregnancies)** *Women supplemented with vitamin C alone or combined with other supplements were at increased risk of giving birth preterm.*

43. **Australian Collaborative Trial of Supplements (ACTS)** (Rumbold et al, 2006) (#1,877 pregnant women); *the rate of low-birth-weight babies was higher and the rate for gestational hypertension was higher for women in the vitamin group.* **Of**

398

concern, there was a downside for the women taking vitamins. *Women in the vitamin group had an increased risk of being hospitalized antenatally for hypertension and having to take antihypertensive medication.* In addition, *a subgroup of women in the vitamin group had a higher frequency of abnormal liver-function tests.*

44. Vitamins in Pre-eclampsia (VIP) Trial Consortium (Poston et al., 2006) (#2,410 women at increased risk for preeclampsia, analayzed 2,395) *Concomitant supplementation with vitamin C and vitamin E does not prevent pre-eclampsia in women at risk, but does **increase the rate of babies born with a low birth weight.***

45. The Melbourne Atherosclerosis Vitamin E Trial (MAVET): a study of high dose vitamin E in smokers. (Magliano et al, 2006) (#409 male and female smokers) *The mean increase in intima-media thickness over time in the vitamin E group was 0.0041 mm/year faster than placebo.*

46. Smoking, alcohol drinking, green tea consumption and the risk of esophageal cancer in Japanese men. (Ishikawa et al, 2006) (#9,008 men in Cohort I and 17,715 men in Cohort 2) *Cigarette smoking, alcohol drinking and green tea consumption were significantly associated with an increased risk of esophageal cancer. The population attributable fractions of esophageal cancer incidence that was attributable to smoking, alcohol drinking and green tea consumption were 72.0%, 48.6%, and 22.1%, respectively.*
CONCLUSIONS: Among the variables studied, *smoking has the largest public health impact on esophageal cancer incidence in Japanese men, followed by alcohol drinking and green tea drinking.*

47. Mortality in Randomized Trials of Antioxidant Supplements for Primary and Secondary Prevention; Systematic Review and Meta-analysis (Bjelakovic et al, 2007) (#232,606 participants);

Conservatively, the supplements increase the likelihood of dying by about 5 percent. When looked at separately, they found that Vitamin A increased death risk by 16 per cent, beta carotene by 7 per cent and Vitamin E by 4 per cent.

48. **Health Professionals Follow-up Study (2007): Effect of vitamins C, E, A and carotenoids and the occurrence of oral pre-malignant lesions** (Maserejian et al, 2007) (#42,340 men enrolled in the Health Professionals Follow-up Study) (#207 found with oral premalignant lesions); *A trend for increased risk of oral pre-malignant lesions was observed with vitamin E, especially among current smokers and with vitamin E supplements. Beta-carotene also increased the risk among current smokers.* However, **dietary vitamin C was significantly associated with a reduced risk of oral premalignant lesions.**

49. **Risk of Mortality with Vitamin E Supplements: The Cache County Study.** (Hayden et al, 2007) *mortality was increased in vitamin E users who had a history of stroke, coronary bypass graft surgery, or myocardial infarction and, independently, in those taking nitrates, warfarin, or diuretics.*

50. **Antioxidant Supplementation Increases the Risk of Skin Cancers in Women but Not in Men.** (Hercberg et al, 2007) (#French adults, 7,876 women and 5,141 men. Total # = 13,017) *In women, the incidence of SC was higher in the antioxidant group* [adjusted hazard ratio (adjusted HR) = 1.68; P = 0.03]. **Conversely, in men, incidence did not differ between the 2 treatment groups.** Despite the small number of events, *the incidence of melanoma was also higher in the antioxidant group for women.*

51. **Antioxidant therapy to prevent preeclampsia: a randomized controlled trial (Spinnato et al. 2007) (#739).** *Among patients without chronic hypertension, there was a slightly higher rate of severe preeclampsia in the study group.*

52. **The effect of vitamin E on blood pressure in individuals with type 2 diabetes: a randomized, double-blind, placebo-controlled trial (Ward et al, 2007) (#58 with type 2 diabetes)** In contrast to our initial hypothesis, *treatment with either alpha- or mixed tocopherols significantly increased BP, pulse pressure and heart rate in individuals with type 2 diabetes.*

53. **National Institutes of Health State-of-the-Science Conference Statement: Multivitamin/Mineral Supplements and Chronic Disease Prevention (NIH State-of-the Science Panel. 2007). reports from RCTs that noted excess lung cancer occurring in asbestos workers and smokers consuming β-carotene.** In addition, *esophageal cancer excess was found with long-term follow-up of older Chinese patients (the Linxian study by Blot et al.) treated with selenium, β-carotene, and vitamin E supplements* (Blot et al, 1993) (NIH State-of-the Science Panel. 2007).

54. **Antioxidant supplements for prevention of mortality in healthy participants and patients with various diseases.** (Bjelakovic, Nikolova, Gludd, Simonetti and Gludd, 2008 Apr) (#232,550 Cochrane Database Syst Rev.) **Overall, the antioxidant supplements had no significant effect on mortality in a random-effects meta-analysis,** but *significantly increased mortality in a fixed-effect model. Vitamin A, beta-carotene, and vitamin E may increase mortality.*

55. **Systematic review: primary and secondary prevention of gastrointestinal cancers with antioxidant supplements.** (Bjelakovic, Nikolova, Simonette and Gludd, 2008 Sept) (#211,818 participants) *Antioxidant supplements had no significant effect on mortality in a random-effects model meta-analysis but significantly increased mortality in a fixed-effect model meta-analysis. CONCLUSIONS: There was no evidence that the studied antioxidant supplements prevented gastrointestinal cancers. On the contrary, they seem to increase overall mortality.*

56. **VITAL (VITamins And Lifestyle) study (2008)** (Slatore et al, 2008) (#77,721 men and women); *Supplemental vitamin E was associated with a small increased risk of lung cancer.*

57. **Vitamin E for Alzheimers and mild cognitive impairment. Cochrane Database Syst Rev. (2008)** (Isaac et al, 2008) (#769 participants); *More participants taking Vitamin E suffered a fall.*

58. **Efficacy of Antioxidant Supplementation in Reducing Primary Cancer Incidence and Mortality: Systematic Review and Meta-analysis** (Bardia et al, 2008) (#104,196 participants) *Beta carotene supplementation was associated with an increase in the incidence of cancer among smokers and with a trend toward increased cancer mortality.*

59. **Vitamin E supplementation may transiently increase tuberculosis risk in males who smoke heavily and have high dietary vitamin C intake.** (Hemila and Kaprio, 2008 Oct) (#29,023 males aged 50-69 years, smoking at baseline, with no tuberculosis) *Among participants who obtained 90 mg/d or more of vitamin C in foods, vitamin E supplementation increased tuberculosis risk by 72%.* **This effect was restricted to participants who smoked heavily.** *Our finding that vitamin E seemed to transiently increase the risk of tuberculosis in those who smoked heavily and had high dietary vitamin C intake should increase caution towards vitamin E supplementation for improving the immune system.*

60. **Vitamin E supplementation and pneumonia risk in males who initiated smoking at an early age: effect modification by body weight and dietary vitamin C.** (Hemila and Kaprio, 2008 Nov) (#21,657 ATBC Study participants who initiated smoking by the age of 20 years) *They had found a 14% higher incidence of pneumonia with vitamin E supplementation in a subgroup of the Alpha-Tocopherol Beta-Carotene Cancer Prevention (ATBC) Study cohort: Vitamin E increased the risk of pneumonia*

in participants with body weight less than 60 kg, and in participants with body weight over 100 kg.

61. **Oral administration of vitamin C decreases muscle mitochondrial biogenesis and hampers training-induced adaptations in endurance performance.** (Gomez-Cabrera et al, 2008) (#14 men) *The administration of vitamin C significantly (P=0.014) hampered endurance capacity.* **Vitamin C supplementation decreases training efficiency because it prevents some cellular adaptations to exercise.**

62. **Oral acetylcysteine does not protect renal function from moderate to high doses of intravenous radiographic contrast** (Boccalandro et al. 2003) **(#106 consecutive patients)** *Patients with* contrast-induced nephropathy (CIN) *and preexisting renal insufficiency had worse clinical outcomes.*

63. **Multivitamin-multimineral supplement use and mammographic breast density.** (Berube et al, 2008) (#Premenopausal (777) and postmenopausal (783) women; total 1,560) *Regular use of multivitamin-multimineral supplements may be associated with higher mean breast density among premenopausal women.*

64. **Antioxidant supplements to prevent or slow down the progression of AMD: a systematic review and meta-analysis.** (Evans, 2008) (#23,099 people were randomized in three trials) *Potential harms of high-dose antioxidant supplementation must be considered. These may include an increased risk of lung cancer in smokers (beta-carotene), heart failure in people with vascular disease or diabetes (vitamin E) and hospitalization for genitourinary conditions (zinc).*

65. **High maternal plasma antioxidant (vit. C) concentrations associated with preterm delivery** (Joshi et al, 2008) **(#140 normotensive women)** *Preterm mothers had higher vitamin C concentrations.*

66. **Dietary antioxidants and the long-term incidence of age-related macular degeneration: the Blue Mountains Eye Study**

(Tan et al. 2008) (#Of 3654 baseline (1992-1994) participants initially 49 years of older, 2454 were reexamined after 5 years, 10 years, or both) *Higher beta-carotene intake was associated with an increased risk of age-related macular degeneration (AMD).*

67. **Vitamins E and C in the prevention of cardiovascular disease in men: the Physicians' Health Study II randomized controlled trial (Sesso et al, 2008) (#14,641 US male physicians)** *vitamin E was associated with an increased risk of hemorrhagic stroke.*

68. **N-acetylcysteine to reduce renal failure after cardiac surgery: a systematic review and meta-analysis** (Naughton et al. 2008) **(#Seven randomized controlled trials (RCTs, n = 1000).** *There was a small, though significant increase in postoperative blood loss among patients treated with NAC.*

69. **Effect of selenium and vitamin E on risk of prostate cancer and other cancers: the Selenium and Vitamin E Cancer Prevention Trial (SELECT) (2009)** (Lippman et al, 2009, SELECT) (#35,533 men) **There were statistically nonsignificant increased risks of prostate cancer in the vitamin E group** but not in the selenium + vitamin E group. CONCLUSION: **Selenium or vitamin E, alone or in combination at the doses and formulations used, did not prevent prostate cancer in this population of relatively healthy men.** *The trial was stopped ahead of its original 12 year deadline because of a lack of any noticeable benefit.*

70. **Decision Analysis Supports the Paradigm That Indiscriminate Supplementation of Vitamin E Does More Harm than Good** (Dotan et al, 2009) The researchers examined data from **more than 300,000 subjects in the US, Europe and**

Israel. *This "disappointing" study* warns that **indiscriminate use of high-dose Vitamin E supplementation does more harm than good** and that indiscriminate supplementation of high doses of vitamin E is not beneficial in preventing **CVD.**

71. **Effects of long-term antioxidant supplementation and association of serum antioxidant concentrations with risk of metabolic syndrome in adults (SU.VI.MAX)** (Czernichow et al, 2009) (#5,220 adults) Adults (*n* = 5,220) **Antioxidant supplementation for 7.5 y did not affect the risk of metabolic syndrome (MetS);** *Baseline serum antioxidant concentrations of β-carotene and vitamin C, however, were negatively associated with the risk of MetS.*

72. **Total and Cancer Mortality After Supplementation With Vitamins and Minerals: 10 year Follow-up of the Linxian General Population Nutrition Intervention Trial.** (Qiao et al, 2009) (#29,584 adult participants) *esophageal cancer deaths increased 14% among those aged 55 years or older. Vitamin A and zinc supplementation was associated with increased total and stroke mortality.*

73. **Vitamin A and retinol intakes and the risk of fractures among participants of the Women's Health Initiative Observational Study.** (Caire-Juvera et al. 2009) (#75,747 women from the Women's Health Initiative Observational Study) *a modest increase in total fracture risk with high vitamin A and retinol intakes was observed in the low vitamin D-intake group.*

74. **Modification of the effect of vitamin E supplementation on the mortality of male smokers by age and dietary vitamin C.** (Hemila and Kaprio, 2009 Apr) (#29,133) The Alpha-Tocopherol, Beta-Carotene Cancer Prevention (ATBC) Study; *Among participants with a dietary vitamin C intake above the median of 90 mg/day, vitamin E increased mortality among those aged 50-62 years by 19%.*

75. **Antioxidants prevent health-promoting effects of physical exercise in humans.** (Ristow et al, 2009) (#39 healthy young men) *Exercise increased parameters of insulin sensitivity (GIR and plasma adiponectin) only in the absence of antioxidants in both previously untrained (P < 0.001) and pretrained (P < 0.001) individuals.* **Supplementation with antioxidants may preclude these health-promoting effects of exercise in humans.**

76. **Long-term use of beta-carotene, retinol, lycopene, and lutein supplements and lung cancer risk: results from the VITamins And Lifestyle (VITAL) study.** (Satia et al, 2009) (#77,126 (VITAL) *Longer duration of use of individual beta-carotene, retinol, and lutein supplements* (but not total 10-year average dose) *was associated with statistically significantly elevated risk of total lung cancer and histologic cell types.*

77. **Vitamin and mineral use and risk of prostate cancer: the case-control surveillance study.** (Zhang et al. 2009) (#1,706 prostate cancer cases and 2,404 matched controls). *Men who used zinc for ten years or more, either in a multivitamin or as a supplement, had an approximately two-fold increased risk of prostate cancer. The finding that long-term zinc intake from multivitamins or single supplements was associated with a doubling in risk of prostate cancer adds to the growing evidence for an unfavorable effect of zinc on prostate cancer carcinogenesis.*

78. **Folic acid and risk of prostate cancer: results from a randomized clinical trial.** (Figueiredo et al, 2009) (643 randomly assigned men). *the estimated probability of being diagnosed with prostate cancer over a 10-year period was 9.7% in the folic acid group and 3.3% in the placebo group.*

79. **Green tea consumption and risk of stomach cancer: a meta-analysis of epidemiologic studies** (Myung, Int J Cancer. et al, 2009) **(#13 epidemiologic studies)** *In the meta-analyses of the recent cohort studies, the highest green tea consumption*

was shown to significantly increase stomach cancer risk using the crude data, but no significant association between them was seen when using the adjusted data.

80. **Green tea (Camellia sinensis) for the prevention of cancer. Cochrane Database Syst Rev. 2009 Jul 8;(3):CD005004).** (Boehm et al, 2009) (#Fifty-one studies with more than 1.6 million participants were included) **there was limited to moderate evidence that the consumption of green tea reduced the risk of lung cancer, especially in men, and** *urinary bladder cancer or that it could even increase the risk of the latter.*

81. **Does antioxidant vitamin supplementation protect against muscle damage? (McGinley et al. 2009).** *antioxidant supplementation may actually interfere with the cellular signaling functions of ROS, thereby adversely affecting muscle performance. Since the potential for long-term harm does exist, the casual use of high doses of antioxidants by athletes and others should perhaps be curtailed.*

82. **Vitamin C supplements and the risk of age-related cataract: a population-based prospective cohort study in women.** (Rautiainen et al, 2010) (#24,593 women) *Among women aged 65 y and over, vitamin C supplement use increased the risk of cataract by 38% (95% CI: 12%, 69%). Vitamin C use among hormone replacement therapy users compared with that among nonusers of supplements or of hormone replacement therapy was associated with a 56% increased risk of cataract. The use of vitamin C supplements may be associated with higher risk of age-related cataract among women.*

83. **Associations of cataract with antioxidant enzymes and other risk factors: the French Age-Related Eye Diseases (POLA) Prospective Study.** (Delcourt et al, 2003) (#1,947 survivors) *Consistent with the baseline analysis, the results of this prospective study suggest that* **antioxidant enzymes might be implicated in the etiology of cataract.**

84. Micronutrient concentrations and subclinical atherosclerosis in adults with HIV. (Falcone et al, 2010) (#298 Nutrition for Healthy Living participants) *elevated serum vitamin E concentrations are associated with abnormal markers of atherosclerosis and may increase the risk of cardiovascular complications in HIV-infected adults.*

85. Multivitamin use and breast cancer incidence in a prospective cohort of Swedish women. (Larsson et al, 2010) (#35,329 cancer-free women) *Multivitamin use was associated with a statistically significant increased risk of breast cancer. Use of multivitamins was linked to a statistically significant 19 per cent increased risk of breast cancer* (after adjusting for lifestyle and risk factors like weight, diet, smoking, exercise, and family history of breast cancer.

86. Effects of vitamin on stroke subtypes: meta-analysis of randomized controlled trials. (Schurks et al. 2010) (#118,765) In this meta-analysis, *vitamin E increased the risk for hemorrhagic stroke by 22% and reduced the risk of ischemic stroke by 10%.*

87. Differential effects of concomitant use of vitamins C and E on trophoblast apoptosis and autophagy between normoxia and hypoxia-reoxygenation (Hung et al, 2010) *Concomitant supplementation of vitamins C and E during pregnancy has been reportedly associated with low birth weight, the premature rupture of membranes and fetal loss or perinatal death in women at risk for preeclampsia. Our results indicate that concomitant administration of vitamins C and E has differential effects on the changes of apoptosis, autophagy and the expression of Bcl-2 family of proteins in the trophoblasts between normoxia and HR. These changes may probably lead to the impairment of placental function and suboptimal growth of the fetus.*

88. **Quercetin and Ferulic Acid Aggravate Renal Carcinoma in Long-Term Diabetic Victims.** (Chiu-Lan Hsieh et al, 2010). Conclusively, the phytoantioxidants *quercetin and ferulic acid are able to aggravate, if not induce, nephrocarcinoma in mice.*

89. **The protective effects of nutritional antioxidant therapy on Ehrlich solid tumor-bearing mice depend on the type of antioxidant therapy chosen: histology, genotoxicity and hematology evaluations.** (Miranda-Vilela AL, et al, 2011) **Antioxidant administrations before tumor inoculation effectively inhibited its growth in the three experimental protocols,** but *administrations after the tumor's appearance accelerated tumor growth and favored metastases.* **Continuous administration of pequi oil inhibited the tumor's growth,** while the same protocol with *vitamins E and C accelerated it (tumor growth), favoring metastasis and increasing oxidative stress on erythrocytes.*

90. **Prenatal exposure to flavonoids: Implication for cancer risk** (Vanhees et al, 2011) *In vitro exposure to genistein/quercetin induced higher numbers of MII rearrangements in bone marrow cells of Atm-ΔSRI mutant mice compared with wt mice. Prenatal exposure to flavonoids associated with higher frequencies of MII rearrangements and a slight increase in the incidence of malignancies in DNA repair-deficient mice.* These data suggest that *prenatal exposure to both genistein and quercetin supplements could increase the risk on MII rearrangements especially in the presence of compromised DNA repair.*

Thus, ninety studies have shown possible harmful endpoints to antioxidant vitamin use. Total human participants is over 15,000,000.

HARMFUL EFFECTS OF ANTIOXIDANTS
- Summary

Ninety three entries:

1. - isotretinoin is **associated with significant adverse systemic effects. (adverse mucocutaneous effects and serum triglyceride elevations)** (Tangrea et al, 1992, 1993)

2. - ATBC (alpha tocopherol and beta carotene) study found **50% increase in hemorrhagic stroke deaths** among vitamin E group; **11% increase in ischemic heart disease deaths among β-carotene group; 18% increase in lung cancer among β-carotene group.** (Heinonen et al, 1994)

3. - CARET (beta carotene and retinol) study found **28% increase in lung cancer; 26% increase in CVD** (nonsignificant); **17% increase in total mortality** among treatment group. This study was stopped 21 months earlier than planned. (Omenn et al, 1996)

4. - **Positive association between retinol and advanced prostate cancer.** (Anderssen et al. 1996)

5. - **beta-Carotene supplementation was associated with increased lung cancer risk. beta-Carotene supplementation at pharmacologic levels may modestly increase lung cancer incidence in cigarette smokers.** (Albanes et al, 1996)

6. - **Vitamin E associated with prediabetic changes in glucose metabolism** (Skrha et al. 1997)

7. - ATBC study found **significantly more deaths from fatal coronary heart disease in the beta-carotene and combined alpha-tocopherol and beta-carotene groups.** (Rapola et al, 1997)

8. - the **Atherosclerosis Risk In Communities (ARIC)** group found individuals with **the highest carotid IMT to have lower levels of plasma carotenoids but higher alpha-tocopherol and retinol levels compared to controls** (Iribarren et al, 1997)

9. - **vitamin E decreased platelet function.** (Calzada et al, 1997)

10. - **Probucol has been pulled off the market due harmful effects and the likelihood of cardiac arrhythmias.** (Tardif et al, 1997)

11. - **Vitamin E positively associated with productive cough. Beta carotene associated with wheezing** (Grievink et al. 1998)

12. - **Vitamin C associated with increased severity of diabetic retinopathy. For those taking insulin, beta carotene and vitamin E was associated with increased severity of diabetic retinopathy** (Mayer-Davis et al. 1998)

13. - the **possibility of increased risk of breast cancer in women taking folic acid supplements throughout pregnancy. The risk of death from breast cancer was much higher in women who had received high doses of the supplement.** (Giovannucci et al, 1998)

14. - **acetyl cysteine or ascorbate and alpha tocopherol together induced further DNA damage to human sperm.** (Hughes et al, 1998)

15. - **Vitamin E intake showed a positive association with productive cough. The intake of beta-carotene had a positive association with wheeze. (Grievink et al. 1998)**

16. - An increase over time in vitamin C intake from the first to ninth deciles was associated with a risk for increased severity of diabetic retinopathy (DR). Increased intake of vitamin

E was associated with increased severity of DR among those not taking insulin. Among those taking insulin, increased intake of beta-carotene was associated with a risk for severity of diabetic retinopathy (DR). (Mayer-Davis et al. 1998)

17. - homozygous familial hypercholesterolemia, intima-to-media thickness (FH IMT study) increased with vitamin E supplements. (Raal et al, 1999)

18. - sperm motility was significantly decreased by antioxidants vitamin C and alpha-tocopherol. (Donnelly et al, Fertil Steril. 1999)

19. - both ascorbate and alpha-tocopherol in combination to sperm preparation medium actually induced DNA damage and intensified the damage induced by H_2O_2. (Donnelly et al, Mutagenesis. 1999)

20. - Alpha tocopherol increased risk of subarachnoid hemorrhage 50% and beta carotene increased risk of intracerebral hemorrhage 62% and alpha tocopherol increased risk of fatal subarachnoid hemorrhage 181% (Leppala et al. 2000)

21. - Multivitamin and combination use had minimal effect on cancer mortality overall, although mortality from all cancers combined was increased among male current smokers who used multivitamins alone or in combination with vitamin A, C, or E. (Watkins et al, 2000)

22. - point estimates suggested a possible decrease in second head and neck cancer risk but a possible increase in lung cancer risk. (Mayne et al, 2001)

23. - When used in combination with simvastatin/niacin, antioxidants negated the benefit of the latter on plasma lipid profile and stenosis progression. (Brown et al, 2001)

24. - **Other subjects, with normal or higher selenium levels, did not benefit and may have an increased risk for prostate cancer. Vitamin E supplements in higher doses (> or =100 IU) were also associated with a higher risk of aggressive or fatal prostate cancer in nonsmokers from a past prospective study.** (Moyad et al. 2002)

25. - In the Women's Angiographic Vitamin and Estrogen Study (WAVE), **postmenopausal women with coronary disease on hormone replacement therapy given vitamin E plus vitamin C had an unexpected significantly higher all-cause mortality rate and a trend for an increased cardiovascular mortality rate.** (Waters et al, 2002)

26. - **Breast cancer–specific survival (i.e., patients censored only at death from breast cancer) and disease-free survival were shorter in the nutrient-supplemented group.** (Lesperance et al, 2002)

27. - **increasing retinol became negatively associated with skeletal health at intakes not far beyond the recommended daily allowance (RDA), intakes reached predominately by supplement users. the Rancho Bernardo Study.** (Promislow et al, 2002)

28. - **women in the highest quintile of total vitamin A intake (≥3000 µg/d of retinol equivalents [RE]) had a significantly elevated relative risk (RR) of hip fracture.** (Feskanich et al, 2002)

29. - **Men with high plasma gamma-tocopherol levels had an increased risk of nonfatal and fatal myocardial infarction (MI).** (Hak et al. 2003)

30. - When beta-carotene was combined with retinol, data from a single study showed that there was a statistically significant, **increased risk of lung cancer incidence and mortality** in people with risk

factors for lung cancer who took both vitamins. (Caraballoso et al., 2003)

31. - **For participants who smoked cigarettes and also drank more than one alcoholic drink per day, beta carotene doubled the risk of adenoma recurrence.** (Baron et al, 2003)

32. - **The risk of fracture was highest among men with the highest levels of serum retinol. The risk of fracture was further increased within the highest quintile for serum retinol. Men with retinol levels in the 99th percentile (>103.12 μg per deciliter [3.60 μmol per liter]) had an overall risk of fracture that exceeded the risk among men with lower levels by a factor of seven.** (Michealsson et al, 2003)

33. - **selenium supplementation was associated with statistically significantly elevated risk of squamous cell carcinoma** and of total nonmelanoma skin cancer. (Duffield-Lillico, 2003)

34. - **antioxidant enzymes might be implicated in the etiology of cataract.** (Delcourt et al, 2003) (POLA study)

35. - **Patients with** contrast-induced nephropathy (CIN) **and pre-existing renal insufficiency had worse clinical outcomes with acetyl cysteine (NAC).** (Boccalandro et al. 2003)

36. - **Early vitamin supplementation (multivitamins) is associated with increased risk for asthma in black children and food allergies in exclusively formula-fed children** (Milner et al. 2004)

37. - **A high vitamin C intake from supplements is associated with an increased risk of cardiovascular disease mortality in postmenopausal women with diabetes.** (Iowa Women's Health study) (Lee et al, 2004)

38. - antioxidant supplements significantly increased mortality. Beta-carotene and vitamin A and beta-carotene and vitamin E significantly increased mortality. When the selenium trials were excluded, both analyses showed a statistically significant increase in mortality, which was particularly strong in patients taking beta carotene and vitamin A. (Cochrane Database Syst Rev.) (Bjelakovic et al, 2004)

39. - ß-Carotene seemed to increase the post-trial risk of first-ever non-fatal MI. (Thornwall et al., 2004)

40. - high doses of vitamin E increased mortality. (Miller et al., 2004)

41. - antioxidant therapy (1,000 mg/day of vitamin C + 800 IU/day of vitamin E) was associated with improvement of coronary atherosclerosis in diabetic women with two copies of the haptoglobin 1 gene but worsening of coronary atherosclerosis in those with two copies of the haptoglobin 2 gene. (Levy et al, 2004)

42. - Women supplemented with vitamin C alone or combined with other supplements were at increased risk of giving birth preterm. (Rumbold et al, Apr 18, 2005. CD004072)

43. - Vitamin E increased the risk of epistaxis (Lee et al. 2005)

44. - Regular multivitamin use was associated with a small increase in prostate cancer death rates. (Stevens et al, 2005)

45. - the trial in Tanzania found evidence that vitamin A supplementation increased the risk of mother to child transfer (MTCT) of HIV. (Wiysonge et al, 2005)

46. - Another subgroup finding in HOPE-TOO was a vitamin E–associated increased risk of heart failure incidence that

appeared in a secondary end point analysis in the 4.5-year report and persisted in the 7-year extended follow-up, as did the risk of hospitalization for heart failure. Patients in the vitamin E group had a higher risk of heart failure and hospitalization for heart failure. (Lonn et al, 2005)

47. - patients receiving alpha-tocopherol supplements had a higher rate of second primary cancers during the supplementation period. alpha-Tocopherol supplementation produced unexpected adverse effects on the occurrence of second primary cancers and on cancer-free survival. Patients taking an antioxidant were 1.65 times more likely to suffer a return of their original cancer during the three years they were on the supplement. The risk was highest among those taking only vitamin E (1.86 times higher). (Bairati et al, 2005 Apr 6)

48. - supplementation with beta-carotene was discontinued because of ethical concerns. Quality of life was not improved by the supplementation. The rate of local recurrence of the head and neck tumor tended to be higher in the supplement arm of the trial. supplementation with beta-carotene was discontinued because of ethical concerns. Quality of life was not improved by the supplementation. The rate of local recurrence of the head and neck tumor tended to be higher in the supplement arm of the trial. This trial suggests that use of high doses of antioxidants as adjuvant therapy might compromise radiation treatment efficacy. (Bairati et al, 2005 Aug 20)

49. - the rate of low-birth-weight babies was higher and the rate for gestational hypertension was higher for women in the vitamin group. Women in the vitamin group had an increased risk of being hospitalized antenatally for hypertension and having to take antihypertensive medication. In addition, a subgroup of women in the vitamin group had a

higher frequency of abnormal liver-function tests. (Rumbold et al, 2006)

50. - Concomitant supplementation with vitamin C and vitamin E does not prevent pre-eclampsia in women at risk, but does **increase the rate of babies born with a low birth weight.** (Poston et al., 2006)

51. - **The mean increase in intima-media thickness over time in the vitamin E group was 0.0041 mm/year faster than placebo.** (Magliano et al, 2006)

52. - **Cigarette smoking, alcohol drinking and green tea consumption were significantly associated with an increased risk of esophageal cancer. The population attributable fractions of esophageal cancer incidence that was attributable to smoking, alcohol drinking and green tea consumption were 72.0%, 48.6%, and 22.1%, respectively.** (Ishikawa et al, 2006)

53. - Conservatively, **the supplements increase the likelihood of dying by about 5 percent. When looked at separately, they found that Vitamin A increased death risk by 16 per cent, beta carotene by 7 per cent and Vitamin E by 4 per cent.** (Bjelakovic et al, 2007)

54. - **Among patients without chronic hypertension, there was a slightly higher rate of severe preeclampsia in the study (vitamin C and E) group**. (Spinnato et al. 2007)

55. - **Treatment with either alpha- or mixed tocopherols significantly increased BP, pulse pressure and heart rate in individuals with type 2 diabetes** (Ward et al, 2007)

56. - **A trend for increased risk of oral pre-malignant lesions was observed with vitamin E, especially among current smokers and with vitamin E supplements. Beta-carotene**

also increased the risk among current smokers. (Maserejian et al, 2007)

57. - mortality was increased in vitamin E users who had a history of stroke, coronary bypass graft surgery, or myocardial infarction and, independently, in those taking nitrates, warfarin or diuretics. (Hayden et al, 2007)

58. - **In women, the incidence of skin cancer (SC) was higher in the antioxidant group and the incidence of melanoma was also four-fold higher in the antioxidant group for women.** (Hercberg et al, 2007, VITAL)

59. - **esophageal cancer excess was found with long-term follow-up of older Chinese patients (the Linxian study by Blot et al.) treated with selenium, β-carotene, and vitamin E supplements** (Blot et al, 1993) (NIH State-of-the Science Panel. 2007)

60. - **Preterm mothers had higher vitamin C concentrations.** (Joshi et al. 2008)

61. - **Higher beta-carotene intake was associated with an increased risk of age-related macular degeneration (AMD).** (Tan et al. 2008)

62. - **Vitamin E was associated with an increased risk of hemorrhagic stroke** (Sesso et al, 2008)

63. - antioxidant supplements **significantly increased mortality in a fixed-effect model. Vitamin A, beta-carotene, and vitamin E may increase mortality.** (Bjelakovic, Nikolova, Gludd, Simonetti and Gludd, 2008 Apr)

64. - **Antioxidant supplements had no significant effect on mortality in a random-effects model meta-analysis** but

significantly increased mortality in a fixed-effect model meta-analysis. CONCLUSIONS: There was no evidence that the studied antioxidant supplements prevented gastrointestinal cancers. On the contrary, they seem to increase overall mortality. (Bjelakovic, Nikolova, Simonette and Gludd, 2008 Sept)

65. - Supplemental vitamin E was associated with a small increased risk of lung cancer. (Slatore et al, 2008)

66. - More Alzheimer's participants taking Vitamin E suffered a fall. (Isaac et al, 2008)

67. - Beta carotene supplementation was associated with an increase in the incidence of cancer among smokers and with a trend toward increased cancer mortality. (Bardia et al, 2008)

68. - Among participants who obtained 90 mg/d or more of vitamin C in foods, vitamin E supplementation increased tuberculosis risk by 72%. Our finding that vitamin E seemed to transiently increase the risk of tuberculosis in those who smoked heavily and had high dietary vitamin C intake should increase caution towards vitamin E supplementation for improving the immune system. (Hemila and Kaprio, 2008 Oct)

69. - They had found a 14% higher incidence of pneumonia with vitamin E supplementation in a subgroup of the Alpha-Tocopherol Beta-Carotene Cancer Prevention (ATBC) Study cohort: Vitamin E increased the risk of pneumonia in participants with body weight less than 60 kg, and in participants with body weight over 100 kg. (Hemila and Kaprio, 2008 Nov)

70. - The administration of vitamin C significantly (P=0.014) hampered endurance capacity. (Gomez-Cabrera et al, 2008)

71.- Regular use of multivitamin-multimineral supplements may be associated with higher mean breast density among premenopausal women. (Berube et al, 2008)

72. - There was a small, though significant **increase in postoperative blood loss after cardiac surgery among patients treated with NAC.** (Naughton et al, 2008)

73.- **There were statistically nonsignificant increased risks of prostate cancer in the vitamin E group** but not in the selenium + vitamin E group. **The trial was stopped ahead of its original 12 year deadline because of a lack of any noticeable benefit.** (Lippman et al, 2009, SELECT)

74.- **Antioxidant supplement use is associated with melanoma risk in light of recently published data from the Supplementation in Vitamins and Mineral Antioxidants (SUVIMAX) study, which reported a 4-fold higher melanoma risk in women.** (Asgari et al, 2009) referred to data from the SUVIMAX study and not Asgari's study.

75. - Antioxidant supplementation may actually interfere with the cellular signaling functions of ROS, thereby adversely affecting muscle performance. Since the potential for long-term harm does exist, the casual use of high doses of antioxidants by athletes and others should perhaps be curtailed. (McGinley et al. 2009)

76. - **Men who used zinc for ten years or more, either in a multivitamin or as a supplement, had an approximately two-fold increased risk of prostate cancer.** (Zhang et al. 2009)

77. - **Prostate cancer over a 10-year period was 9.7% in the folic acid group and 3.3% in the placebo group.** (Figueiredo et al, 2009)

78. - **This "disappointing" study** warns that **indiscriminate use of high-dose Vitamin E supplementation does more harm than good.** (Dotan et al, 2009)

79. - **esophageal cancer deaths increased 14% among those aged 55 years or older. Vitamin A and zinc supplementation was associated with increased total and stroke mortality.** (Qiao et al, 2009)

80. - **a modest increase in total fracture risk with high vitamin A and retinol intakes was observed in the low vitamin D-intake group.** (Caire-Juvera et al. 2009) (**Women's Health Initiative Observational Study**)

81. - **Among participants with a dietary vitamin C intake above the median of 90 mg/day, vitamin E increased mortality among those aged 50-62 years by 19%.** (Hemila and Kaprio, 2009 Apr)

82. - **Exercise increased parameters of insulin sensitivity (GIR and plasma adiponectin) only in the absence of antioxidants in both previously untrained (P < 0.001) and pretrained (P < 0.001) individuals.** (Ristow et al, 2009)

83. - **Longer duration of use of individual beta-carotene, retinol, and lutein supplements** (but not total 10-year average dose) **was associated with statistically significantly elevated risk of total lung cancer and histologic cell types. (VITAL) study.** (Satia et al, 2009)

84. - **In the meta-analyses of the recent cohort studies, the highest green tea consumption was shown to significantly increase stomach cancer risk using the crude data, but no significant association between them was seen when using the adjusted data.** (Myung, Int J Cancer. et al, 2009)

85. - **there was limited to moderate evidence that the consumption of green tea reduced the risk of lung cancer, especially in men, and urinary bladder cancer or that it could even increase the risk of the latter.** (Boehm et al, 2009)

86. - **Among women aged 65 y and over, vitamin C supplement use increased the risk of cataract by 38% (95% CI: 12%, 69%). Vitamin C use among hormone replacement therapy users compared with that among nonusers of supplements or of hormone replacement therapy was associated with a 56% increased risk of cataract.** (Rautiainen et al, 2010)

87. - **Vitamin E increased the risk for hemorrhagic stroke by 22%** and reduced the risk of ischemic stroke by 10%. (**indiscriminate widespread use of vitamin E should be cautioned against**) (Schurks et al. 2010)

88. - **Concomitant supplementation of vitamins C and E during pregnancy has been reportedly associated with low birth weight, the premature rupture of membranes and fetal loss or perinatal death in women at risk for preeclampsia. These changes may probably lead to the impairment of placental function and suboptimal growth of the fetus** (Hung et al, 2010)

89. - **elevated serum vitamin E concentrations are associated with abnormal markers of atherosclerosis and may increase the risk of cardiovascular complications in HIV-infected adults.** (Falcone et al, 2010)

90. - **Multivitamin use was associated with a statistically significant increased risk of breast cancer.** Use of **multivitamins was linked to a statistically significant 19 per cent increased risk of breast cancer.** (Larsson et al, 2010)

91. - **quercetin and ferulic acid are able to aggravate, if not induce, nephrocarcinoma in mice.** (Chiu-Lan Hsieh et al, 2010)

92. - **antioxidant administrations after the tumor's appearance accelerated tumor growth and favored metastases.** Continuous administration of pequi oil inhibited the tumor's growth, while the same protocol with **vitamins E and C accelerated it (tumor growth),** favoring metastasis and increasing oxidative stress on erythrocytes. (Miranda-Vilela AL, et al, 2011)

93. - In vitro exposure to **genistein/quercetin induced higher numbers of MII rearrangements in bone marrow cells of Atm-ΔSRI mutant mice compared with wt mice. Prenatal exposure to flavonoids associated with higher frequencies of MII rearrangements and a slight increase in the incidence of malignancies in DNA repair-deficient mice.** These data suggest that **prenatal exposure to both genistein and quercetin supplements could increase the risk on MII rearrangements** especially in the presence of compromised DNA repair. (Vanhees et al, 2011)

Adverse consequences of antioxidant vitamins and increased risk of death

Any agent which causes an increased risk of the following can contribute to the number of dead bodies. They need repeating.

Antioxidant vitamins can lead to increased risk of:

- **lung cancer, 18%, 28%;**
- **lung cancer incidence and mortality;**
- **breast cancer;**
- **prostate cancer;**
- **multivitamins doubled risk of fatal prostate cancer**

- head and neck cancer;
- skin cancer;
- melanoma;
- oral pre-malignant lesions, especially in smokers;
- total mortality, 17%
- likelihood of dying by about 5%;
- esophageal cancer deaths 14%;
- doubled adenoma (polyp) recurrence in smokers and drinkers;
- had a higher rate of second primary cancers during the supplementation period
- hemorrhagic stroke deaths, 50%;
- increased total and stroke mortality;
- supplements increase the likelihood of dying by about 5%
- vitamin A increased death risk by 16%,
- beta carotene by 7%,
- vitamin E by 4 %;
- significant increase in all-cause mortality and a slight increase in cardiovascular death;_
- mortality was increased in vitamin E users who had a history of stroke, coronary bypass graft surgery, or myocardial infarction and, independently, in those taking nitrates, warfarin, or diuretics;
- participants with a dietary vitamin C intake above the median of 90 mg/day, vitamin E increased mortality among those aged 50-62 years by 19%
- increased post-trial risk of first-ever non-fatal Myocardial Infarction (heart attack);
- ischemic heart disease deaths, 11%;
- significantly more deaths from fatal coronary heart disease;
- 2.3 times risk of death from stroke and 2 times risk of dying from coronary artery disease in diabetic post-menopausal women;
- higher rates of heart failure and hospitalizations for heart failure

- total fracture;
- hip fracture;
- osteoporotic hip fractures in men and women;
- accelerate the progression of retinitis pigmentosa (eye disease);
- increased risk of age-related macular degeneration;
- higher risk of age-related cataract among women;
- among women aged ≥65 y, vitamin C supplement use increased the risk of cataract by 38%;
- vitamin C use among hormone replacement therapy users was associated with a 56% increased risk of cataract
- increased risk of Mother-to-child transmission of HIV;
- increased rate of low-birth-weight babies;
- increased rate for gestational hypertension;
- increased in having to take antihypertensive medication antenatally (after birth)
- suffered falling more often;
- E supplementation increased tuberculosis 72% and pneumonia risk 14%;
- vitamin E results in loss of quality-adjusted life years
- increase in the incidence of cancer among smokers and increased cancer mortality
- -Multivitamin/Minerals associated with doubling in the risk of fatal prostate cancer
- increase in intima-media thickness over time in smokers
- rate of local recurrence of the head and neck tumor tended to be higher
- higher rate of second primary head and neck cancers
- worsening of coronary atherosclerosis in those with two copies of the haptoglobin 2 gene
- promote the clogging of arteries;
- accelerated thickening of the walls
- higher mortality rate in men
- adverse mucocutaneous effects and serum triglyceride elevations
- cardiac arrhythmias (probucol)

- negated the benefit of cholesterol lowering drugs (statin plus niacin)
- negatively associated with skeletal health
- 2 studies suggested poorer survival with concurrent administration of antioxidants and cytotoxic therapy in non-metastatic breast cancer
- C hampered endurance capacity
- multivitamins associated with higher mean breast density in premenopausal women
- multivitamins statistically significant increased risk 19 % of breast cancer
- supplementing the general public with vitamin E results in loss of quality-adjusted life years (loss of about four months)
- decreased sperm motility
- induced sperm DNA damage

www.ingramcontent.com/pod-product-compliance
Lightning Source LLC
Chambersburg PA
CBHW071353170526
45165CB00001B/24